STECK-VAUGHN

Access
Mathematics

Harcourt Achieve
Rigby • Steck-Vaughn

www.HarcourtAchieve.com
1.800.531.5015

ACKNOWLEDGMENTS

Editorial Director: Ellen Northcutt

Senior Editor: Donna Townsend

Associate Design Director: Joyce Spicer

Senior Designer: Jim Cauthron

Photo Researcher: Nicole Mlakar

Editorial Development: Learning Unlimited, Oak Park, IL

Photography Credits: p. 10 © Reuters NewMedia Inc./CORBIS; p. 64 © Stewart Cohen/Getty Images; p. 70 © LWA-Stephen Welstead/CORBIS; p. 112 © Mark Gibson Photography; p. 118 © Spencer Grant/Picture Quest; p. 154 © EyeWire/Getty Royalty Free; p. 160 © James Leynse/CORBIS; p. 222 © Julian Cotton/Image State.

ISBN 0-7398-8929-X

Printed in the United States of America.

1 2 3 4 5 6 7 8 9 10 073 09 08 07 06 05 04

CONTENTS

How to Use This Book

This book presents basic mathematics skills and concepts in the context of everyday, real-life applications. The four units are divided into lessons. Each lesson presents a mathematical concept, along with the computation skills you need to solve problems related to each concept. Each lesson provides practice sections, and ends with a practice review. Use the Answers and Explanations section at the back of the book to check your computation skills.

Pretest and Posttest

The Pretest is a self-check to see which skills you already know. When you complete all the questions in the Pretest, check your work in the Answers and Explanations section in the back of the book. Then fill out the Pretest Evaluation Chart. This chart tells you where each skill is taught in this book. When you complete this book, you will take a Posttest to see how much your math skills have improved.

Units

Unit 1: Whole Numbers. This unit covers whole number concepts such as place value, order of operations, rounding, equations, and expressions, as well as addition, subtraction, multiplication, and division of whole numbers.

Unit 2: Fractions. Fraction concepts and using the four operations with fractions are presented in this unit.

Unit 3: Decimals. This unit covers decimal concepts and using the four operations with decimals.

Unit 4: Ratios, Proportions, and Percents. In this unit you will learn the concept of ratios and how to solve proportions. You will also use the percent and interest formulas. This unit also covers the concept of probability.

Unit Reviews and Mini-Tests

Unit Reviews let you see how well you have learned the skills and concepts in each unit. Each Unit Review also includes a Math Extension Activity that provides an opportunity for further practice with the skills presented in the unit. Mini-Tests follow each Unit Review. These timed tests allow you to practice answering questions based on the content of the unit.

Math at Work

Math at Work is a two-page feature included in each unit. Each Math at Work feature introduces a specific job, describes the math skills the job requires, and includes a math activity related to it. It also gives information about other jobs in the same career area.

Real-Life Applications

All the math skills and concepts are applied to problems in real-life applications in the following areas:

- Workplace Math—math used on the job, in the office, and in business situations, with such items as order forms, timecards, and pay stubs; and in computing sales discounts, commissions, and sale prices.
- Measurement and Geometry—working with both standard and metric units to find perimeter, area, and volume, and reading maps and scales.
- Numbers and Operations—applying percent, ratios, and rates.
- Data Analysis—working with graphs, charts, and tables, and computing the mean and median and probability.
- Algebra Topics—writing and solving equations and using formulas.

Problem-Solving Strategies

You will learn many helpful strategies that are used to solve math problems. As you work with these strategies, use the following four-step plan to solve word problems.

Read Read the problem to determine what you need to find out. Identify the information you need to solve the problem.

Plan Plan how you will solve the problem. Decide which operation or operations you will use to solve the problem.

Solve Solve the problem by doing all the computations necessary to find the answer.

Check Check your answer by reading the problem again. Ask yourself, *Does my answer make sense?* Also check your computations to make sure you did them correctly.

Calculator Handbook

A Calculator Handbook is included after the Posttest. It presents step-by-step instructions for applying the four operations—addition, subtraction, multiplication, and division—to whole numbers and decimals, and for solving exponent, square root, and percent problems using the CASIO *fx-260SOLAR* calculator.

Setting Goals

A goal is something you aim for, something you want to achieve. What is your long-term goal for using this book? You may want to improve your test scores or perhaps you just want to learn more about math and how to use it in your life or on the job. These are large goals that may take some time to accomplish.

Write your long-term goal for mathematics.

____

____

This section of the book will help you to think about how you already use math and then to set some goals for what you would like to learn in this book. These short-term goals will be stepping-stones to the long-term goal that you wrote.

Check each activity that you do. Add more activities.

I use mathematics in my everyday life to

- ______ write checks and figure out my bank balance
- ______ budget my money
- ______ decide whether I have enough money to make a purchase
- ______ get to appointments on time
- ______ other ______________________________

____

____

List your experiences with learning and using mathematics.

What I've Liked	What I Haven't Liked
______________	______________
______________	______________
______________	______________
______________	______________
______________	______________

Setting Goals

Think about your mathematics goals.

1. I decided to improve my mathematics skills when I ______________________

____

____

2. My mathematics goals include (check as many as you like)

- ☐ improving my skills in adding and subtracting
- ☐ improving my skills in multiplying and dividing
- ☐ solving word problems
- ☐ using fractions
- ☐ using decimals
- ☐ reading graphs
- ☐ writing ratios
- ☐ using proportion to solve problems
- ☐ finding percents
- ☐ understanding probability
- ☐ other ______________________________________

____

____

3. I will meet my long-term goal for mathematics when I am able to

____

____

____

Keep track of your goals.

As you work through this book, turn back often to this page. Add more goals at any time you wish.

Learn about the skills you have.

Complete the Pretest that begins on the next page. It will help you learn more about your strengths and weaknesses in mathematics. You may wish to change some items in your list of goals after you have taken the Pretest and completed the Pretest Evaluation Chart on page 9.

PRETEST

Use this Pretest before you begin Unit 1. Don't worry if you can't easily answer all the questions. The Pretest will help you determine which skills you are already strong in and which skills you need to practice.

Read and answer the questions that follow. Check your answers on pages 254–256. Then enter your scores on the chart on page 9. Use the chart to figure out which skills to work on and where to find those skills in this book.

Write the value of the underlined digit in words.

1. $6,\underline{3}02,450$ ______________________

2. $5.34\underline{7}9$ ______________________

Compare each pair of numbers. Write >, <, or = between the two numbers.

3. 46,023 ______ 46,203

4. 0.76 ______ 0.456

Write the correct answers.

5. Round 534,103 to the nearest ten thousand. ____________

6. Round 3.725 to the nearest hundredth. ______

7. Write the mixed number that names the shaded portions.

______

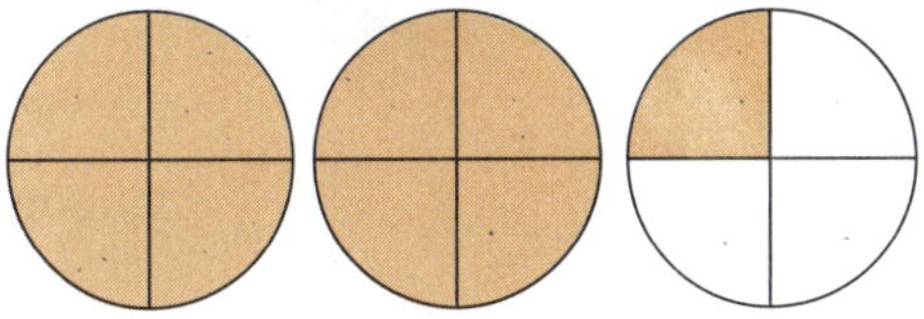

Solve. Reduce answers to lowest terms.

8. $\begin{array}{r} 600 \\ -328 \\ \hline \end{array}$

9. $\begin{array}{r} 6.7 \\ -3.846 \\ \hline \end{array}$

10. $\begin{array}{r} 743 \\ \times 608 \\ \hline \end{array}$

11. $4\overline{)\$12.28}$

12. $\begin{array}{r} 4\frac{7}{8} \\ +2\frac{1}{3} \\ \hline \end{array}$

13. $\begin{array}{r} 6\frac{2}{5} \\ -4\frac{2}{3} \\ \hline \end{array}$

14. $11 + (2 + 1) \times 6 \div 2 =$

15. $42 + 376 + 57 =$

Go on to the next page.

16. $3.27 + 26.4 + 8.531 =$

17. $\$5.08 - \$2.99 =$

18. $0.054 \times 0.3 =$

19. $4\frac{2}{5} \times 3\frac{3}{4} =$

20. $7{,}658 \div 42 =$

21. $31.2 \div 0.06 =$

22. $3\frac{1}{2} \div 4\frac{3}{8} =$

23. $7^2 =$

24. Change 375% to a decimal.

25. Change 0.07 to a percent.

26. Change 80% to a fraction.

27. Change $\frac{3}{4}$ to a percent.

28. What is 8% of 150?

29. What percent of \$72 is \$18?

30. 54 is 150% of what number?

31. $\frac{8}{20} = \frac{?}{35}$

Circle the best answer for each question.

32. Handy Hardware had 144 gallons of interior paint and 96 gallons of exterior paint in stock. During the week, 48 gallons of interior paint were sold. Which is the correct expression to find the number of gallons of interior paint still in stock?

(1) $96 - 48$
(2) $144 + 48$
(3) $144 - 48$
(4) $144 - 96$
(5) $144 + 96 - 48$

33. Carol is building a rectangular dog kennel that is 12 feet long and 6 feet wide. What is the perimeter of the kennel in feet?

(1) 18
(2) 36
(3) 72
(4) 144
(5) 180

34. Darius wants to carpet a living room that measures 20 feet on each side. Which is the correct expression to find the area of the room in square feet?

(1) 20×9
(2) $20 \div 9$
(3) 20^2
(4) $20^2 \div 9$
(5) $20 + 20 + 20 + 20$

35. On Friday, Village Deli sold 79 sandwiches on rye, 36 sandwiches on white, and 52 sandwiches on whole wheat bread. Which is the best estimate of the total number of sandwiches sold on Friday?

(1) 150
(2) 160
(3) 170
(4) 180
(5) Not enough information is given.

36. A package contains 8 hamburger buns. Maria bought 16 packages of buns for a picnic. Which is the correct expression to find the number of buns she bought?

(1) $8 \div 16$
(2) $8 + 16$
(3) $16 - 8$
(4) 16×8
(5) $16 \div 8$

37. The spinner shown has 6 equal sections. What is the probability that the wheel will stop on a 2?

(1) $\frac{1}{6}$ or $16\frac{2}{3}\%$
(2) $\frac{1}{3}$ or $33\frac{1}{3}\%$
(3) $\frac{1}{2}$ or 50%
(4) 1 or 100%
(5) 2 or 200%

38. Jesse has a recipe that calls for $2\frac{1}{4}$ cups of chicken broth. He wants to make one third of this recipe. How many cups of broth should Jesse use?

(1) $\frac{4}{27}$
(2) $\frac{3}{4}$
(3) $2\frac{1}{4}$
(4) $5\frac{1}{4}$
(5) $6\frac{3}{4}$

39. Joan's Market has onions on sale at 3 pounds for \$1.56. A customer bought 5 pounds. Which is the correct expression to find the cost of 5 pounds of onions?

(1) $\frac{3}{5} = \frac{?}{\$1.56}$
(2) $\frac{3}{\$1.56} = \frac{5}{?}$
(3) $\frac{3}{\$1.56} = \frac{?}{5}$
(4) $\frac{5}{3} = \frac{\$1.56}{?}$
(5) $\frac{5}{\$1.56} = \frac{3}{?}$

Go on to the next page.

40. Jermaine works 40 hours per week as a clerk at Central Auto Supply. He spends 15% of his time restocking shelves. Which is the correct expression to find the number of hours he spends restocking shelves each week?

(1) 15 × 40%
(2) 15 ÷ 40%
(3) 15% ÷ 40
(4) 40 × 15%
(5) 40 ÷ 15%

41. The vet told Andy to put his dog on a diet. During the first week of the diet, Andy's dog lost 0.2 kilogram. How many grams did the dog lose?

(1) 0.002
(2) 0.02
(3) 2
(4) 200
(5) 2,000

42. The Food Mart had ground beef on sale for $2.48 per pound. Joy bought 4 pounds of ground beef. Which is the correct expression to find the total cost of the ground beef?

(1) $2.48 × 0.4
(2) $2.48 ÷ 0.4
(3) 4 ÷ $2.48
(4) $2.48 × 4
(5) $2.48 ÷ 4

43. Ricardo's gross pay is $420.00 each week. He had $25.20 deducted for his credit union savings account. What percent of Ricardo's gross pay is deducted for savings?

(1) 0.6%
(2) 6%
(3) $16\frac{2}{3}\%$
(4) 20%
(5) 25%

44. Maggie bought a jacket at Discount Fashions. The price tag is shown here. Which is the correct expression to find the original price of the jacket?

(1) $18.60 × 30%
(2) $18.60 × 70%
(3) $18.60 ÷ 30%
(4) 30% × $18.60
(5) 30% ÷ $18.60

45. Which expression best describes how to find the width of the rectangle?

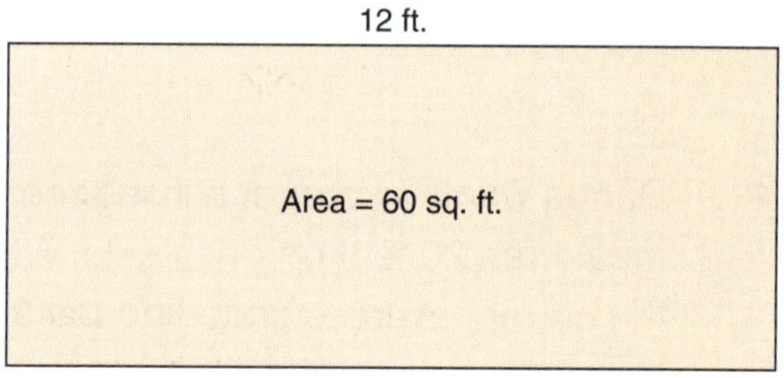

(1) 12 + 60
(2) 60 − 12
(3) 12 × 60
(4) 60 ÷ 12
(5) Not enough information is given.

46. Abdul just started work at a fast-food restaurant. His current wage is $5.25 per hour. After three months, his new wage will be $5.46 per hour. What is the percent of increase in his wages?

(1) 3%
(2) 4%
(3) 20%
(4) 21%
(5) 96%

Write your answers in the space provided. Show your work.

47. What is the volume in cubic inches of the shipping carton shown?

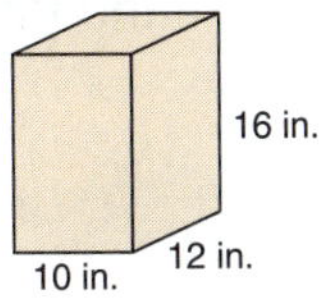

48. Marcus drives a delivery truck for a bread company. His total route is $36\frac{9}{10}$ miles long. He drives $19\frac{2}{5}$ miles in the morning and finishes the route in the afternoon. Estimate the number of miles Marcus drives in the afternoon.

Question 49 refers to the following table.

Fuel Economy in Miles per Gallon

	Car A	Car B	Car C
Highway	32	23	27
In Town	21	14	18

49. How many miles can car B travel on the highway on 15 gallons of gas?

50. Miranda needs a board that is $4\frac{1}{2}$ feet long for a shelf. She has a board that is 75 inches long. How many feet are in 75 inches?

Question 51 refers to the following timecard.

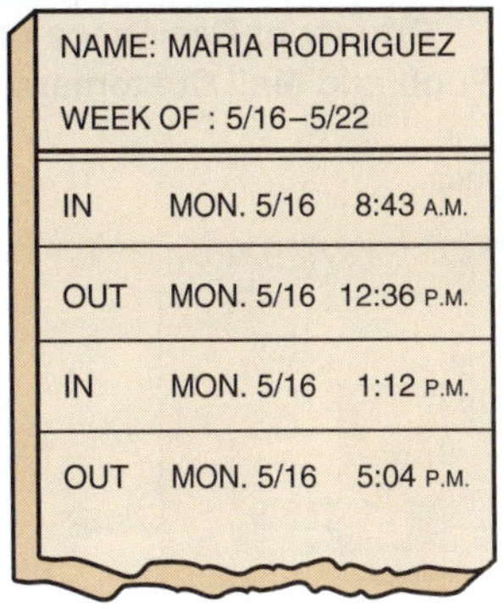

51. Approximately how many hours did Maria work before taking a lunch break?

52. David and Grace are playing cards. David has the following ten cards in his hand. If Grace takes one of his cards, what is the chance that she will take a 5 or higher?

53. Tawanna works at a gift shop. Her customer is buying a statue for $32.50 and a card for $1.75. Sales tax on this purchase is $1.71. The customer gives Tawanna $40.00 in cash. How much change should Tawanna give her customer?

Go on to the next page.

Question 54 refers to the following graph.

County of Residence of Woodland Mall Customers, 6/23

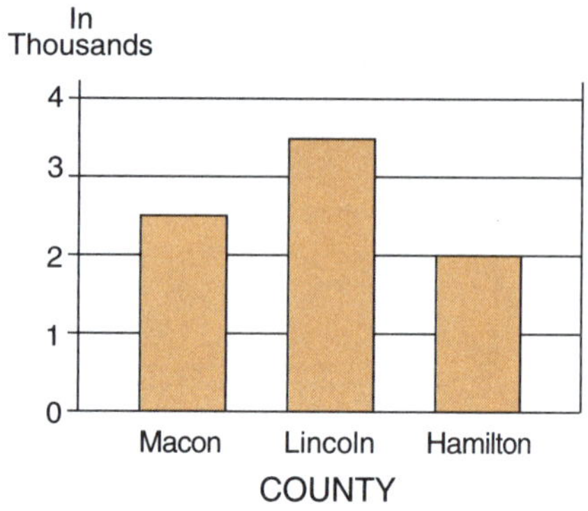

54. What is the approximate difference in the number of mall customers who live in Lincoln County and the number who live in Hamilton County?

55. The weights of three packages are $\frac{5}{8}$, $\frac{3}{4}$, and $\frac{9}{16}$ pound. Write the weights in order from least to greatest.

Questions 56 and 57 refer to the following data.

The points scored by a football team during 4 preseason games were: 37, 18, 21, and 24.

56. What is the mean of these data?

57. What is the median of these data?

58. Of the 36 employees at Home Products, Inc., 20 work in the factory. What is the ratio in lowest terms of those who work in the factory compared to all employees?

59. Mai borrowed $3,000 from her parents to buy a new car. She will pay this back in 5 years at 5% interest. How much interest will she pay?

Question 60 refers to the following number line.

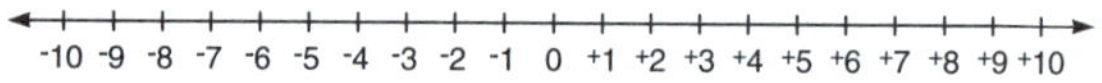

60. Before going to bed, Joe read that the temperature was 2 degrees. The next morning he heard that the temperature had dropped 7 degrees overnight. What was the temperature in the morning?

Pretest Evaluation Chart

Check your answers on pages 254–256. Circle the number of each question that you answered correctly on the Pretest. Count the number of questions in each row that you answered correctly. Write the number in the Total Correct space in each row. For example, in the *Whole Numbers* row, write the number correct in the blank before *out of 21*. Complete this process for the remaining rows. Then add the four totals to get your total correct for the whole Pretest.

Skill Area	Computation	Data Analysis	Measurement	Total Correct
Whole Numbers (Pages 10–69)	1, 3, 5, 8, 10, 14, 15, 20, 23, 32, 35, 36	**54, 56, 57**	33, 34, 45, **47, 49, 60**	_____ out of 21
Fractions (Pages 70–117)	7, 12, 13, 19, 22, 48, 55		38, 50, **51**	_____ out of 10
Decimals (Pages 118–159)	2, 4, 6, 9, 11, 16, 17, 18, 21, 53		41, 42	_____ out of 12
Ratios, Proportions, and Percents (Pages 160-227)	24, 25, 26, 27, 28, 29, 30, 31, 40, 43, 44, 46, 58, 59	**37, 52**	39	_____ out of 17

Total Correct for Pretest _____ out of 60

Boldfaced numbers indicate questions based on charts, diagrams, graphs, and maps.

UNIT 1

Whole Numbers

Whole numbers are the first numbers we learn and the numbers we use all of our lives—0, 1, 2, 3, 4, 5, 6, 7, 8, and 9. We use them for counting and numbering everything around us. We put them together to write addresses and phone numbers, to keep score in our favorite sport or game, and to read the dates on a calendar. We add, subtract, multiply, and divide them to solve problems. Below are examples of very simple ways we use whole numbers.

Write the numbers in your street address. ______________________

Write a sports score. ______________________

Write today's date. ______________________

Thinking About Whole Numbers

You may not realize how often you say, read, or think about numbers as you go about your daily life. Think about your recent activities.

Check the box for each activity you have done recently.

- ☐ Did you pay cash for a purchase?
- ☐ Did you count the change you got back?
- ☐ Did you write down an address or telephone number?
- ☐ Did you keep score in a game or sporting event?
- ☐ Did you estimate how long it would take you to get somewhere?
- ☐ Did you read advertisements to look for good prices?

Write some other activities where you used numbers.

____

____

____

____

____

Previewing the Unit

In this unit, you will learn:

- what whole numbers mean in different situations
- how to perform basic operations with whole numbers
- how to use whole numbers to measure the space inside shapes and objects
- how to perform whole number operations on a calculator

LESSON 1

Number Sense

Place Value

Our number system uses ten **digits:** 0, 1, 2, 3, 4, 5, 6, 7, 8, and 9. Writing one or more digits in a row forms a **whole number.** The number 7 is a one-digit number, and the number 154 is a three-digit number. The number 5,000 is a four-digit number even though three of the digits are zeros.

The **place-value** chart below shows the first ten place values in our whole number system. The value of a digit depends on its place in the number. The value of the places increases as you move to the left.

billions	hundred millions	ten millions	millions	hundred thousands	ten thousands	thousands	hundreds	tens	ones
			5,	2	1	4,	0	8	4

The number 5,214,084 is written on the place-value chart. From the chart, you can see that this whole number has 5 millions, 2 hundred thousands, 1 ten thousand, 4 thousands, 0 (or no) hundreds, 8 tens, and 4 ones. The 4 in the thousands place has a greater value than the 4 in the ones place.

TIP
When you write numbers, place commas every three digits, counting from the right. For example, 32,548,901.

Example Write the place value for the 4 in 54,201.

Step 1 Find the number being discussed: the 4 in 54,201.

Step 2 Find the place value of the digit being discussed. Look at the place-value chart if necessary. The 4 is the fourth digit from the right. Thus, the 4 is in the thousands place.

PRACTICE

Write the place value of each underlined digit. Refer to the chart above if necessary.

1. 9,650 ______tens______
2. 972 ____________
3. 28,730 ____________
4. 826,110 ____________
5. 7,535,000 ____________
6. 57,425 ____________
7. 9,853,483 ____________
8. 37,654,321 ____________
9. 56,839 ____________
10. 348,653 ____________

 Check your answers on page 256.

Reading and Writing Whole Numbers

To write a check, you must write the amount in both words and digits.

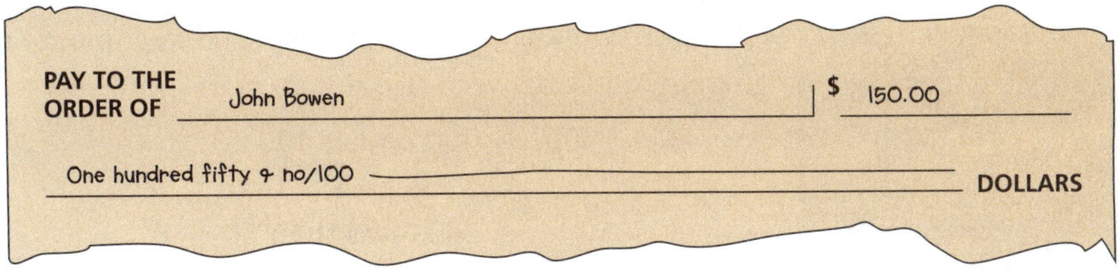

Use the following rules to read and write whole numbers:

> **TIP**
> Do not say the word *and* when reading whole numbers.
> Read 1,523 as "one thousand, five hundred twenty-three."

Rule 1 When writing a number using digits, use zero as a placeholder.

Example "Two hundred four" is written **204.** The zero in the tens place has no value, but it is needed to hold the tens place.

Rule 2 To read a number, read each group of digits from left to right, stopping at each comma. Then say the name of the group to the left of the comma.

Example 4,819,520 is read "**four million, eight hundred nineteen thousand, five hundred twenty.**"

Rule 3 When writing big numbers in words, place a comma after the word *million* and the word *thousand.* Do not place a comma after the word *hundred.* This puts a comma in the same place with the words as with the digits.

Example 5,150,425 is read "**five million, one hundred fifty thousand, four hundred twenty-five.**"

PRACTICE

Write each number in words.

1. 756 seven hundred fifty-six
2. 43,018 ____
3. 115,200 ____
4. 5,400,012 ____

Write each number using digits. Then enter it in a calculator.

5. one thousand, seven hundred eight 1,708
6. two hundred fifty thousand, nine hundred eleven ____
7. twelve thousand, sixteen ____
8. nine million, fourteen thousand, five hundred sixty ____

For more practice, see page 241 of the Calculator Handbook.

Check your answers on page 257.

Comparing Whole Numbers

Would you rather work for $9 or $7 an hour? To answer the question, you would compare the two numbers and pick $9, the greater amount.

The symbols =, >, and < are used to compare numbers. You can write these symbols between two numbers to show how the numbers compare.

100 = 100	100 **equals** 100
$9 > $7	$9 **is greater than** $7
10 < 50	10 **is less than** 50

TIP
Think of the symbols > and < as arrows that always point to the smaller number.
12 < 20 and
20 > 12

Use these rules to compare whole numbers:

Rule 1 A number with more digits is greater than a number with fewer digits.

Example Compare 8,500 and 920.

Step 1 Count the number of digits in one number. There are four digits in 8,500.

Step 2 Count the number of digits in the other number. There are three digits in 920.

8,500 > 920 because 8,500 has more digits than 920. We read this as "8,500 is greater than 920."

Rule 2 If the numbers have the same number of digits, work from left to right and compare each place value until the digits in a place value are different.

Example Compare 6,410 and 6,481.

Step 1 Compare the digits in the thousands place. 6,410 6,481 → same

Step 2 Compare the digits in the hundreds place. 6,410 6,481 → same

Step 3 Compare the digits in the tens place. 6,410 6,481 → different

6,410 < 6,481 because 1 is less than 8. We read this as "6,410 is less than 6,481."

PRACTICE

Compare each pair of numbers. Write >, <, or =.

1. 4,700 __>__ 740
2. 38,000 _____ 38,500
3. 179 _____ 179
4. 210,580 _____ 210,480
5. 1,000,000 _____ 10,000,000
6. 496 _____ 4,690
7. 13,415 _____ 13,415
8. 802,165 _____ 803,980
9. 5,000 _____ 50,000
10. 1,345 _____ 1,435
11. 10,334 _____ 10,334
12. 479 _____ 476
13. 340,635 _____ 340,835
14. 5,010 _____ 5,001
15. 682,489 _____ 682,489
16. 4,609 _____ 4,906

 Check your answers on page 257.

Rounding

Rounded numbers are easier to remember and use than exact amounts. They can be used to estimate answers or to check your calculations. Whenever you see the word *about,* you know the amounts are estimates.

> **TIP**
> Rounded numbers often end in one or more zeros.
> 40; 500; and 10,000 are all rounded numbers.

Example Miwa earns $34,765 a year as a computer operator. Rounded to the nearest thousand dollars, how much does she earn per year?

Step 1 Look at 34,765. This number falls between 34,000 and 35,000, but which is it closer to? You need to round to the nearest thousand, so underline the digit in that place. In this case, the 4 is in the thousands place. $34,765

Step 2 Look at the digit to the right of the underlined digit: 7 $34,765

Step 3 If this digit is 5 or more, add 1 to the underlined digit. If the digit is less than 5, do not change the underlined digit. Since 7 is greater than 5, add 1: 4 + 1 = 5 $35,_ _ _

Step 4 Change all the digits to the right of the underlined digit to zeros. **$35,000**

Example Round 4,514 to the nearest hundred.

Step 1 Underline the digit in the place you want to round the number: the 5 is in the hundreds place 4,514

Step 2 Look at the digit to the right of the underlined digit. 4,514

Step 3 Decide if the underlined digit needs to change. Since 1 is less than 5, the underlined digit does not change. 4,5_ _

Step 4 Change digits to the right of the underlined digit to zeros. **4,500**

PRACTICE

Round each number to the given place value.

1. Round 58 to the nearest ten. 60
2. Round 1,723 to the nearest hundred. ____________
3. Round 6,509 to the nearest thousand. ____________
4. Round 861 to the nearest hundred. ____________
5. Round 19,580 to the nearest thousand. ____________
6. Round 209,320 to the nearest ten thousand. ____________
7. Round 64,299 to the nearest thousand. ____________
8. Round 5,256,000 to the nearest hundred thousand. ____________

Check your answers on page 257.

PROBLEM SOLVING

Estimation

Estimation is a useful tool when you work with numbers. An *estimate* gives you a general idea of a value. You can estimate to find an approximate answer or to make sure an answer is reasonable. Rounding is one method of estimation.

TIP
Some situations need only an approximate answer or estimate. Other situations, like how much money to charge someone, need an exact answer.

Example Samuel works at an office supply store. A customer wants to know if 15 packages of pastel-colored copy paper will cost less than $60. Each package sells for $4.99. Excluding tax, is $60 enough money?

Step 1 Ask "Is an exact answer needed?" No, the customer wants a general idea of how much the paper will cost.

Step 2 To estimate, round to the nearest dollar: $4.99 rounded to the nearest dollar is $5.00.

Step 3 Perform the needed operations. Multiply the number of packages by the estimated cost per package: $15 \times \$5 = \$75$. **No, $60 is not enough money.**

Example Bridget is a cashier at a department store. A customer is buying a vest for $24, a cotton turtleneck for $16, a denim skirt for $19, and a blouse for $29. How much do the items cost before tax?

Step 1 Is an exact answer needed? Yes. Since Bridget is a cashier, she needs to find the exact total so that she can add the correct tax and the customer can pay for the purchases.

Step 2 Find the total cost before tax.
$\$24 + \$16 + \$19 + \$29 = \$88$ **The items total $88.**

PRACTICE

Find an estimate and an exact answer for each problem.

	Estimate	Exact
1. $11 + $15 + $23 (round to the nearest $5)	______	______
2. 58 feet + 13 feet + 28 feet	______	______

Based on the problem, decide to find either an estimate or an exact answer.

3. Chin recommended that his boss buy the following equipment to protect the office computer and to back up information. He listed a surge protector for $19, a back-up disk drive for $119, and an extra back-up disk for $16. About how much money would the items on Chin's list cost?

4. Elory has a pet-sitting service. She charges $15 for one hour, per pet. Last week she sat 1 hour each with 18 cats and 1 hour each with 22 dogs. How much money did Elory charge customers last week?

 Check your answers on page 257.

ALGEBRA TOPIC

Number Line

All whole numbers—positive numbers, negative numbers, and zero—are called **integers.**

Integers that are greater than zero are called **positive numbers.** Integers that are less than zero are called **negative numbers.** Zero is neither positive nor negative. A positive number can be written with or without a positive sign (+). Negative numbers must always have a minus sign (−) in front of the number.

A **number line** shows the relationship of the positive and negative numbers and zero. On a number line, positive numbers are to the right of zero and negative numbers are to the left of zero.

TIP
To solve problems using a number line, count to the right to find an increase or gain. Count to the left to find a decrease or loss.

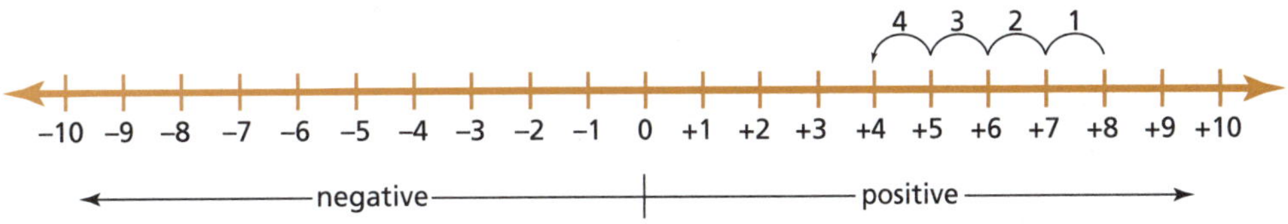

Example Darius is a farmer. During cold weather he checks the temperature hourly. Before dinner, he saw that the temperature was 8 degrees. During the next hour, the temperature fell 4 degrees. What was the temperature at that time?

Step 1 Find the number on the number line: +8.

Step 2 Move 4 units to the left since the temperature fell (decreased) 4 degrees.

The temperature at that time was **+4 degrees.**

PRACTICE

Write your answers in the blanks. Refer to the number line above if necessary.

1. 3 degrees with a drop of 12 degrees −9
2. 0 yards with a gain of 4 yards and a loss of 10 yards ______
3. $3 and $5 more ______
4. $0 with a loss of $8 and a loss of $3 ______
5. deduct 10 points from 12 points ______
6. −7 percent with an increase of 10 percent ______
7. When Elena left for work in the morning, the temperature was −6 degrees. By lunchtime the temperature had increased 5 degrees. What was the temperature when Elena left work to go to lunch? ______
8. In a high school football game, the home team lost 4 yards on its first play and gained 6 yards on its second play. In all, how many yards has the home team lost or gained so far? ______

Check your answers on page 257.

Operating Your Calculator

Calculators are important tools. We use them at home, at work, and in school. Throughout the lessons in this book, you will be directed to the Calculator Handbook at the back of the book to learn how to use the CASIO *fx-260SOLAR*.

Key Reference

The illustration shows the CASIO *fx-260SOLAR* and the features you will need to do the work in this book.

The CASIO *fx-260SOLAR* Calculator

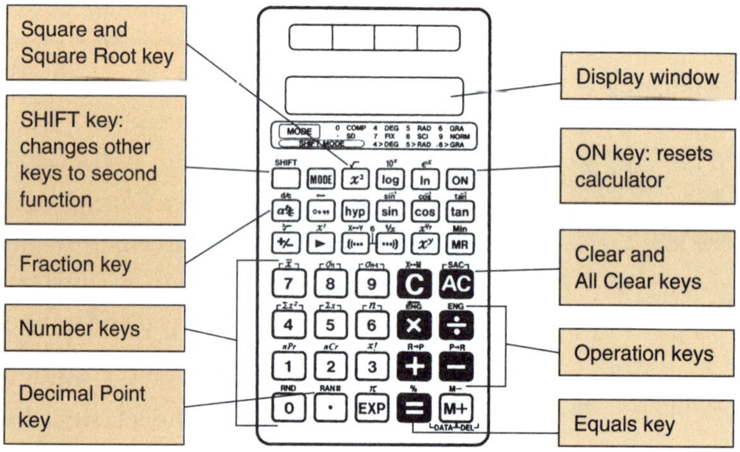

TIP
Commas are not used with calculators.
1,200 is entered as 1200.

Example Enter 1,965 on your calculator.

Step 1 Press the all clear key (AC) to erase any previous entries.

Step 2 Press the appropriate digit keys 1965

Your calculator display should look like this: 1965.

Adding and Subtracting on a Calculator

The (+) key is used to add numbers. Use the (−) key to subtract numbers. To display the answer, press the (=) key.

Refer to pages 241–242 in the Calculator Handbook for more information on adding and subtracting with calculators.

Example Ariel brings home $1,500 a month. She pays $435 a month for her mortgage and $450 for other bills. What is the difference in Ariel's monthly take-home pay and her monthly payments?

Step 1 Find the total of Ariel's monthly payments.

(AC) 435 (+) 450 (=) 885

Step 2 Subtract Ariel's monthly payments from her salary.

(AC) 1500 (−) 885 (=) 615 or **$615**

Multiplying and Dividing on a Calculator

The ⌈×⌉ key is used to multiply numbers. Use the ⌈÷⌉ key to divide numbers. To clear the display, press the ⌈AC⌉ key.

Example Michael bought a package of laser paper for $7.99. How much would four packages cost?

Step 1 Press ⌈AC⌉ to erase any previous entries.

Step 2 Multiply the cost for each package by the number of packages. Be sure to enter the decimal point.

7.99 ⌈×⌉ 4 ⌈=⌉ 31.96, which is $31.96.

PRACTICE

A. For each number, write what you would see on the calculator display.

1. 780 ___780___
2. 1,653 ________
3. 10,241 ________
4. $18.92 ________

B. Show how to calculate each problem. Then, use your calculator to find the answer.

5. 15 + 89 ___⌈AC⌉ 15 + 89 = 104___
6. 73 × 46 ________
7. 2,187 ÷ 3 ________
8. $1,406.19 − $94.37 ________

C. Use your calculator to solve these problems.

9. Stephanie ordered the following holiday gifts for company clients:

Five-box fruit gift	$31.95
Fruit basket	$43.95
Chocolate truffles	$18.95

What is the total cost of the order? ________

10. Roberto said he would bring soda pop to the class holiday party. One 2-liter bottle costs $1.05. How much will Roberto spend if he buys 6 bottles?

____

Check your answers on page 257.

LESSON 1

Lesson 1 Review

Write the place value of each underlined digit.

1. $2,\underline{8}34$ ________________
2. $3\underline{2}8,954$ ________________
3. $\underline{7},927,480$ ________________
4. $8,\underline{0}51,939$ ________________

Write each number in words.

5. 28,302 ____
6. 1,076,500 ____

Write each number using digits.

7. forty-two thousand, fifty-seven ________________
8. three million, four hundred thousand, five hundred ninety

____

Compare each pair of numbers. Write >, <, or = to make each expression true.

9. 5,680 ________ 856
10. 32,457 ________ 32,457
11. 82,346 ________ 82,546
12. 790,300 ________ 709,300

Round each number to the given place value.

13. Round 43 to the nearest ten. ________________
14. Round 2,453 to the nearest hundred. ________________
15. Round 307,216 to the nearest ten thousand. ________________
16. Round 4,293,785 to the nearest million. ________________

Use your calculator to solve these problems.

17. 1,023 [−] 645 [+] 270 [=] ________
18. 36,450 [÷] 3 [×] 12 [=] ________
19. 2,010 [+] 45 [−] 70 [=] ________
20. 5,388 [×] 2 [−] 375 [=] ________

Circle the best answer for each question.

Questions 21 through 23 refer to the following chart.

Week 1	12,435
Week 2	14,526
Week 3	12,345
Week 4	14,814
Week 5	13,706

21. Which statement is false?
    (1) Fewer parts were produced during week 5 than during week 4.
    (2) Fewer parts were produced during week 1 than during week 3.
    (3) More parts were produced during week 4 than during week 2.
    (4) More parts were produced during week 5 than during week 1.
    (5) Fewer parts were produced during week 2 than during week 4.

22. During which week was the greatest number of parts produced?
    (1) week 1
    (2) week 2
    (3) week 3
    (4) week 4
    (5) week 5

23. About 14,000 parts were produced during one week. The manager had rounded the number to the nearest thousand. Which week was she talking about?
    (1) week 1
    (2) week 2
    (3) week 3
    (4) week 4
    (5) week 5

24. The number of cars sold in a state was 1,030,402. Select the expression below that shows in words the number of cars sold.
    (1) one million, three thousand, four hundred two
    (2) one million, thirty thousand, four hundred two
    (3) one hundred thirty thousand, forty-two
    (4) one hundred thirty thousand, four hundred two
    (5) one million, three thousand, four hundred twenty

25. Marsha spent $2,264 last year for day care. Rounded to the nearest hundred dollars, how much did she spend for day care?
    (1) $2,000
    (2) $2,200
    (3) $2,260
    (4) $2,300
    (5) $3,000

26. When Kevin left work, the temperature was 8 degrees. By the time he went to sleep, the temperature had dropped 10 degrees. What was the temperature when Kevin went to sleep?
    (1) 18 degrees
    (2) 8 degrees
    (3) 2 degrees
    (4) −2 degrees
    (5) −10 degrees

Check your answers on pages 257–258.

LESSON 2

Adding and Subtracting Whole Numbers

Adding Whole Numbers

Addition means putting numbers together to find a total. The total is called the **sum.** You write an addition problem with a plus sign (+). Addition problems can be written in rows or in columns.

If the numbers in a column add up to more than 10, you regroup the number of tens to the tens column. Do the same with hundreds and so on.

TIP
To check addition, add again from bottom to top.

$$\begin{array}{r} 125 \\ 12 \\ +291 \\ \hline 428 \end{array}$$

Example Add: 291 + 12 + 125

Step 1 Line up the numbers so that the digits in each column have the same place value.

$$\begin{array}{r} 291 \\ 12 \\ +125 \\ \hline \end{array}$$

Step 2 Start with the ones column and add the numbers. Working from right to left, add the numbers in each column.

$$\begin{array}{r} 29\mathbf{1} \\ 1\mathbf{2} \\ +12\mathbf{5} \\ \hline 8 \end{array}$$

Step 3 If the numbers in a column add up to more than 10, regroup the number of tens to the tens column. Do the same with hundreds and so on.

$$\begin{array}{r} {}^{1} \\ 2\mathbf{9}1 \\ \mathbf{1}2 \\ +1\mathbf{2}5 \\ \hline 28 \end{array} \qquad \begin{array}{r} {}^{1} \\ \mathbf{2}91 \\ 12 \\ +\mathbf{1}25 \\ \hline 428 \end{array}$$

Many times you will see addition problems written in rows rather than columns. It is often helpful to rewrite the addition problem in a column to help you line up the digits.

Example 153 + 148 Rewrite: $\begin{array}{r} 153 \\ +148 \\ \hline \end{array}$ Add: $\begin{array}{r} {}^{1\,1} \\ 153 \\ +148 \\ \hline \mathbf{301} \end{array}$

Adding Whole Numbers on a Calculator

To add numbers on a calculator, enter each number, use the plus key between each number, then use the equals key to get the answer.

Add: 4,643 and 7,982.

[AC] 4643 [+] 7982 [=] **12625.**

For more practice adding with the calculator, see Calculator Handbook, page 320.

✔ Sal checks his work by adding the numbers again, in reverse order.

PRACTICE

A. Add the numbers below. Check your answers. The first one is done for you.

1. $\begin{array}{r} 254 \\ +325 \\ \hline 579 \end{array}$ Check $\begin{array}{r} 325 \\ +254 \\ \hline 579 \end{array}$

2. $\begin{array}{r} 476 \\ +\ 18 \\ \hline \end{array}$

3. $\begin{array}{r} 387 \\ +264 \\ \hline \end{array}$

4. $\begin{array}{r} 473 \\ +182 \\ \hline \end{array}$

5. $\begin{array}{r} 148 \\ 327 \\ +232 \\ \hline \end{array}$

6. $\begin{array}{r} 135 \\ 456 \\ +784 \\ \hline \end{array}$

7. $\begin{array}{r} 564 \\ 125 \\ +\ 68 \\ \hline \end{array}$

8. $\begin{array}{r} 817 \\ 76 \\ +453 \\ \hline \end{array}$

B. Rewrite the problems below. Line up the numbers. Solve and check. The first one is started for you.

9. 38 + 157 = $\begin{array}{r} 38 \\ +157 \\ \hline \end{array}$

10. 450 + 326 + 188 =

11. 3,947 + 18,889 + 232 =

12. 142 + 7 + 802 + 3 =

C. Solve the problems below with a calculator. Refer to page 241 in the Calculator Handbook for additional information.

13. Rachel ordered computer components for her department. She ordered the following:

Cable (1)	$ 39
Monitor (1)	229
Disks (1 box)	25
Software package	499

What was the total cost of the components?

Calculator Answer: ______________

Checked Answer: ______________

When working with calculators, be careful to enter digits in the correct order.

14. Sandra is in the third grade. The teacher encouraged the 26 students in the class to read more. When the students had read 100 books, they won a class party. By which week did they meet their goal? ______________

Week 1	13 books
Week 2	12 books
Week 3	21 books
Week 4	19 books
Week 5	16 books
Week 6	18 books
Week 7	17 books
Week 8	15 books
Week 9	16 books
Week 10	16 books

Check your answers on pages 258–259.

Subtracting Whole Numbers

Subtraction is taking an amount away from another amount. Subtract when you need to find the **difference** or to make a comparison. The minus sign (−) tells you to subtract. Like addition problems, subtraction problems can be written in rows or in columns.

Example Subtract 62 from 145.

Step 1 Line up the numbers. Make sure the smaller number is on the bottom.

$$\begin{array}{r} 145 \\ -62 \\ \hline \end{array}$$

Step 2 Start with the ones column and subtract. Working to the left, subtract the numbers in each column.

$$\begin{array}{r} 145 \\ -\ 62 \\ \hline 3 \end{array}$$

Step 3 Regroup whenever the digit being subtracted is greater than the digit above it.

$$\begin{array}{r} {\scriptstyle 0\,14} \\ \not{1}\not{4}5 \\ -\ 62 \\ \hline 83 \end{array}$$

If a subtraction problem has one or more zeros in it, you can regroup from the first digit to the left that is not a zero.

TIP
To check subtraction, add your answer and the number that was subtracted.

$$\begin{array}{r} 300 \\ -148 \\ \hline 152 \end{array} \qquad \begin{array}{r} 152 \\ +148 \\ \hline 300 \end{array}$$

Example Subtract: 300 − 148

Step 1 Start with the ones column. You need to regroup from the hundreds column.

$$\begin{array}{r} {\scriptstyle 2\,10\ } \\ \not{3}\not{0}0 \\ -148 \\ \hline \end{array}$$

Step 2 Next, regroup from the tens column.

$$\begin{array}{r} {\scriptstyle 9\,10} \\ {\scriptstyle 2\,\not{1}\not{0}\ } \\ \not{3}\not{0}0 \\ -148 \\ \hline \end{array}$$

Step 3 Work from right to left and subtract each column.

$$\begin{array}{r} {\scriptstyle 2\,9\,10} \\ \not{3}\not{0}\not{0} \\ -148 \\ \hline \mathbf{152} \end{array}$$

Subtracting Whole Numbers on a Calculator

To subtract numbers on a calculator, enter each number. Then use the minus key between each number. Use the equals key to get the answer.

Subtract: 408 − 299

[AC] 408 [−] 299 [=] 109

PRACTICE

A. **Subtract the numbers below. Check your answers using addition.**

1. 48 − 25 = 23 Check: 23 + 25 = 48	2. 957 − 304	3. 899 − 425	4. 786 − 43
5. 53 − 28	6. 426 − 271	7. 329 − 150	8. 843 − 376
9. 500 − 167	10. 715 − 328	11. 906 − 29	12. 520 − 482

B. **Rewrite the problems below. Line up the numbers. Solve and check. The first one is started for you.**

13. 827 − 254 =   827 − 254

14. 817 − 499 =

15. 700 − 543 =

16. 684 − 486 =

C. **Solve the problems below with a calculator. Refer to page 242 in the Calculator Handbook for additional information.**

17. Jaime was balancing his checkbook. His starting balance was $650. He wrote checks for $125, $89, $57, and $21. What was Jaime's ending balance?

Calculator Answer: ____________________

Checked Answer: ____________________

18. Gayle works at a bookstore. She needs to order a total of 120 books from different publishers. She ordered 43 books on Monday, 54 books on Tuesday, and 12 books on Wednesday. How many more books does she need to order?

Calculator Answer: ____________________

Checked Answer: ____________________

Check your answers on page 259.

MEASUREMENT AND GEOMETRY

Perimeter

The distance around the edge of something is called the **perimeter.** Suppose you wanted to find how much wood you would need to frame a picture. First, you would measure the four sides of the picture. Then you would add the measurements. The sum of the sides is the perimeter of the picture.

Many things have the shape of a rectangle. In a rectangle, the opposite sides are equal in length.

Example Pete and Bev Ramos want to find out how many feet of fencing they will need to fence the yard to keep in their dog.

The Ramos's yard, shown in the drawing below, is a rectangle. The yard is 200 feet long and 120 feet wide. To find the perimeter of the yard, they need to know the measurement of each side.

What is the perimeter, in feet, of the Ramos's yard?
(1) 720
(2) 640
(3) 560
(4) 520
(5) 360

Answer (2) is the correct choice. Since opposite sides of a rectangle are equal, each length is 200 feet and each width is 120 feet. Pete and Bev need to find the total, or sum, of all the sides, so they will **add all four lengths** together:

200 + 200 + 120 + 120 = 640 feet, so the perimeter of the yard **is 640 feet.**

Example Pete and Bev want a 10-foot gate in the side of the yard, which is 120 feet wide. Which expression shows the amount of fencing needed to fence in the yard if they install the gate?
(1) 640 − 10
(2) 640 + 10
(3) 10 − 640
(4) 120 + 120 + 10 + 10
(5) 120 + 120 + 120 − 10

Answer (1) is correct. Pete and Bev need to deduct, or **subtract,** the width of the gate (10 feet) from the perimeter of the yard.

640 − 10 = 630 feet, so they need **630 feet of fencing** for the entire yard.

PRACTICE

Circle the best answer for each question.

1. Joe is planning a new exhibit at the zoo for a rectangular space that is 40 feet long and 30 feet wide. What is the perimeter, in feet, of the exhibit?
   (1) 10
   (2) 70
   (3) 100
   (4) 140
   (5) 160

2. Shaunte works at The Frame Place. She is framing a rectangular painting for a customer. The size of the painting is shown below.

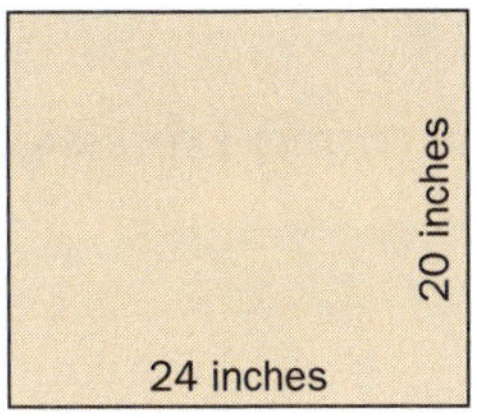

   Which expression shows how to find the amount of framing she needs?
   (1) 24 – 24
   (2) 24 – 20
   (3) 20 + 20 + 24
   (4) 20 + 24 + 24
   (5) 20 + 20 + 24 + 24

3. The landscaping service where Sandy works is putting a fence across a front yard. The yard is 80 feet wide. Sandy needs to leave a 3-foot opening for the sidewalk. Which is the correct expression to find how much fencing she needs?
   (1) 80 + 3
   (2) 80 − 3
   (3) 80 + 80 + 3
   (4) 80 + 80 − 3
   (5) 80 + 80 + 3 + 3

4. Lee is building a rectangular dog run that is 30 feet long and 6 feet wide. What is the perimeter, in feet, of the dog run?
   (1) 24
   (2) 36
   (3) 48
   (4) 66
   (5) 72

5. Jackie is planning to add a border to the walls of her bedroom. The room is 10 feet wide and 14 feet long. To find how much border she needs, Jackie has to figure out the perimeter of her room. Which is the correct expression to find the perimeter of the bedroom?
   (1) 10 + 10 + 14 + 14
   (2) 10 + 10 − 14
   (3) 10 + 14 + 14
   (4) 10 + 14
   (5) 10 − 14

6. Roberto wants to put a railing along the two shorter sides and one longer side of a rectangular deck. The deck is 10 feet wide and 18 feet long. Which is the correct expression to find how many feet of railing he will need?
   (1) 10 + 18
   (2) 10 + 10 + 18
   (3) 10 + 18 + 18
   (4) 10 + 10 + 18 + 18
   (5) 10 + 10 + 10 + 18

7. **Communicate** Explain why you can find the perimeter of a rectangular room by measuring only one length and one width. Write your explanation in one or two sentences. Make a drawing to illustrate your point.

Check your answers on pages 259–260.

PROBLEM SOLVING

Extra or Not Enough Information

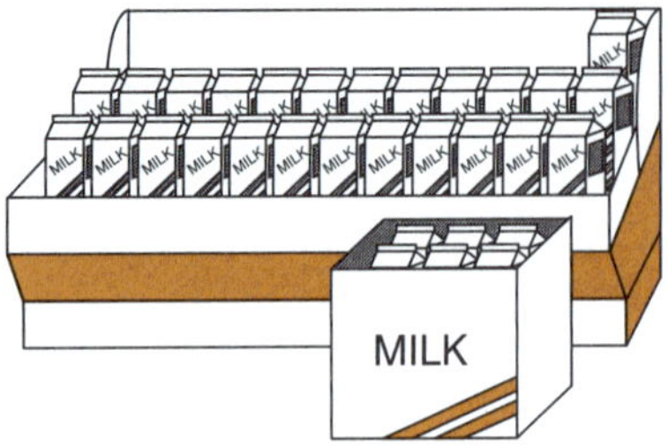

Example Lila is in charge of keeping the dairy cases filled at Jons Grocery. The milk case holds 60 half-gallon cartons. When the store opens, Lila takes out any cartons that are out-of-date. Then she adds new cartons to fill the case.

On Monday Lila pulled 14 outdated cartons at 8:00 A.M. and added 38 new cartons. At 10:00 A.M. she added 12 cartons. At noon she put out 19 new cartons, and at 2:00 P.M. she added 18 more. At the end of the day, Lila has to figure out how many new cartons she put in the case.

This problem contains a lot of **extra information.** You don't need all of the details to solve the problem.

Think about what the question asks. Which is the correct expression to find the number of new cartons?

(1) 60 + 14 + 38 + 12 + 19 + 18
(2) 60 + 38 + 12 + 19 + 18
(3) 38 + 12 + 19 + 18
(4) 14 + 38 + 12 + 19 + 18
(5) 14 + 38 + 12 + 19

Answer (3) is the correct choice. Lila needs to find the **total number of new cartons** she put in the case: **38 + 12 + 19 + 18 = 87 cartons,** so Lila put **87 new cartons** in the case. You need only the numbers of **new cartons** to solve the problem. You don't need to know how many cartons the case will hold, 60, or how many cartons were outdated, 14.

Sometimes **not enough information** is given to solve a problem. These types of questions are similar to some situations in real life. Sometimes, you just don't have all the information you need to solve a problem.

Example At 2:00 P.M. Lila refilled the milk case by adding 18 more cartons. If the milk case holds 60 cartons, how many new cartons will Lila need to add on Tuesday morning?

(1) 18
(2) 42
(3) 60
(4) 78
(5) Not enough information is given.

Answer **(5)** is correct. You don't know how much milk was sold by the end of the day on Monday, so you can't answer the question.

PRACTICE

Circle the best answer for each question.

1. Paula's take-home pay is $1,200 a month. She pays $450 a month for rent and $225 a month for food. Which is the correct expression to find how much more she pays for rent than for food a month?
   (1) $1,200 − $450
   (2) $1,200 − $225
   (3) $450 + $225
   (4) $450 − $225
   (5) Not enough information is given.

2. Rasheed delivers bottled water three times a week to an office building. He delivers 132 bottles on Wednesdays and 76 bottles on Fridays. What is the total number of bottles Rasheed delivers in a week?

   (1) 56
   (2) 208
   (3) 211
   (4) 624
   (5) Not enough information is given.

3. Danielle drove 45 miles in 1 hour. Her car gets 18 miles per gallon of gasoline. Which expression shows how many miles Danielle could drive in 5 hours at that rate?
   (1) 45×18
   (2) 45×5
   (3) $45 \div 5$
   (4) $45 + 18$
   (5) Not enough information is given.

4. Kunio is a driver for a recycling company. He collected 56 boxes of newspapers Monday morning and 17 more boxes that afternoon. On Tuesday morning Kunio collected 49 boxes of newspapers and 21 more that afternoon. Which is the correct expression to find the total number of boxes Kunio picked up on Monday?
   (1) 49 + 21
   (2) 56 + 49
   (3) 56 + 17
   (4) 17 + 21
   (5) Not enough information is given.

5. The shoe store where Eva works had a 3-day sale. By the end of the sale, there were 27 pairs of sandals and 38 pairs of tennis shoes left in stock. How many pairs of sandals were sold during the sale?
   (1) 27
   (2) 38
   (3) 65
   (4) 81
   (5) Not enough information is given.

6. Sandra is a records clerk. This week she handled requests for 67 marriage licenses, 73 birth certificates, and 81 death certificates. Estimate the total number of requests she handled this week.
   (1) 200
   (2) 220
   (3) 240
   (4) 260
   (5) 280

7. **Communicate** Think of a situation where you would purchase more than one of a certain item. Write a sentence or two describing what information you would need to find the total cost of this purchase.

Check your answers on page 260.

DATA ANALYSIS

Using a Bar Graph

A **bar graph** displays data in horizontal or vertical bars. Bar graphs are used to compare data when only one aspect of the data is numbers. In the example, the data being compared is the number of subscribers in the different sections of town. There are three parts to a bar graph: the **title,** the horizontal and vertical **axis lines,** and a **scale** that is used to show number values.

TIP
Read across from the top of the bar to the mark on the scale. The number by the mark is the value of the bar.

Example Tatsu is a clerk at the *Tilden Times,* a daily newspaper. Her boss has asked her to divide the town into three sections so that she could quickly answer questions such as the following:

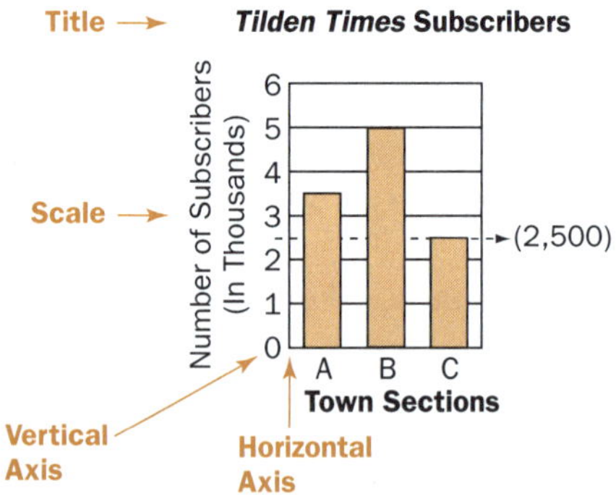

1. Which section of the town has the most subscribers?
2. About how many more subscribers live in the biggest section than in the smallest section?

To answer question **1,** find the tallest bar, which represents the greatest number of subscribers. The bar for Section B is the tallest. It reaches the 5 mark and represents **5,000 subscribers.**

For question **2,** find the biggest and smallest section. The biggest section is Section B with 5,000 subscribers. The bar for Section C is the smallest. It represents about 2,500 subscribers. Subtract to find the difference. **Section B, the largest section, has about 2,500 more subscribers than Section C, the smallest.**

Drawing a Bar Graph

Example Use the information in the table to begin building the bar graph in Part B on the next page.

Tall Buildings in the United States

Building	Height (Stories)
Amoco Building	80
Chrysler Building	77
Empire State Building	102
First Interstate World Center	75
Sears Tower	110

Step 1 Based on the information you will be graphing, write a title for the graph.

Step 2 Think about the information that is being listed on the vertical axis. Write a label for the vertical axis.

Step 3 Observe the pattern used for the scale on the vertical axis. Complete the scale following the same pattern.

PRACTICE

A. Use the bar graph at the right to answer the following questions.

1. Which type of car sold the least? 4-door
2. What was the most popular car sold? ____________
3. How many minivans were sold? ____________
4. What is the total of the two least popular types of cars sold? ____________
5. How many more Sports Vehicles were sold than 2-door cars? ____________
6. How many total sales did Country Motors have? ____________

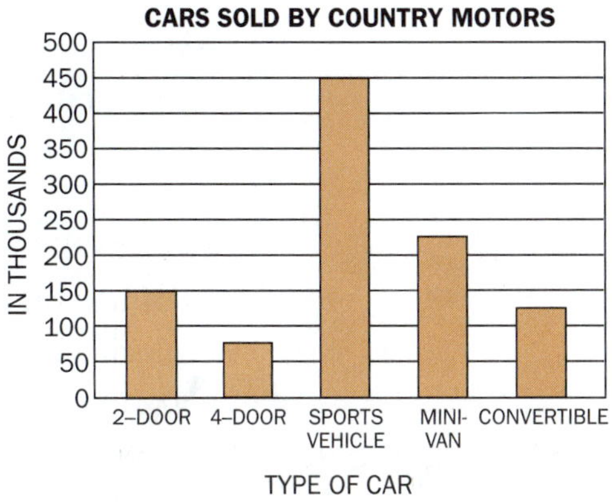

B. Complete the bar graph below using the information from the table in the second example on page 30.

7.

C. Use the bar graph you drew in Part B to answer these questions.

8. What label did you use for the vertical axis? ____________
9. What pattern did you use for the vertical axis scale? ____________
10. What title did you give your bar graph? ____________
11. Which building is the tallest? How many stories does it have? ____________
12. Which building has the fewest stories? ____________
13. How many more stories does the Sears Tower have than the Amoco Building? ____________

Check your answers on pages 260–261.

WORKPLACE MATH

Using Order Forms

Catalogs and order forms are commonly found in workplaces. Order forms organize important information so that an order can be successfully completed for both the person placing the order and the company filling the order. Order forms contain information such as addresses, item codes, quantities, and prices. If you do business over the Internet, you will be asked to complete an online order form.

Example Kelly is an office manager at a factory. The company orders all of its office equipment and cleaning supplies from mail-order companies. Kelly is placing an order for new computer accessories and supplies for one of the departments. Since she will need to get approval for the purchase, Kelly needs to figure out the total amount.

Item Number	Qty.	Description	Item Price	Total
414-527	1	Color computer monitor	299.99	299.99
306-291	5	3.5" diskettes	10.99	54.95
414-271	1	Color scanner	199.99	199.99
414-644	1	Laser printer	699.99	699.99
306-702	5	Laser paper	6.99	34.95
–				
			Merchandise Total	1,289.87
			Shipping/Handling	
			TOTAL	

Shipping and Handling

Merchandise Total	Add	Merchandise Total	Add
Up to $25	$4.95	$100.01 to $200	$10.95
$25.01 to $50	$6.95	$200.01 to $300	$12.95
$50.01 to $100	$8.95	$300.01 and over	$14.95

Based on the merchandise total, how much will Kelly's company need to pay in shipping and handling charges for this order?

(1) $6.95
(2) $8.95
(3) $10.95
(4) $12.95
(5) $14.95

Answer **(5)** is correct. Use the shipping and handling chart to determine the Merchandise Total range for this order. The merchandise total is $1,289.87, which is greater than $300.01. For any order over $300.01, the shipping and handling charge is **$14.95.**

PRACTICE

Circle the best answer for each question based on the order form on page 32.

1. Which item matches the item number 414-644?
   (1) monitor
   (2) diskettes
   (3) scanner
   (4) laser paper
   (5) laser printer

2. Mail-order companies often put out different versions of their catalog. The first three digits of an item number let a company know in which catalog the item was listed. Which three items listed on the order form are from the same catalog?
   (1) monitor, diskettes, scanner
   (2) diskettes, scanner, printer
   (3) monitor, scanner, printer
   (4) diskettes, scanner, printer
   (5) monitor, scanner, laser paper

3. The total for the laser paper is \$34.95. Which expression shows how Kelly found that amount?
   (1) 5 × \$10.99
   (2) 5 × \$6.99
   (3) 1 × \$34.95
   (4) 5 × \$4.95
   (5) 3 × \$10.99

4. Which of the following represents the best estimate of the cost of the diskettes and laser paper?
   (1) \$11 + \$7
   (2) \$55 + \$7
   (3) \$55 + \$200
   (4) \$55 + \$35
   (5) \$11 + \$35

5. When Kelly receives the order, she will need to make sure that everything she ordered was sent. The column labeled *Qty. (Quantity)* lists the number of items ordered. Which expression shows the number of items that should be delivered?
   (1) 1 + 5 + 5
   (2) 1 + 5 + 1 + 5
   (3) 1 + 5 + 1 + 1
   (4) 1 + 5 + 1 + 1 + 5
   (5) 1 + 5 + 1

6. Kelly compared prices with another mail-order company. She finds that she can purchase a similar laser printer for \$459.99. Which expression shows how much Kelly could save if she bought the less expensive printer?
   (1) \$699.99 ÷ \$459.99
   (2) \$699.99 + \$459.99
   (3) \$699.99 × \$459.99
   (4) \$459.99 ÷ \$699.99
   (5) \$699.99 − \$459.99

7. **Reason** Write a sentence to explain the following: If diskettes cost \$10.99 each, why is the total amount for the diskettes \$54.95?

Check your answers on page 261.

LESSON 2

Lesson 2 Review

Add or subtract.

1. $\begin{array}{r} 237 \\ +496 \\ \hline \end{array}$

2. $\begin{array}{r} 547 \\ -385 \\ \hline \end{array}$

3. $\begin{array}{r} \$295 \\ +\ 487 \\ \hline \end{array}$

4. $\begin{array}{r} 600 \\ -321 \\ \hline \end{array}$

5. $\begin{array}{r} \$607 \\ -\ 295 \\ \hline \end{array}$

6. $\begin{array}{r} 503 \\ 58 \\ +367 \\ \hline \end{array}$

7. $\begin{array}{r} \$227 \\ 317 \\ +\ 145 \\ \hline \end{array}$

8. $\begin{array}{r} \$993 \\ -\ 294 \\ \hline \end{array}$

9. $\begin{array}{r} 762 \\ -684 \\ \hline \end{array}$

10. $\begin{array}{r} 836 \\ +414 \\ \hline \end{array}$

11. $3,342 + $467 =

12. 52,304 − 7,099 =

Solve each word problem. Show your work.

13. Ed works in the mailroom of a large company. This morning he needs to make deliveries to 238 people. He makes deliveries to 149 people before his coffee break. How many people does he have to make deliveries to after his coffee break?

14. On Saturday the Perez family drove to Tall Pines Park where they will camp. They drove 243 miles before lunch and 179 miles after lunch. How many miles did the Perez family drive on Saturday?

15. Gloria delivers newspapers. She has 153 customers who receive the paper every day, 26 who get it on weekends only, and 42 who receive it on Sundays only. How many total customers does Gloria have?

16. Carolyn works in the theater box office. She sold 83 tickets for the 5:00 show and 127 tickets for the 7:30 show. How many more tickets did she sell for the 7:30 show?

Circle the best answer for each question.

17. Sue is putting a wallpaper border around her rectangular living room. The living room is 20 feet long and 14 feet wide. Which is the correct expression to find the amount of border she will need?
(1) 20 + 14
(2) 20 − 14
(3) 20 + 20
(4) 20 + 20 + 20 + 14
(5) 20 + 20 + 14 + 14

18. Last month Mark's utility bills included $28 for electricity, $53 for the telephone, and $11 for water. What is the total of Mark's utility bills last month?
(1) $39
(2) $70
(3) $81
(4) $92
(5) $102

19. Bill weighs 208 pounds. His doctor said he should weigh only 185 pounds. Which is the correct expression to find how much weight Bill should lose?
(1) 185 − 208
(2) 208 − 185
(3) 208 + 185
(4) 208 + 208 − 185
(5) 208 − 185 + 185

20. Joshua is taking a course on office procedures. Tuition is $348, and the cost of books and supplies is $104. What is the total cost for Joshua to take this course?
(1) $244
(2) $342
(3) $418
(4) $452
(5) $888

Questions 21 through 23 refer to the following graph.

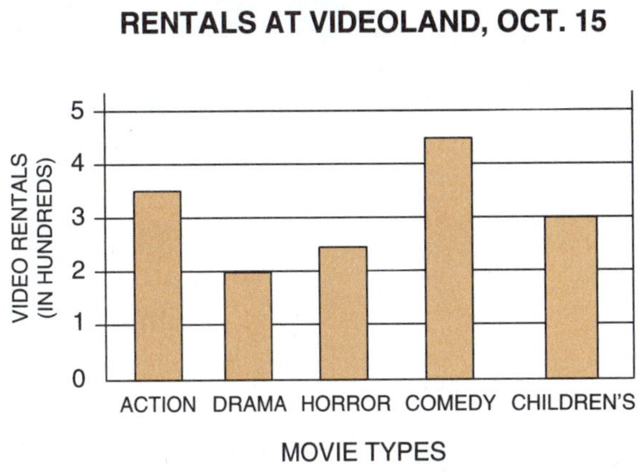

21. Which type of movie had the greatest number of rentals?
(1) action
(2) drama
(3) comedy
(4) horror
(5) children's

22. What is the approximate difference between the movie type that had the most rentals and the movie type that had the fewest rentals?
(1) 100
(2) 150
(3) 200
(4) 250
(5) 650

23. How many total movie rentals were there on October 15?
(1) 1,600
(2) 1,550
(3) 1,500
(4) 1,150
(5) 1,000

Check your answers on pages 261–262.

LESSON 3

Multiplying and Dividing Whole Numbers

Multiplying Whole Numbers

Multiplying, like adding, helps you find a total. When you need to add the same number many times, use **multiplication**. The times sign, (×), tells you to multiply. Multiplication is also shown by using parentheses: (3)(5) means 3 × 5.

> **TIP**
> To check multiplication, you can multiply the numbers in the reverse order.
> 15
> ×62

Example Multiply 62 by 15.

Step 1 Multiply each digit in the top number by each digit in the bottom number. Work from right to left. Be sure to line up the digits according to place value. If a column is 10 or more, you need to regroup. Since the 1 in the number 15 is in the tens place, you are really multiplying by 10. So, put a 0 in the ones place.

```
  1
  62        62
 ×15      × 15
 310       310
          +620
```

Step 2 Add the results.

```
   62
 × 15
  310
 +620
  930
```

Example Multiply 169 by 205.

Step 1 Multiply as normal. Multiply each digit in 169 by the 5 in 205.

```
  3 4
  169
 ×205
  845
```

Step 2 Multiply by 0. Since 0 times any number is 0, write 0 directly below the 0 in 205. You do not need to write a row of zeros.

```
  169
 ×205
  845
   0
```

Step 3 Multiply by 2. Begin directly under the 2, to the left of the 0 you just wrote. Notice how the answer is the same as if you had written the row of zeros.

```
   1 1
   169        169
  ×205       ×205
   845        845
 3380         00
 34645     338
           34645
```

Example Keesha is a salesperson at Frank's Furniture. Yesterday she sold 3 dining room sets at $1,129 each. Use a calculator to find the total sales.

[AC] 1129 [×] 3 [=] 3387

For more practice multiplying with the calculator, see Calculator Handbook, page 243.

PRACTICE

A. **Multiply. Show your work. The first one is done for you.**

1. $\begin{array}{r} ^{2} \\ 172 \\ \times\ \ 4 \\ \hline 688 \end{array}$
2. $\begin{array}{r} 234 \\ \times\ \ 2 \\ \hline \end{array}$
3. $\begin{array}{r} 384 \\ \times\ \ 6 \\ \hline \end{array}$
4. $\begin{array}{r} 73 \\ \times 18 \\ \hline \end{array}$
5. $\begin{array}{r} 83 \\ \times 24 \\ \hline \end{array}$
6. $\begin{array}{r} 259 \\ \times\ 47 \\ \hline \end{array}$
7. $\begin{array}{r} \$475 \\ \times\ \ \ 9 \\ \hline \end{array}$
8. $\begin{array}{r} 64 \\ \times 300 \\ \hline \end{array}$
9. $\begin{array}{r} 837 \\ \times 402 \\ \hline \end{array}$
10. $\begin{array}{r} 126 \\ \times 280 \\ \hline \end{array}$

B. **Rewrite the problems below. Line up the numbers. Solve and check. The first one is started for you.**

11. $389 × 7 = $\begin{array}{r} \$389 \\ \times\ \ \ 7 \\ \hline \end{array}$
12. 48 × 40 =
13. 527 × 63 =
14. 936 × 508 =

C. **Solve the problems below with a calculator. Refer to page 248 in the Calculator Handbook for additional information.**

15. Javier works in a warehouse. He knows that 1 carton can hold 55 of Product 1 or 40 of Product 2. Javier counted the number of cartons holding each product.

Product 1	198 cartons
Product 2	57 cartons

What was the total number of both products in the packed cartons?

Calculator Answer: ______________

Checked Answer: ______________

16. Natina makes wreaths to sell at craft fairs. She uses 8 yards of ribbon for each wreath. If she makes 45 wreaths for one craft fair and 48 wreaths for another, how many yards of ribbon does Natina need? ______

17. Greenhall Corp. held an awards banquet for 125 people at Westside Cafe. If dinner for each person costs $12, how much did the banquet cost? ______

18. Leal Furniture Manufacturers can make 280 chairs each week. If the factory operates 51 weeks this year, how many chairs can be made? ______

19. Lobodo Publishers shipped 84 boxes of a new auto repair manual today. Each box holds 24 books. How many books were shipped today? ______

20. Robert drives a gasoline tank truck. Today he delivered gas to 14 different gas stations. The tank at each station holds 350 gallons. How many gallons of gasoline did Robert deliver? ______

✔ Check your work by entering the numbers again.

Check your answers on page 263.

Dividing Whole Numbers

Use **division** to figure out how many times one number goes into or divides another number. Division problems can be written using the division sign, (÷). To solve the problem, you can write it using a division bracket, $\overline{)\ \ }$.

```
35 ÷ 7 = 5
     5
   7)35
    -35
      0
```

TIP
To check division, multiply your answer by the number you divided by. If you divided correctly, the result should be the same as the number you divided.

Example Divide 2,352 by 12.

Step 1 How many times does 12 **divide into** 23? 12 × 1 = 12 and 12 × 2 = 24. Choose the answer that is closer to 23 without going over 23. Write 1 in the answer space over 23. **Multiply** 12 × 1 = 12. Write 12 under 23 and **subtract. Bring down** the next digit: 5.

```
     196
 12)2,352
   -12
     115
    -108
       72
      -72
        0
```

Step 2 How many times does 12 divide into 115? 12 × 9 = 108. Write 9 in the answer space over the 5. Multiply and subtract. Bring down the next digit: 2.

Step 3 How many times does 12 divide into 72? 12 × 6 = 72. Write 6 in the answer space over the 2. Multiply and subtract.

Step 4 Check your answer. Multiply 196 by 12. The result is the number you divided, 2,352.

```
   196
 ×  12
   392
 +196
 2,352
```

Dividing Money

Divide as you would with whole numbers. Then, put the decimal point and dollar sign in your answer above the decimal and dollar sign in the problem.

```
Divide: 3)$12.60

Solve:      $4.20
          3)$12.60
            -12
               6
             -60
              00

Check:  $4.20
        ×   3
       $12.60
```

PRACTICE

Divide and check. Show your work. The first one is done for you.

```
1.     54          54
    8)432        ×  8
     -40          432
       32
      -32
        0

2. 7)$1,505

3. 6)5,478

4. 36)5,940
```

Remainders

TIP
The letter *r* means remainder.
59 r3 means 59 with a remainder of 3.

Example Divide 475 by 8.

Step 1 How many times does 8 divide into 47? 5 times. Write 5 in the answer space above 47. Multiply, subtract, and bring down the next digit.

Step 2 How many times does 8 divide into 75? 9 times. Write 9 in the answer space above 5. Multiply and subtract.

```
   59 r3
8)475
 -40
   75
  -72
    3
```

When there are no more numbers to bring down and you have an amount left over, you have a **remainder.**

Step 3 Write the remainder in the answer space.

Step 4 Check your answer. Multiply 59 by 8,then add the remainder 3.

```
  7
  59
×  8
 472
+  3
 475
```

To divide numbers on a calculator, enter each number. Use the **division key,** ÷, between numbers and the **equals key** to get the answer.

Example Divide $2,596 by 15. Be sure to enter the number being divided first.

AC 2596 ÷ 15 = **173.0667**

Notice in your answer there are digits to the right of the decimal point. The digits 0667 are your remainder. When doing division with money, often you are asked to round your answer to the nearest cent or the nearest dollar. The answer rounded to the nearest cent is $173.07 and to the nearest dollar is $173.

PRACTICE

Divide and check. Show your work. For money amounts, round your answer to the nearest cent. The first one is done for you.

1.
```
   49 r1
6)295        49
 -24       ×  6
   55       294
  -54      +  1
    1       295
```

2. 24)2,099

3. 9)843

4. 5)1,234

5. 18)1,739

6. 62)$2,170

7. 8)375

8. 27)$1,269

Check your answers on page 264.

Zeros in the Answer

Example Divide 490 by 7.

Step 1 How many times does 7 divide into 49? 7 times. Write 7 in the answer space over 49. Multiply, subtract, and bring down the next digit.

```
   70
7)490
 -49
   00
```

TIP
0 ÷ any number = 0

Step 2 How many times does 7 divide into 00 (0)? 0 ÷ 7 = 0, or 0 times. Write 0 in the answer space over 0.

Step 3 Check your answer.

```
  70
×  7
 490
```

Example Divide 4,872 by 24.

Step 1 How many times does 24 divide into 48? 2 times. Write 2 in the answer space over 48. Multiply, subtract, and bring down the next digit.

```
     203
24)4,872
  -4 8
     72
    -72
      0
```

Step 2 How many times does 24 divide into 7? 0 times. The number 24 cannot divide into 7, so write 0 in the answer space over 7. Bring down the next digit.

Step 3 How many times does 24 divide into 72? 3 times. Write 3 in the answer space over 2. Multiply and subtract.

Step 4 Check your answer.

```
  203
×  24
  812
+4 06
4,872
```

PRACTICE

A. **Divide. Watch for zeros in the answers. Check your answers. The first one is done for you.**

1.
```
   50
9)450     50
 -45    ×  9
    0    450
```
2. $26\overline{)2,626}$
3. $8\overline{)\$320}$
4. $6\overline{)1,824}$

B. Divide. Watch for zeros. Check your answers. The first one is done for you.

5. 5)280 = 56

```
    56      Check
5)280         56
 −25        ×  5
   30        280
  −30
    0
```

6. 27)5,103

7. 19)5,852

8. 8)$2,208

9. 6)4,800

10. 13)6,695

11. 4)$2,320

12. 10)4,550

C. Divide. Watch for remainders. Check your answers. The first one is done for you.

13. 6)465 = 77 r3

```
    77 r3   Check
6)465          77
 −42         ×  6
   45         462
  −42        +  3
    3         465
```

14. 7)3,429

15. 9)5,688

16. 21)1,953

17. 8)1,252

18. 46)4,955

19. 4)$3,668

20. 5)1,734

D. Solve the problems below with a calculator. Refer to page 244 in the Calculator Handbook for additional information.

21. Brian works at Frank's Foods. He is packing cans of corn in boxes. The cans fit 24 to a box. How many boxes does Brian need for 1,632 cans?

Calculator Answer: ____________

Checked Answer: ____________

22. Melanie earns $17,460 a year. How much does she make a month? (Hint: Think how many months are in a year.)

____

23. Adrienne can enter 80 words a minute on her computer. How long will it take her to enter an 18,480-word report?

____

24. Sara works for a music library. One week she listened to 840 minutes of music. How many hours of music did she hear? (Hint: 60 min = 1 hr)

____

Check your answers on pages 264–266.

ALGEBRA TOPIC

Writing and Solving Equations

An **equation** is two mathematical expressions, or statements, with an equals sign (=) in between. Some equations are all numerals, such as $6 + 4 = 10$. But some equations contain unknown numbers, or **variables,** represented by letters such as $y + 4 = 10$. (Any letter can be used.)

Writing Equations

Example Write an equation for the following statement: A number increased by nine equals eighteen.

Step 1 Identify the key words that tell you what operation to use. The words *increased by* tell you to *add*.

Step 2 Identify the unknown amount. *A number* is the unknown.

Step 3 Write the equation.

Unknown	add	9	=	18
y	**+**	**9**	**=**	**18**

Example Write an equation for the following question: How many dozen equal sixty cookies? The product of twelve and a number equals sixty.

Step 1 Identify the operation. The word *product* means to multiply.

Step 2 Identify the unknown amount. *A number* is the unknown.

Step 3 Write the equation. Notice that multiplication can be shown without using multiplication signs.

Multiply	12 times the unknown	=	60
(12)	**(y)**	**=**	**60**

Solving Equations

When solving equations, you want to get the unknown number (variable) alone on one side of the equation. To do this, perform the **inverse** (opposite) operations on both sides of the equation.

Example Solve $z + 18 = 34$.

Step 1 The inverse of addition is subtraction. Subtract 18 from both sides.

$$\begin{aligned} z + 18 &= 34 \\ -18 &= -18 \\ \hline z + 0 &= 16 \\ z &= \mathbf{16} \end{aligned}$$

(You do not write the zero in your final answer.)

Step 2 Check your answer using substitution, or putting in the value for z into the equation: $16 + 18 = 34$

Example Solve $w \div 5 = 16$.

Step 1 The inverse of division is multiplication. Since w is being divided by 5, multiply both sides of the equation by 5 to solve for w.

$$\begin{aligned} w \div 5 &= 16 \\ \times 5 &= \times 5 \\ \hline w &= \mathbf{80} \end{aligned}$$

Step 2 Check the answer using substitution. $80 \div 5 = 16$

TIP

- The inverse of addition is subtraction.
- The inverse of subtraction is addition.
- The inverse of multiplication is division.
- The inverse of division is multiplication.

PRACTICE

A. **Write an equation for each statement. Do not solve. The first one is done for you.**

1. A number divided by eight equals four. $n \div 8 = 4$
2. Sixteen more than a number equals twenty. ______
3. Five times a number equals fifty. ______
4. Twenty-seven decreased by a number equals twelve. ______
5. A number divided by six equals five. ______

B. **Solve and check. The first one is started for you.**

6. $a + 12 = 32$ $\quad a + 12 = 32$ $\quad -12 = -12$
7. $d \div 7 = 8$
8. $n - 18 = 56$
9. $d - 47 = 24$
10. $n \div 4 = 15$
11. $9n = 108$
12. $3z = 54$
13. $c + 13 = 13$
14. $y + 19 = 45$
15. $10c = 100$

C. **Write and solve the equation for each problem.**

16. Anne earns d dollars an hour. She worked 45 hours last week and received $540. How much does she earn an hour?

17. Monica weighs p pounds. After she loses 15 pounds, Monica will weigh 125 pounds. How much does Monica weigh?
18. Van is making a fruit salad for his class party. He bought p pounds of melon and 6 pounds of berries. He bought a total of 18 pounds of fruit for the salad. How many pounds of melon did he buy?
19. Luis was taking inventory at the video rental store where he works. He noted that the total number of videos divided by 26 people who rented videos equals an average of three videos per person. What was the total number of rented videos?
20. At work Andrea received a $25 bonus added to her check. The total amount of the check was $560. How much was her check before the bonus?

Check your answers on page 266.

PROBLEM SOLVING

Solving Word Problems

Calculating the answer is only part of the problem-solving process. First you have to figure out how to approach and solve the problem.

Choosing the Correct Operation

Look for the meaning behind the words in a problem. Ask yourself, "How should I approach the problem?" The chart below will help you choose the correct operation.

You should . . .	When you need to . . .
Add	Find a **total,** find **how many in all,** put amounts **together,** or find the **sum.**
Subtract	Find a **difference,** find **how many more, how many less,** or **how much is left.**
Multiply	Add the same number more than once to find a **total** or to find a **product.**
Divide	Break or **separate** an amount into equal parts; find **how many** or **how much** for **each.**

TIP
Use estimation to see if your answer is reasonable.

Example Yuri is a cashier at the Empire Theater. He counted ticket stubs for one show and found out 56 adults and 39 children or senior citizens were at the show. What was the total number of ticket sales for the one show?

There are five steps to solving word problems:

Step 1 Read the problem carefully. Make sure you understand what you are being asked to find. **What was the total number of tickets sold?**

Step 2 Think about what you are trying to find.You are finding **the total number of ticket sales for one show.**

Step 3 Ask yourself how the facts in the problem can help you find the answer. Determine which numbers you need to use. **There are 56 adults and 39 children or senior citizens at the show.**

Step 4 Look for clue words that tell you which operation to use.
Total: use addition. **56 + 39 = 95**

Step 5 Make sure your answer is reasonable. Estimate: 60 + 40 = 100. The total number of tickets is about 100. The answer, 95, is reasonable.

Using the Calculator to Solve Word Problems

Example Spring has a population of 1,437,306. Denton has a population of 731,523. How many more people live in Spring than in Denton?

The clue words are *how many more*. Therefore, you subtract.

[AC] 1437306 [−] 731523 [=] 705783.

PRACTICE

A. Underline clue words in each problem. Write which operation to use. Do not solve the problem. The first one is done for you.

1. Marcus drove 156 miles the first day, 202 miles the second day, and 175 miles the third day. How many miles in all did Marcus drive?
   Addition

2. In a 40-hour work week, Carol makes $542. How much does she earn for each hour she works?

3. Kevin is in a jazz band. Sixty-four people paid $4 to hear him play. What is the total amount of money paid to hear the band?

4. Marilyn is an office manager. She has a budget of $1,000 to buy a new computer for the office. The computer sells for $1,399. How much more money does Marilyn need in the budget?

B. Write the operation needed and solve. Use estimation to check the reasonableness of your answers.

5. Midori is making a fruit basket for a party. She buys 12 pounds of melon and 6 pounds of berries. How many pounds of fruit does she buy?

6. Tastee Burger sold 198 hamburgers on Friday. Of the hamburgers sold, 72 were sold after 7:00 P.M. How many hamburgers sold during the rest of the day?

C. Use the five-step problem solving approach to solve the word problems below. You may use a calculator.

7. Marcy is a cashier at Best Video. Each video costs $17. If her customer has $50 to spend, how many videos can the customer buy?

8. Terrence was reviewing his utility bills. He had the following bills: Electricity $53, water $16, telephone $27, trash pickup $5, gas $11. What is the total amount Terrence owes for utilities?

9. Lelia put on layaway a coat that cost $371. She put $25 down to hold the coat. How much money does Lelia still owe for the coat?

10. Carmen is a records clerk. This week she handled requests for 67 marriage licenses, 73 birth certificates, and 81 death certificates. What was the total number of requests she handled this week?

Check your answers on page 267.

DATA ANALYSIS

Finding Averages

Numbers that people group and study to make decisions are called **data.** Finding the **average** is one way to figure out the usual value of a set of data.

> **TIP**
> The average is also called the **mean.**

Example Stan works at a department store that is open 6 days a week. Stan's boss wants him to find the average number of customers who shop between 4 P.M. and 6 P.M. What would be the best way for Stan to get this information?

Stan decides to count the people who go through the checkout lines within the 2-hour period every day for 1 week. Stan collects the following data:

Monday	106	Thursday	85
Tuesday	94	Friday	100
Wednesday	75	Saturday	104

Step 1 To find an average, first find the total of the numbers in the set of data. 106 + 94 + 75 + 85 + 100 + 104 = 564

Step 2 Next, divide the total by how many numbers are in the data set. There are six numbers in the data set. Divide 564 by 6. The usual (or average) number of shoppers between 4 P.M. and 6 P.M. is 94.

```
   94
6)564
 −54
   24
  −24
    0
```

Finding the Median

The **median** is another kind of typical value. The *median* is the middle number of an ordered set of data.

Example Women's Shoes Unlimited had nine types of shoes on sale in these sizes: 5, 7, 8, 10, 6, 9, 4, 9, 6. What is the median for this set of data?

Step 1 Arrange the data in numerical order, from least to greatest.
4 5 6 6 <u>7</u> 8 9 9 10

Step 2 Since there are nine values in all, the middle value would be the fifth number: four numbers to the left and four numbers to the right. The median, or middle number, is 7.

To find an average on a calculator, enter each number to find a total. Use the [÷] key to divide the total by how many numbers are in the set of data.

Example Find the average for this set of data: 2, 4, 6, 8, 10, 12

[AC] 2 [+] 4 [+] 6 [+] 8 [+] 10 [+] 12 [=] 42

42 [÷] 6 [=] 7

PRACTICE

A. **Find the mean and median for each set of numbers. The first one is done for you.**

1. Hours worked Monday by a group of employees: 8, 8, 3, 9, 7, 6, 8

   Mean:
   8 + 8 + 3 + 9 + 7 + 6 + 8 = 49
   49 ÷ 7 = 7

   Median: 3, 6, 7, <u>8</u>, 8, 8, 9

   Mean ___7___ Median ___8___

2. Number of hot dogs sold for three days: 410, 350, 368

   Mean ______ Median ______

3. Bowling scores for five games: 261, 254, 105, 280, 165

   Mean ______ Median ______

4. Calories eaten each day for a week: 2,450; 2,100; 1,970; 2,430; 2,840; 1,800; 2,860

   Mean ______ Median ______

B. **When there is an even amount of numbers in the set of data, there will be two middle numbers. Find the average of the two middle numbers to find the median. The first one is done for you.**

5. 7, 8, 10, 10, 8, 7, 12, 15 Median ___9___
   7, 7, 8, <u>8</u>, <u>10</u>, 10, 12, 15

   8 + 10 = 18

   $$\begin{array}{r} 9 \\ 2\overline{)18} \\ -18 \\ \hline 0 \end{array}$$

6. 146, 162, 234, 212, 184, 198, 523, 206 Median ______

7. 81, 46, 73, 32, 54, 81, 26, 41 Median ______

8. 1,903; 7,418; 5,651; 4,342; 2,950; 3,420 Median ______

C. **Solve each problem below with a calculator.**

9. On Saturday Angela kept a record of the number of customers she had each hour during the 9 hours her fruit stand was open: 25, 42, 45, 37, 102, 86, 46, 38, 29. What is the average number of customers per hour? ______

10. John is a manager at a movie theater. He recorded the attendance for one Saturday evening. What is the average attendance for that evening? ______

Movie Attendance on Saturday Evening, Nov. 1

Cinema 1	502	Cinema 2	147
Cinema 3	425	Cinema 4	454
Cinema 5	518	Cinema 6	504

Check your answers on pages 267–268.

Area of a Rectangle

Example Donna is putting wood tiling in a rectangular office. The office is 16 feet wide and 20 feet long. She wants to use tiles that measure 1 foot on each side. To find out how many tiles she needs, Donna first needs to find the area of the office.

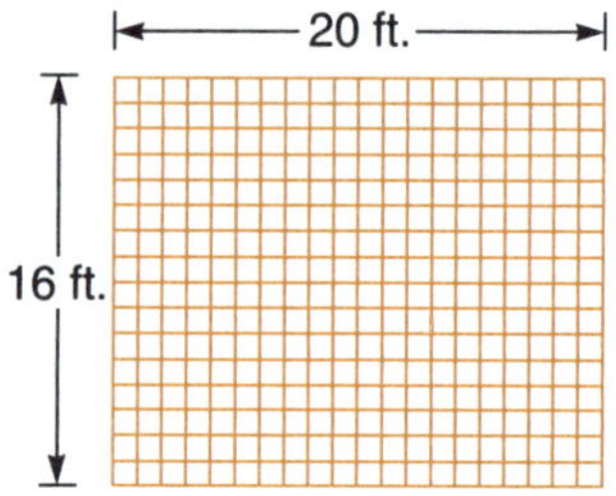

Look at the diagram of the office. There are 16 rows with 20 squares in a row. You can find the area by counting all the squares, or you can add 20 sixteen times. However, the easiest way to find the area is to multiply.

Which is the correct expression to find the area of the office?

(1) $16 + 16 + 20 + 20$
(2) $20 + 20 + 20 + 20$
(3) 20×20
(4) 16×20
(5) 16×16

Answer **(4)** is the correct choice. Donna uses the area formula to find the area of the office. She knows the length, 20 feet, and the width, 16 feet, of the office. She multiplies the length by the width to find the area. **16 × 20 = 320 square feet,** so the area of the office is **320 square feet.** Because the tiles measure 1 square foot, Donna needs **320 tiles** to cover the floor.

Area is the amount of surface something takes up.

The formula for the area of a rectangle or a square is written:

$A = (l)(w)$, which means *Area* = (*length*)(*width*).

Area is measured in square units.

A square foot measures 1 foot on each side.

Example Jason works at Picture Frames. The picture he is framing measures 12 inches by 16 inches. He needs to put glass down on the picture before he frames it.

Which is the correct expression to find the surface area of the picture?

(1) $12 + 16$
(2) 12×16
(3) $12 + 12 + 16 + 16$
(4) 12×12
(5) 16×16

Answer **(2)** is the correct choice. Jason needs to find the surface area of the picture. **12 × 16 = 192 square inches,** so Jason needs **192 square inches of glass.**

PRACTICE

Circle the best answer for each question.

1. Yakov is planning a new exhibit at the zoo for a rectangular space that is 40 feet long and 30 feet wide. Which is the correct expression to find the area, in square feet, of the exhibit?
   (1) 40×30
   (2) $40 + 30$
   (3) $40 + 40 + 30$
   (4) $40 \times 40 \times 30 \times 30$
   (5) $40 + 40 + 30 + 30$

2. The Juarez family is installing carpet in their family room. The size of the family room is shown below. Which expression shows how to find how many square feet of carpeting they need?

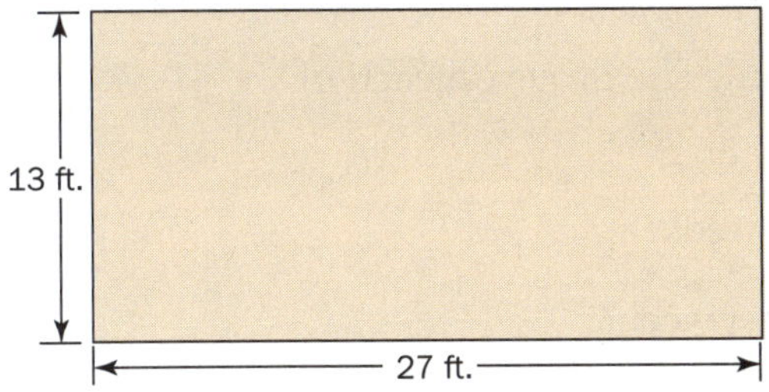

   (1) $13 \times 13 \times 27$
   (2) $27 + 27 + 13 + 13$
   (3) $27 \times 13 \div 2$
   (4) 13×27
   (5) $27 \div 3$

3. Mrs. Juarez is making four curtain panels. She needs 6 feet of fabric for each curtain panel. There are 3 feet in a yard. Which is the correct expression to find the number of yards of fabric Mrs. Juarez needs?
   (1) $6 \times 4 \times 3$
   (2) $6 + 4 + 3$
   (3) $6 \times 4 \div 3$
   (4) $6 + 4$
   (5) $6 \div 3$

4. Akeo is putting new vinyl flooring in the basement of her house. The basement is 42 feet long and 32 feet wide. Which expression shows how many square feet of vinyl flooring Akeo will need?
   (1) $42 + 32$
   (2) $42 + 32 + 42 + 32$
   (3) 42×32
   (4) $42 - 32$
   (5) $42 \div 32$

5. Nancy is going to put lime on her rectangular garden to fertilize the soil. When deciding how much lime she needs, Nancy needs to know the size of the area to be covered. She knows the length of the garden is 12 feet. What is the area, in square feet, of her garden?
   (1) 12
   (2) 24
   (3) 48
   (4) 144
   (5) Not enough information is given.

6. Maria wants to cover the surface of an old coffee table with three-inch square tiles. She measured the tabletop and found that it is 20 inches wide and 40 inches long. What is the area, in square inches, of the tabletop?
   (1) 60
   (2) 63
   (3) 120
   (4) 800
   (5) Not enough information is given.

7. **Reason** Given the formula for the area of a rectangle is $A = (l)(w)$, develop the formula for finding the area of a square.
   (Hint: Let s = one side of the square.)

Check your answers on page 268.

LESSON 3

Lesson 3 Review

Multiply or divide. Show your work.

1. $\begin{array}{r} 456 \\ \times \quad 3 \\ \hline \end{array}$

2. $\begin{array}{r} 83 \\ \times 49 \\ \hline \end{array}$

3. $\begin{array}{r} \$297 \\ \times \quad\; 6 \\ \hline \end{array}$

4. $\begin{array}{r} 178 \\ \times 702 \\ \hline \end{array}$

5. $\begin{array}{r} 653 \\ \times 305 \\ \hline \end{array}$

6. $4\overline{)3,294}$

7. $7\overline{)\$6,440}$

8. $18\overline{)5,794}$

9. $43\overline{)8,657}$

10. $27\overline{)4,131}$

11. $39 \times 800 =$

12. $\$842 \times 5 =$

13. $\$1,640 \div 8 =$

14. $7,463 \div 52 =$

Questions 15 and 16 refer to the following information.

Freddie drives a van for a delivery service. The number of miles he drove on each of the 5 days he worked last week are 62, 28, 57, 65, and 63.

15. What is the mean number of miles Freddie drove last week?

16. What is the median number of miles Freddie drove last week?

Solve the following:

17. Solve for y in this equation: $y - 13 = 29$.

18. The Auto Lube Shop charges $22 for an oil change. This week the shop collected $330 for oil changes. How many customers had an oil change?

Use a calculator to solve the following problems.

19. 723 [×] 25 [=] ______

20. 40 [×] 300 [=] ______

21. 5,520 [÷] 12 [=] ______

22. 3,145 [÷] 5 [=] ______

Circle the best answer for each question.

23. A professional basketball court is 94 feet long and 50 feet wide. Which is the correct expression for finding the area of the court?
(1) 94 + 50
(2) 94 × 50
(3) 94 + 50 + 50
(4) 94 + 94 + 50 + 50
(5) 94 × 94 × 50 × 50

24. Lena earns $8 an hour making deliveries for a pharmacy. She worked 7 hours on Friday, 8 hours on Saturday, and 10 hours on Sunday. How much did Lena earn on Friday?
(1) $56
(2) $64
(3) $80
(4) $200
(5) Not enough information is given.

25. Felicia bought 6 quarts of oil for her car. The total cost was $12. Which is the correct expression for finding the price of one quart of oil?
(1) 6 × $12
(2) 6 ÷ $12
(3) $12 ÷ 6
(4) $12 + 6
(5) $12 × 6

26. Kelvin dug a rectangular garden in his backyard. The garden is 40 feet long. What is the area of the garden in square feet?
(1) 15
(2) 65
(3) 130
(4) 1,000
(5) Not enough information is given.

27. A shoe store has children's shoes on sale for $5 a pair. Martin bought *p* pairs of shoes for his children. He paid $20. Which is the correct expression to find how many pairs of shoes Martin bought?
(1) $20 ÷ 5 = *p*
(2) $5 + *p* = $20
(3) $5 − *p* = $20
(4) $20 − *p* = $5
(5) $20 − $5 = *p*

Use your calculator for question 28 and 29.

28. Perry is a salesperson. The odometer on his car reads 126,000. If his car is 3 years old, about how many miles is Perry driving a year?
(1) 2,000
(2) 4,200
(3) 12,600
(4) 32,000
(5) 42,000

29. The Santos family is taking a driving vacation. Below is a list of cities and the road mileage. How many total miles will they drive each way between Cincinnati and Los Angeles?

Location	Miles
Cincinnati, OH to Chicago, IL	244
Chicago, IL to Denver, CO	996
Denver, CO to Los Angeles, CA	1,059

(1) 1,303
(2) 2,055
(3) 2,289
(4) 2,299
(5) 2,300

Check your answers on pages 268–269.

LESSON 4

Squares, Cubes, and Square Roots

exponent

7^2 means 7 × 7

base number

TIP

To find the value of a number with an exponent, write the expression as a multiplication problem and multiply.

$4^3 = 4 \times 4 \times 4$
$= 64$

Finding Squares and Cubes

An **exponent** tells you how many times a number is to be multiplied by itself. In the expression 7 × 7, you multiply 7 times itself. Another way to write this expression is 7^2. The 7 is the **base** number of the expression. The 2 is the **exponent.**

The expression 7^2 can be read "seven to the second power" or "seven squared." The square of 7 is 49 because 7 × 7 = 49. The **value** of the expression 7^2 is 49.

A number is **cubed** when the number is multiplied by itself three times. Using an exponent, "4 cubed" is written 4^3. The expression 4^3 means 4 × 4 × 4.

Example Find the value of the expression 4^3.

Step 1 Multiply 4 × 4. $4 \times 4 = 16$

Step 2 Multiply the product from Step 1 by 4. $16 \times 4 = 64$

The value of the expression 4^3 **is 64.**

Finding Square Roots

To find the square of a number, multiply that number by itself. To find the **square root** of a number, think: "What number squared equals this number?" The symbol for square root is $\sqrt{}$.

Example What is the square root of 36?

Step 1 Think of what number squared equals 36. $6^2 = 6 \times 6 = 36$

Step 2 Write in square root notation. $\sqrt{36} = 6$

Some calculators assign two functions to the same key. Press SHIFT to use the second function of a key. To find the square root of a number, enter the number, press (SHIFT), and press (x^2). Note the $\sqrt{}$ symbol above the (x^2) key.

For more practice finding the square root with a calculator, see Calculator Handbook page 321.

Example Find the value of $\sqrt{3,136}$ using a calculator.

(AC) 3136 (SHIFT) (x^2) 56

PRACTICE

A. Write each expression as a multiplication problem and multiply. The first one is done for you.

1. 6^3 $6 \times 6 \times 6$
   $= 36 \times 6$
   $= 216$
2. 5^2
3. 1^2
4. 7^3

B. Refer to the table below to find the value of each expression. The first one is started for you.

TABLE OF SQUARES

1^2	=	1	11^2	=	121
2^2	=	4	12^2	=	144
3^2	=	9	13^2	=	169
4^2	=	16	14^2	=	196
5^2	=	25	15^2	=	225
6^2	=	36	16^2	=	256
7^2	=	49	17^2	=	289
8^2	=	64	18^2	=	324
9^2	=	81	19^2	=	361
10^2	=	100	20^2	=	400

5. 8^2 8×8
6. 11^2
7. 18^2
8. 16^2
9. 5^3
10. 8^3
11. 4^3
12. 10^3
13. $\sqrt{121}$
14. $\sqrt{16}$
15. $\sqrt{289}$
16. $\sqrt{196}$

C. Find the value of the expressions below with a calculator. Refer to page 321 in the Calculator Handbook for additional information.

17. $21^2 =$ ______ $31^2 =$ ______
    $22^2 =$ ______ $32^2 =$ ______
    $23^2 =$ ______ $33^2 =$ ______
    $24^2 =$ ______ $34^2 =$ ______
    $25^2 =$ ______ $35^2 =$ ______

18. $\sqrt{441} =$ ______ $\sqrt{3,481} =$ ______
    $\sqrt{625} =$ ______ $\sqrt{3,721} =$ ______
    $\sqrt{841} =$ ______ $\sqrt{4,225} =$ ______
    $\sqrt{1,225} =$ ______ $\sqrt{4,761} =$ ______
    $\sqrt{1,521} =$ ______ $\sqrt{5,041} =$ ______

Check your answers on page 270.

ALGEBRA TOPIC

Using Formulas

Example Linda works for a trucking company. She drives an average of 50 miles per hour. If Linda drives for 3 hours at that rate, how many miles could she drive?

To find how many miles Linda could drive, you can use the **distance formula.** A **formula** is an algebraic equation. To solve a formula, you need to know 2 of the 3 variables.

$d = rt$, or $d = r \times t$, which means distance = rate × time.

Follow these steps when using a formula:

Step 1 Substitute the numbers you know for variables, or letters, in the formula.

Step 2 Do the operations (add, subtract, multiply, or divide) shown in the formula.

Which expression shows how many miles Linda could drive in 3 hours?

(1) $50 + 3$
(2) $50 - 3$
(3) 50×3
(4) $50 \div 3$
(5) $3 \div 50$

Answer **(3)** is the correct choice. The distance formula tells you to multiply the rate (in this case, speed) by the time. Substitute the numbers you know into the formula. $\boldsymbol{d = rt = 50 \times 3 = 150}$ **miles,** so at 50 miles per hour, Linda could drive **150 miles** in 3 hours.

Example Mai-Ling wants to buy flooring for her kitchen. The kitchen is 14 feet long by 10 feet wide. You learned in Lesson 3 that the formula for area of a rectangle is $A = l \times w$, or $A = lw$.

Which is the correct expression to find the area of Mai-Ling's kitchen?

(1) $14 + 10$
(2) $14 \div 10$
(3) $14 - 10$
(4) 14×10
(5) $10 - 14$

Answer **(4)** is correct. The area formula tells you to multiply the length by the width. Substitute the numbers you know into the formula.
$\boldsymbol{A = lw = 10 \times 14 = 140}$ **square feet,** so the area of the kitchen is **140 square feet.**

PRACTICE

Circle the best answer for each question.

1. Richard is putting wood trim along the walls of his living room as shown in the picture.

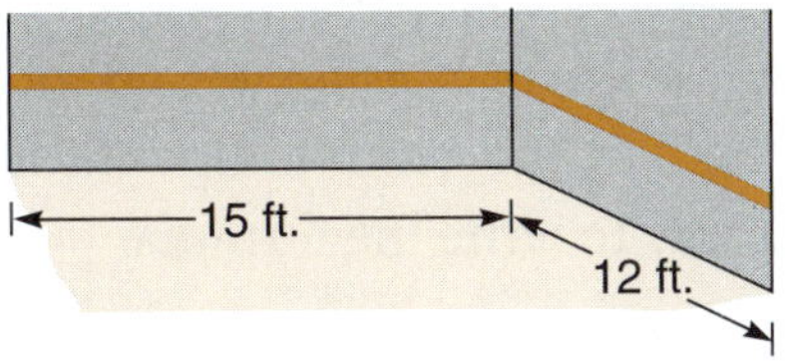

The formula for perimeter (P) can be written as $P = 2 \times l + 2 \times w$. Which is the correct expression to find how long the trim will be?

(1) $2 \times 15 \times 2 \times 12$
(2) $2 \times 15 - 2 \times 12$
(3) $2 \times 15 + 2 \times 12$
(4) $2 + 15 \times 2 + 12$
(5) $2 \times 15 \times 2 + 12$

2. Lilia is a salesperson for Advanced Technologies. On average she drives 5 hours a day. If she drives at a rate of 45 miles per hour, which expression finds how many miles Lilia drives in one day? ($d = rt$)

(1) 45×5
(2) $45 + 5$
(3) $45 - 5$
(4) $45 \div 5$
(5) $5 \div 45$

3. John's customer wants to put a 1-foot-wide strip of contrasting color carpet around the edges of her living room. The room measures 20 feet on each side. The formula for the perimeter (P) of a square is $P = 4s$, or $P = 4 \times s$. Which is the correct expression to find the perimeter of the room?

(1) 20^4
(2) 4×20
(3) $20 + 20$
(4) 2×20
(5) $20 + 20 + 20 + 20$

4. Kathleen is training for a two-day 150-mile bicycle ride to benefit the children's hospital. She wants to ride six hours each day. How many miles does she need to ride each hour to finish the ride in 12 hours? (Use $r = d \div t$.)

(1) 12×150
(2) $12 \div 150$
(3) $150 \div 12$
(4) 150×12
(5) Not enough information is given.

5. Kathy has been driving for 7 hours. She left at 9:00 and put 12 gallons of gas in her car. How many miles has she driven so far?

(1) 19
(2) 28
(3) 54
(4) 756
(5) Not enough information is given.

6. Emily needs a rectangular storage shed with at least 100 square feet of floor space. Of the following dimensions, which would have a floor space of at least 100 square feet? Use the formula: $A = lw$.

(1) length 14 feet, width 5 feet
(2) length 13 feet, width 7 feet
(3) length 12 feet, width 8 feet
(4) length 11 feet, width 10 feet
(5) length 11 feet, width 9 feet

7. **Connect** Recall the distance formula is $d = rt$. Use what you know about solving equations to rewrite the distance formula to find t (time), given d (distance) and r (rate).

Check your answers on page 270.

Area of a Square

Example John works for Culver Carpets. On Tuesday John's boss sent him to meet Sara Vega. She wants to carpet her office. John measured her office. He found that each side is 24 feet long. John knows her office is a square.

A square is a rectangle that has four sides of equal length.

Recall in Lesson 3 you used the formula *Area* = *length* × *width* to find the area of a rectangle. You can also multiply the length by the width to find the area of a square: $A = s \times s$. But there is another way to write the formula.

$A = s^2$, where s is one side of the square.

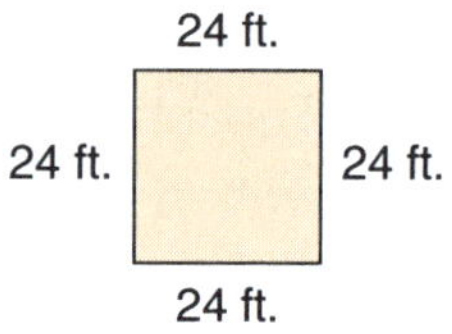

The raised 2 (s^2) tells you how many times to multiply the side by itself. So $A = s^2$ is the same as $A = s \times s$.

Which is the correct expression to find the area of Sara Vega's office?

(1) 24 + 24
(2) 24 + 24 + 24 + 24
(3) 24 × 24
(4) 24 × 2
(5) 24 × 4

Answer **(3)** is correct. John needs to find the area of the office, so he multiplies the side by itself. **$24 \times 24 = 24^2 = 576$ square feet,** so the office area is **576 square feet.**

Example Carpeting is sold by the square yard. John knows the number of square feet needed (576 square feet). He needs to change the square feet to square yards.

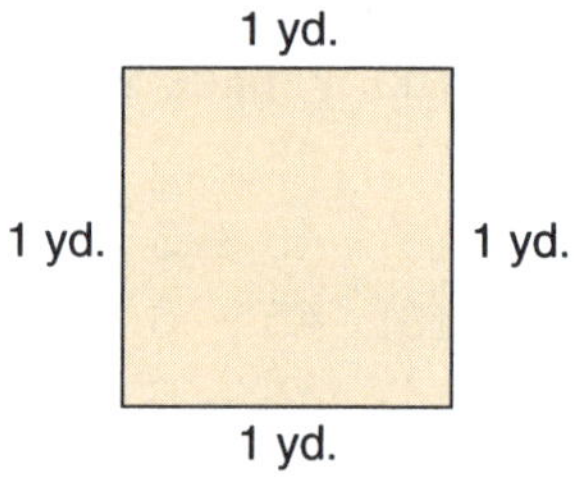

1 square yard

A **square yard** is a square that measures 1 yard on each side. Also, 1 yard = 3 feet. A square that measures 3 feet on each side has 9 square feet, so 1 square yard = 9 square feet.

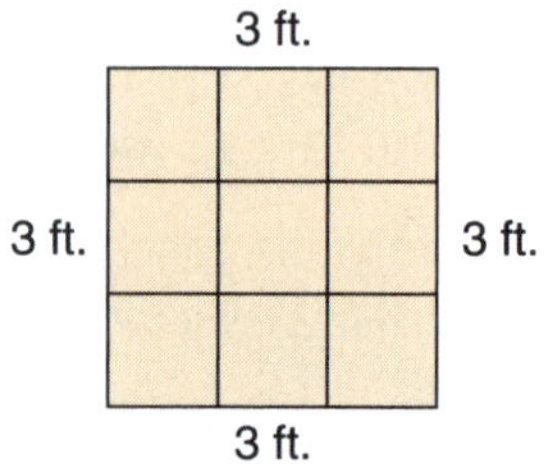

9 square feet
1 sq. yd. = 9 sq. ft.

Which is the correct expression to find the area of the office in square yards?

(1) 576 + 9
(2) 576 × 9
(3) 576 ÷ 9
(4) 576 − 9
(5) 9 ÷ 576

Answer **(3)** is correct. John needs to divide the number of square feet by 9. **576 ÷ 9 = 64 square yards,** so the area of the office is **64 square yards.**

PRACTICE

Circle the best answer for each question.

1. Ms. Vega wants to carpet another room that measures 15 feet on each side. John needs to find the area of the room. Which is the correct expression to find the area of the room in square feet?

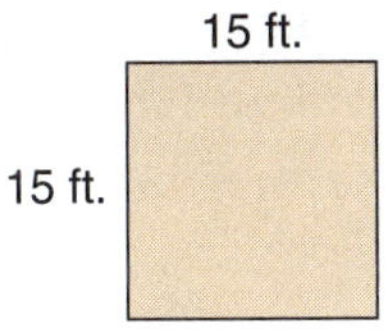

(1) 15×9
(2) $15 + 15 + 15 + 15$
(3) $9 \div 5$
(4) 15^2
(5) $15 \div 9$

2. John figures out that the area of the room is 225 square feet. He needs to find the area in square yards. Which is the correct expression to change 225 square feet to square yards?

(1) 225×9
(2) $9 \div 225$
(3) 9^2
(4) 225^2
(5) $225 \div 9$

3. Stanley wants to use a tarp to cover an area that measures 18 feet on each side. In square feet, which is the correct expression to find the minimum area that the tarp needs to cover?

(1) 18^2
(2) $18 + 18$
(3) 18×9
(4) $18 \div 9$
(5) $18 + 18 + 18 + 18$

4. Silvia has been hired to cover the floor of a kitchen and dining area with tile. The floor measures 12 feet on each side. Which is the correct expression to find the area of the floor in square feet?

(1) $12^2 \div 9$
(2) $12^2 \times 9$
(3) $12 + 12 + 12 + 12$
(4) 12^2
(5) $12 + 12 \div 9$

5. Ramon wants to order bathroom carpet from a catalog. The carpet is sold by the square yard. He finds that the area of his bathroom is 36 square feet. Which is the correct expression to change 36 square feet to square yards?

(1) $36 \div 4$
(2) $36 \div 9$
(3) 36×9
(4) $36 \div 3$
(5) Not enough information is given.

6. Irene wants to tile a hallway. The area of the hallway is 18 square feet. Four tiles are needed to cover one square foot. Which is the correct expression to find how many tiles Irene needs?

(1) $4 \div 18$
(2) $18 + 4$
(3) $18 \div 4$
(4) 4×18
(5) $18 + 18$

18 sq. ft.

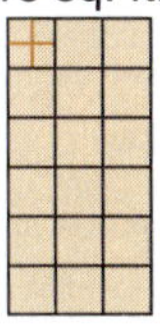

7. **Communicate** Explain why you can find the area of a square room by measuring only one side. Write your explanation in a sentence or two. Include a drawing with your explanation.

Check your answers on page 270.

MEASUREMENT AND GEOMETRY

Volume of Rectangular Solids and Cubes

Example Celia works for a shipping company. She figures out how many cubic feet each shipment will be. In the diagram, the shipment is 12 feet by 8 feet by 11 feet. To find out the cubic feet of this shipment, Celia needs to find its volume.

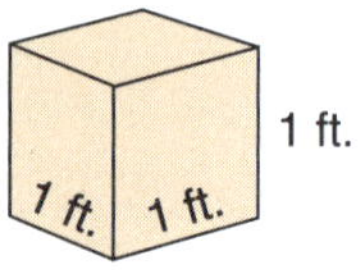

A cubic foot is a volume that measures 1 foot on each side. For example, the box is 1 foot long, 1 foot wide, and 1 foot high. The volume of the box is 1 cubic foot.

Volume measures the space inside an object. The space is measured in cubic units such as cubic inches or cubic feet. The formula for the volume of a rectangular solid is written:

$V = l \times w \times h$, or $V = lwh$, which means
Volume = length × width × height.

Which is the correct expression to find the volume of the shipment in the example above?

(1) $12 + 8 + 11$
(2) $12 \times 8 \times 11$
(3) 12×8
(4) 12×11
(5) 12×8

Answer **(2)** is correct. To find the volume of a rectangular solid, multiply the length by the width by the height: $V = 12 \times 8 \times 11 = 1{,}056$ **cubic feet.**

Example Another shipment is in the shape of a cube. Each side measures 8 feet. Celia needs to find the volume of the cube before loading it.

You can multiply the length by the width by the height to find the **volume** of a **cube,** but there is another way to write the formula.

$V = s^3$, which means
Volume = side × side × side.

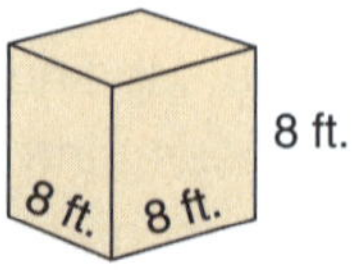

Which is the correct expression to find the volume for the cube?

(1) $8 + 8$
(2) $8 + 8 + 8$
(3) 8×8
(4) $8 \times 8 \times 8$
(5) $8 + 1$

Answer **(4)** is correct. Since the shipment is a cube, all sides have the same measurement. To find the volume of a cube, multiply the length of the side by itself three times: $V = 8 \times 8 \times 8 = 512$ **cubic feet.**

PRACTICE

Circle the best answer for each question.

1. The food compartment of a refrigerator is 3 feet long, 2 feet wide, and 5 feet high. Which is the correct expression to find the volume of the food compartment?
   (1) $3 + 2 + 5$
   (2) $3 \times 2 + 5$
   (3) $3 \times 2 \times 5$
   (4) $3 + 2 \times 5$
   (5) 3×5

2. Tony is going to buy sand for his children's sandbox. The measurements for the sandbox are shown below.

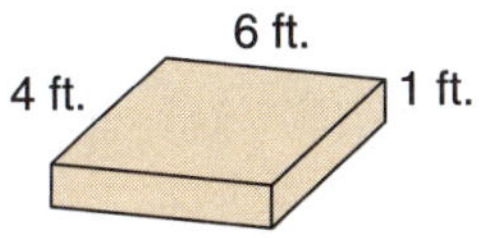

   Which expression shows how to find the volume of the sandbox?
   (1) 6×4
   (2) $6 \times 4 \times 1$
   (3) $6 + 4 + 1$
   (4) 6×1
   (5) 4×1

3. Joann works for a packing and mailing service. A customer brings in a gift-wrapped box that measures 7 inches on each side. Which is the correct expression to find the volume of the box?
   (1) 7^1
   (2) 7^2
   (3) 7^3
   (4) 7^4
   (5) 7^5

4. Josefina is shipping a carton that is 12 inches long, 10 inches wide, and 8 inches tall. Which expression shows how to find the volume of the shipping carton?
   (1) 10×8
   (2) $12 + 10$
   (3) 12×10
   (4) $12 \times 10 \times 8$
   (5) $12 + 10 + 8$

5. Juan is shipping a carton that is 28 inches long and 40 inches wide. In cubic inches, what is the volume of the carton?
   (1) 12
   (2) 40
   (3) 68
   (4) 1,120
   (5) Not enough information is given.

6. Gail has filled the delivery truck with 1,056 cubic feet of cartons. The inside of the truck measures 16 feet long, 8 feet wide, and 14 feet high. How many cubic feet of space is left on the truck?
   (1) 320
   (2) 736
   (3) 1,376
   (4) 1,792
   (5) 2,528

7. **Reason** Explain why you can find the volume of a cube by measuring only one side. Write your explanation in a sentence or two. Include a drawing with your explanation.

Check your answers on page 271.

Filling In the Standard Grid

Not all of the problems on the GED Mathematics Test have multiple-choice answers. Some problems require answers be filled in on **standard grids.** To correctly fill in a grid, use the information below.

- Start your answer in any column, as long as the answer fits.
- Write your answer in the boxes on the top row.
- Use the digits you wrote on the top row as a guide to filling in the correct circles.
- Fill in only one circle in each column.
- Leave any unused columns blank.

When you solve problems, you write some answers with a comma (12,350) or a symbol ($12). However, when you use the standard grid, you will fill in only numbers, no commas or symbols.

TIP

To make sure you have enough room, start writing your answer in the left column in a grid.

Example Marcus is saving for a new CD player. He saved $30 in September, $35 in October, and $40 in November. How much has he saved?

Step 1 Add to find the total amount of money that Marcus saved. **$30 + $35 + $40 = $105**

Step 2 Write the numbers in the answer in the top row of the grid.

Step 3 Fill in the matching circle below each digit. Leave the unused columns blank.

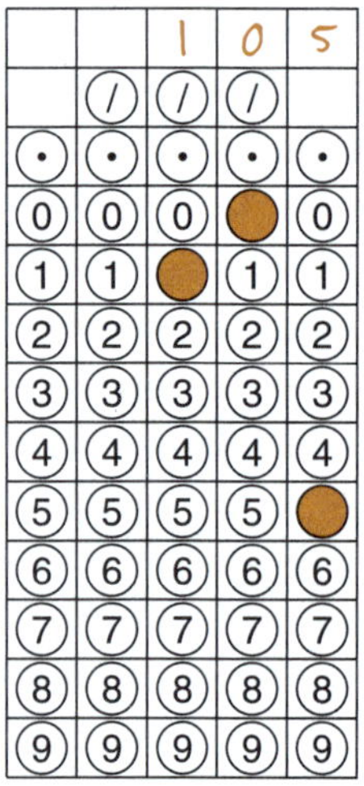

Example Lina is moving to a bigger apartment. Her old apartment has 1,079 square feet, and the new one has 1,352 square feet. How much bigger is the new apartment in square feet?

Step 1 Subtract to find the answer: 1,352 − 1,079 = 273

Step 2 Write the answer in the top row.

Step 3 Fill in the circle below your answer. Leave the remaining columns blank.

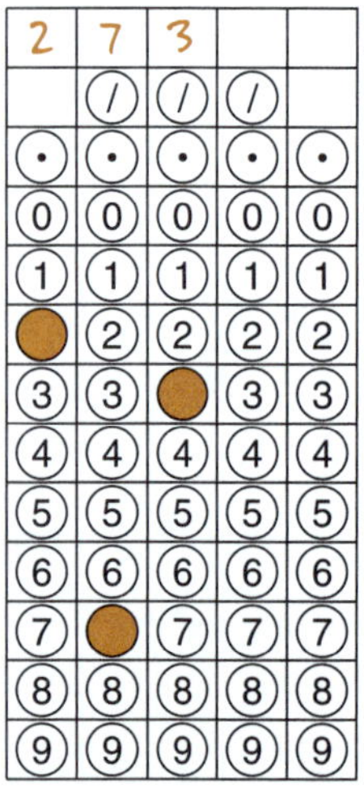

PRACTICE

Solve each problem. Then fill in your answer on the grids.

1. Jennifer is buying a new television. She signs a contract to make 18 payments of $25 each. How many dollars will she pay for the television?

	/	/	/	
.	.	.	.	.
0	0	0	0	0
1	1	1	1	1
2	2	2	2	2
3	3	3	3	3
4	4	4	4	4
5	5	5	5	5
6	6	6	6	6
7	7	7	7	7
8	8	8	8	8
9	9	9	9	9

2. Nita drove 322 miles on Thursday and 247 miles on Friday. How many miles did she drive in all for the two days?

	/	/	/	
.	.	.	.	.
0	0	0	0	0
1	1	1	1	1
2	2	2	2	2
3	3	3	3	3
4	4	4	4	4
5	5	5	5	5
6	6	6	6	6
7	7	7	7	7
8	8	8	8	8
9	9	9	9	9

3. Four friends share an apartment. They each pay $348 per month. How many dollars in rent do they pay for the apartment in all?

	/	/	/	
.	.	.	.	.
0	0	0	0	0
1	1	1	1	1
2	2	2	2	2
3	3	3	3	3
4	4	4	4	4
5	5	5	5	5
6	6	6	6	6
7	7	7	7	7
8	8	8	8	8
9	9	9	9	9

4. Ned needs 250 fluid ounces of paint thinner for a project. He already has three cans of paint thinner. If each can holds 32 fluid ounces, how many more fluid ounces does he need?

	/	/	/	
.	.	.	.	.
0	0	0	0	0
1	1	1	1	1
2	2	2	2	2
3	3	3	3	3
4	4	4	4	4
5	5	5	5	5
6	6	6	6	6
7	7	7	7	7
8	8	8	8	8
9	9	9	9	9

Check your answers on page 271.

LESSON 4

Lesson 4 Review
Lesson 4 Review

Lesson 4 Review

Write each expression as a multiplication problem and multiply.

1. 8^2
2. 1^3
3. 3^3
4. 9^2

Find the value of each expression.

5. 17^2
6. 15^2
7. $\sqrt{25}$
8. $\sqrt{49}$

Use your calculator to find the value of each expression.

9. 215^2 = ______
10. 19^3 = ______
11. $\sqrt{324}$ = ______
12. $\sqrt{1{,}089}$ = ______

Solve each problem. Show your work.

13. The Mahers' living room measures 20 feet on each side. Use the formula $A = s^2$ to find the area of the living room.

14. A rectangular storage bin is 15 feet long, 12 feet wide, and 5 feet deep. Use the volume formula, $V = l \times w \times h$, to find the volume of the bin.

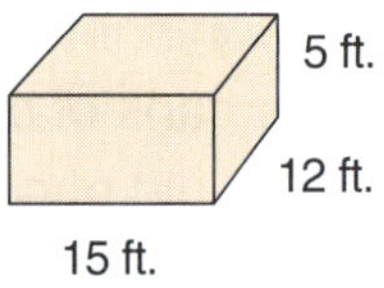

15. Willis installs pools. Today he is beginning to install an in-ground pool that is 30 feet long, 6 feet wide, and 4 feet deep. What is the volume, in cubic feet, of the dirt that he must remove before he installs the pool?

16. Marsha bought 9 square yards of carpet. The carpet cost $14 per square yard. What was the total cost of the carpet?

17. Viola sells carpet. She measures a customer's family room and finds its area is 576 square feet. What is the area in square yards?

18. A trash hauler places large trash bins at apartment complexes. Each bin is 12 feet long, 8 feet wide, and 6 feet high. What is the volume of each bin in cubic feet?

Mark your answer in the circles in the grid.

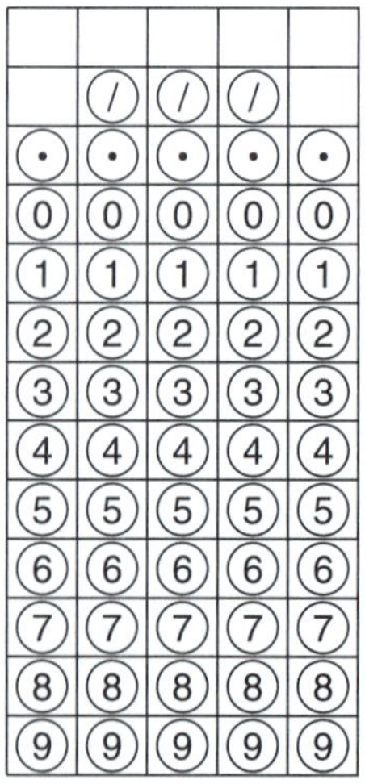

Circle the best answer for each question.

19. Which is the correct expression to find the volume of the cube shown in cubic inches?

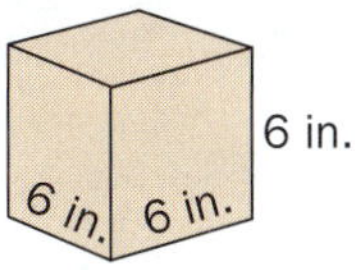

(1) $6 + 6 + 6$
(2) $6 \times 6 \times 6$
(3) $6 \times 6 + 20$
(4) $6 \times 6 \times 6 \div 6$
(5) $6 + 6 + 6 \div 6$

20. Dave orders vinyl flooring for his kitchen, which measures 15 feet on each side. Which is the correct expression to find the area of the kitchen in square feet?
(1) $15 + 15$
(2) $15^2 \div 9$
(3) $15 \times 15 \times 15$
(4) $15 + 15 + 15 + 15$
(5) 15^2

21. Dennis drives a delivery van. The cargo area of the van is 8 feet long, 5 feet wide, and 4 feet high. What is the volume of the cargo area in cubic feet?
(1) 17
(2) 40
(3) 52
(4) 105
(5) 160

22. Dwayne measures the conference room of his office building and finds it is 42 feet on each side. Which is the correct expression to find the area in square yards?
(1) 42^2
(2) $42^2 \times 9$
(3) $42^2 \div 9$
(4) $42 \div 9$
(5) 42×4

23. Marcus needs to buy bricks for a patio he is building. The patio will measure 10 feet on each side. What will be the area of the patio in square feet?
(1) 20
(2) 40
(3) 100
(4) 200
(5) 1,000

24. Joan needs to fill a rectangular carton with foam pieces. The carton is 18 inches long, 12 inches wide, and 3 inches high. Which is the correct expression to find the volume of the carton in cubic inches?
(1) $18 \times 12 \times 3$
(2) $18 + 12 + 3$
(3) $18 \times 12 + 3$
(4) $18 + 12 \times 3$
(5) $18 \times 12 \div 3$

25. Pat measures the hallways and bathroom in her house so she can buy new vinyl flooring. She finds a total area of 108 square feet. What is the area in square yards?

Mark your answer in the circles in the grid.

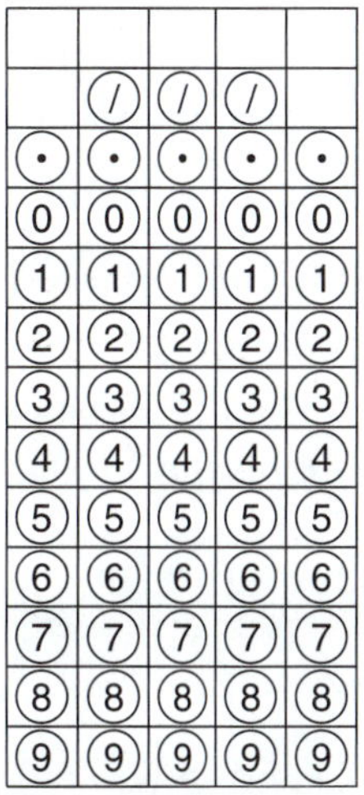

Check your answers on page 272.

Math at Work

Transportation: Truck Driver

Do you enjoy driving and seeing new places? If so, a job in the transportation industry may interest you. Because these types of jobs involve moving goods or people from place to place, truck drivers must have good map reading and math skills.

Every day truck drivers and other transportation workers use many of the math skills you've studied in this section. Drivers must add, subtract, multiply, and divide whole numbers. They often round numbers to make them easier to work with. Planning their routes and judging the time it will take to drive the distances requires drivers to rely heavily on good map reading skills.

Look at the Some Careers in Transportation chart.

- Do any of the careers interest you? If so, which ones?
- What information would you need to find out more about those careers? On a separate piece of paper, write some questions that you would like answered. You can find out more information about those careers in the *Occupational Outlook Handbook* at the library or online.

Some Careers in Transportation

Bus Driver
transports people from one location to another on set routes

Chauffeur
drives people to a desired location in a car or limousine

Delivery Driver
takes packages and goods to specific locations

Taxi Driver
drives one person or small groups of people to their desired locations, usually in a car or van

Truck Driver
transports goods from one destination to another, checking equipment and cargo

Transportation workers must have good map reading skills. Use the map below to answer the questions that follow.

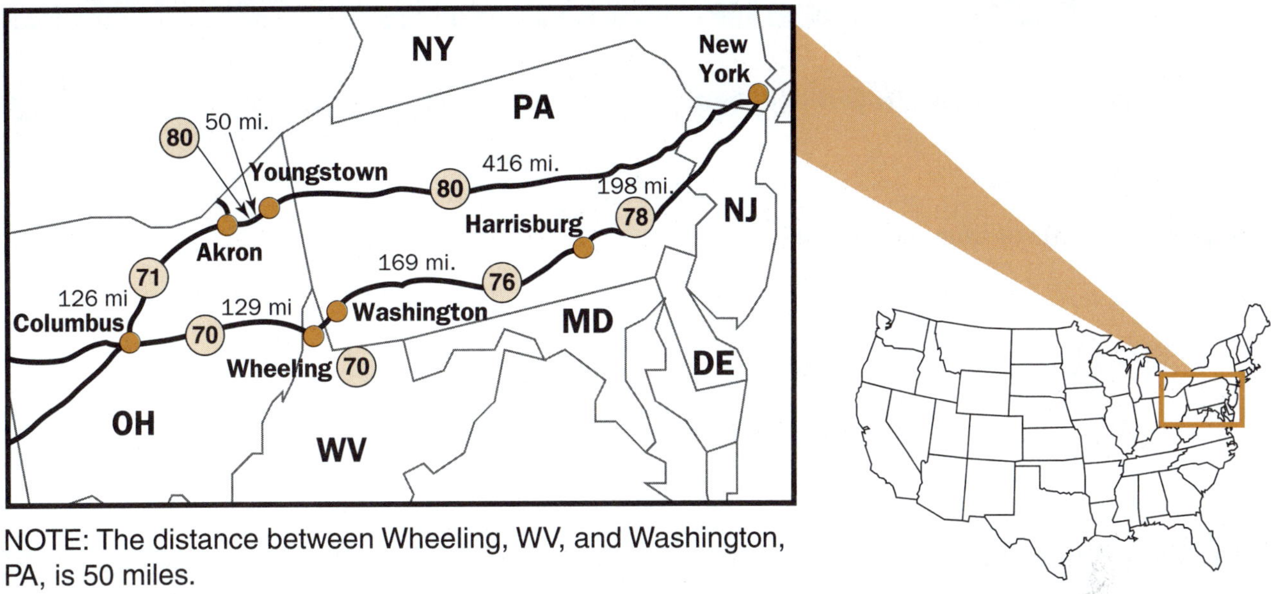

NOTE: The distance between Wheeling, WV, and Washington, PA, is 50 miles.

Mike drives a tractor-trailer truck. His job requires him to haul fruits and vegetables from Ohio to New York City. Because his cargo is perishable, he must be very careful about how long it takes him to make his runs.

Example According to the mileage written on the map, how far would Mike have to drive from Youngstown to Akron, Ohio?

Read the mileage on the map to see that it is **50 mi.** (miles) from Youngstown to Akron.

Use a calculator to answer the following questions.

1. According to the interstate map above, how many miles is it between Harrisburg, Pennsylvania, and New York City?
   (1) 50
   (2) 129
   (3) 169
   (4) 198
   (5) 416

2. There are two routes between Columbus, Ohio, and New York City.
   a. Find the miles round trip from Columbus to New York City on each route.
   b. Compare the routes. Which is shorter?
   c. Mike needs to get this cargo to New York City as quickly as possible. If there is more than one hour of delays on Route 76, should he take the northern or southern route? Explain.

Check your answers on page 273.

Unit 1 Review Whole Numbers

Write the place value of each underlined digit.

1. 234,957 ________________ 2. 6,030,495 ________________

Compare each pair of numbers. Write >, <, or =.

3. 417 __________ 1,740 4. 54,972 __________ 54,927

Follow the directions for each problem.

5. Round 4,375,429 to the nearest ten thousand. ____________________
6. Write 23,012 in words. ____

Solve. Show your work.

7. $x + 8 = 19$ 8. $4x = 20$

9. $\begin{array}{r} 426 \\ 84 \\ +253 \\ \hline \end{array}$ 10. $\begin{array}{r} \$248 \\ \times \quad 4 \\ \hline \end{array}$ 11. $\begin{array}{r} \$954 \\ -\ 378 \\ \hline \end{array}$ 12. $27\overline{)785}$

13. $\begin{array}{r} 726 \\ \times 308 \\ \hline \end{array}$ 14. $4^3 =$ 15. $\begin{array}{r} 500 \\ -243 \\ \hline \end{array}$ 16. $\sqrt{64} =$

17. $2,504 ÷ 8 = 18. 3,976 ÷ 13 =

19. The heights of the starting players on a basketball team are 80 inches, 76 inches, 75 inches, 80 inches, and 79 inches. What is the mean height of the players?

20. The temperatures for one week in Chicago were 35, 42, 37, 31, 24, 21, and 18 degrees. What was the median temperature for that week?

Use your calculator to find the value of each expression.

21. 4,203 [×] 3,056 [=] __________ 22. 160,208 [÷] 2,356 [=] __________

23. 83 [x^2] __________ 24. 55,225 [SHIFT] [x^2] __________

Circle the best answer for each question.

25. Martin is building a fence around his rectangular vegetable garden. The garden is 48 feet long and 20 feet wide. Which is the correct expression to find the perimeter of the garden?

(1) 48×20
(2) 48^2
(3) $48^2 + 20^2$
(4) $48 + 20$
(5) $48 + 48 + 20 + 20$

26. On Friday Lorene spent $57 for groceries, $13 for dry cleaning, and $21 for an oil change. Estimate the total amount of money Lorene spent on Friday.

(1) $70
(2) $80
(3) $90
(4) $100
(5) $110

27. Roberto earns $14 an hour as a cook. He worked 38 hours last week and 43 hours this week. Which is the correct expression to find how much Roberto earned this week?

(1) $14 × 38
(2) $14 × 43
(3) 38 + 43
(4) $14 × 38 + 43
(5) $14 + 38 + 43

<u>Question 28</u> refers to the following graph.

28. Find the approximate difference between the number of sweatshirts sold during the month with the greatest number of sales and the number of sweatshirts sold during the month with the least number of sales.

(1) 100
(2) 250
(3) 300
(4) 350
(5) 400

29. Carol is a carpet salesperson. She measures a bedroom and finds each side is 14 feet long. What is the area of this room in square feet?

Mark your answer in the circles in the grid.

Math Extension

Look through today's newspaper. Find ten different features of the paper that use whole numbers. Make a list of the features and give an example of how the whole number is used. For example, Classified Ads—Apartment for Rent, 3 bedrooms, 2 baths.

Check your answers on pages 273–274.

Mini-Test • Unit 1

Directions: This is a 15-minute practice test. After 15 minutes, mark the last number you finished. Then complete the test and check your answers. If most of your answers were correct but you did not finish, try to work faster next time.

Part I Directions: Choose the one best answer to each question. You MAY use your calculator.

1. Tony earns $12 per hour as a bank teller. He works between 30 and 40 hours per week. Last month, he worked 128 hours. How much did he earn last month?

   (1) $360
   (2) $480
   (3) $1,440
   (4) $1,536
   (5) $1,920

2. A baseball player hit 29, 34, 47, 18, 22, and 24 home runs during his six-year career. How many home runs did he average per year?

   (1) 26
   (2) 29
   (3) 34
   (4) 53
   (5) 174

3. Stephanie needs $3,600 for a down payment on a car. She already has $1,362 in a savings account. If she deposits $250 from her paycheck, how much more will she need for the down payment?

   (1) $1,988
   (2) $2,238
   (3) $2,488
   (4) $3,350
   (5) $5,212

4. Donna wants to make a frame for the mirror shown below. How many inches of molding does she need?

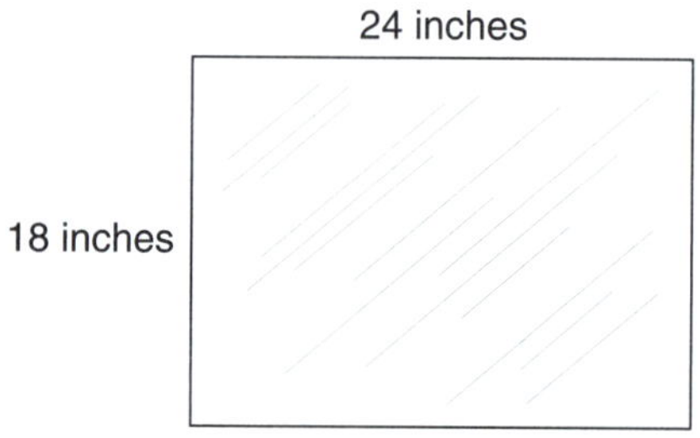

   (1) 60
   (2) 64
   (3) 84
   (4) 324
   (5) 432

5. At a city park, 312 children signed up for the winter basketball league. If the park puts 12 children on each team, how many teams will there be?

   Mark your answers in the circles in the grid.

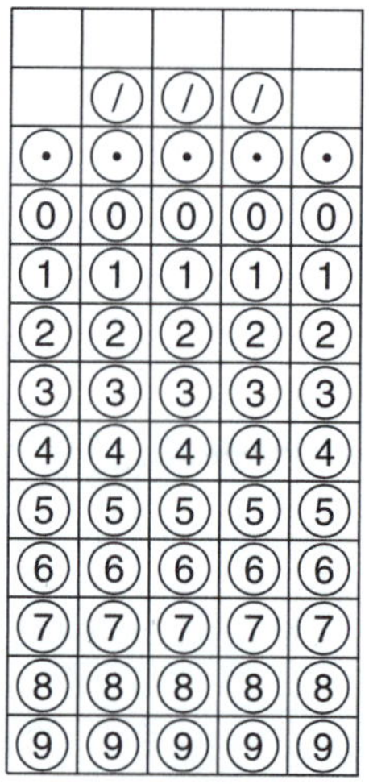

Part II **Directions:** **Choose the one best answer to each question. You MAY NOT use your calculator.**

Question 6 refers to the following graph.

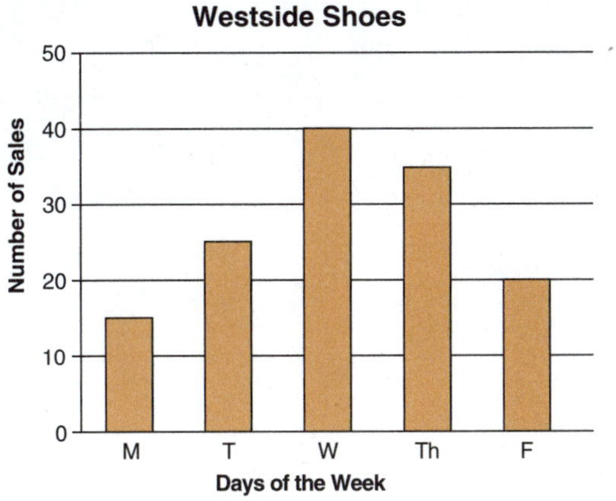

6. How many more sales were made on Thursday than on the day with the fewest sales?
   (1) 5
   (2) 10
   (3) 15
   (4) 20
   (5) 25

7. A rectangular tabletop is 84 inches long and 40 inches wide. Which is the correct expression to find the area in square inches?
   (1) $84 + 84 + 40 + 40$
   (2) 84^2
   (3) 40×4
   (4) $84^2 + 40^2$
   (5) 84×40

8. Which of the following shows the correct expression for finding the value of 6^3?
   (1) 6×3
   (2) $3 \times 3 \times 3 \times 3 \times 3 \times 3$
   (3) $6 \times 6 \times 6$
   (4) $6 + 6 + 6$
   (5) $3 + 3 + 3 + 3 + 3 + 3$

9. A clothing store sells pants for five different prices: \$26, \$52, \$46, \$24, and \$32. What is the median price of the amounts?
   (1) \$26
   (2) \$32
   (3) \$36
   (4) \$39
   (5) \$52

10. A softball team has four pitchers and eight position players. The entry fee for a tournament is \$420. If the players split the fee equally, which is the correct expression to find out how much each player should pay?
   (1) $\$420 \div (4 + 8)$
   (2) $(\$420 \div 4) + (\$420 \div 8)$
   (3) $(4 + 8) \times \$420$
   (4) $(4 + 8) \div \$420$
   (5) $(\$420 \times 4) + (\$420 \times 8)$

11. John finds that he can make 34 cookies from a single batch of a recipe. If he makes six batches, how many cookies can he make?

   Mark your answer in the circles in the grid.

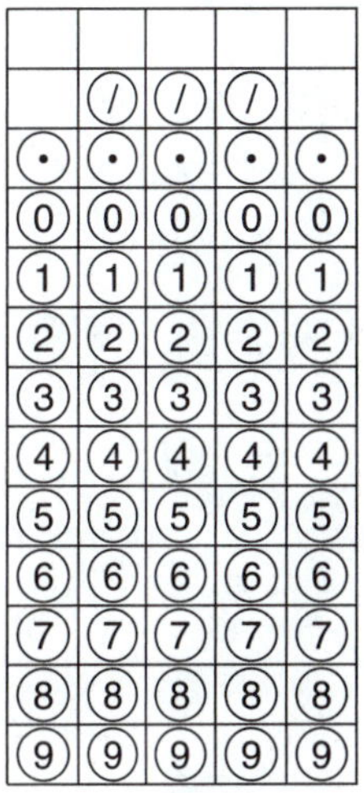

Check your answers on pages 274–275.

UNIT 2

Fractions

Fractions

Fractions and mixed numbers are often used to measure things. A **fraction** shows part of a whole, such as $\frac{1}{2}$ inch. A **mixed number** includes a whole number and a fraction, such as $1\frac{1}{2}$ inches. You may use fractions for many activities, such as cooking, or measuring and cutting lumber. After taking measurements, you may need to add, subtract, multiply, or divide fractions or mixed numbers to solve problems.

Write two fractions you might find on a measuring cup.

____________ ____________

Write two mixed numbers you might get from measuring with a ruler.

____________ ____________

Thinking About Fractions and Mixed Numbers

You may not realize how often you say, read, or use fractions and mixed numbers as you go about your daily life. Think about your recent activities.

Check the box for each activity you have done recently.

- ☐ Did you use a measuring cup?
- ☐ Did you use a scale?
- ☐ Did you fill out a scorecard at a game?
- ☐ Did you use a recipe?
- ☐ Did you cut a pie or cake into sections?
- ☐ Did you use a measuring tape or ruler?
- ☐ Did you read a road sign with fractions or mixed numbers?

Write some other activities where you used fractions or mixed numbers.

____

____

____

Previewing the Unit

In this unit, you will learn:

- definitions for proper and improper fractions and mixed numbers
- how to change improper fractions to mixed numbers
- how to find equivalent fractions and compare fractions
- how to reduce to lowest terms and raise to higher terms
- how to find common denominators
- how to add, subtract, multiply, and divide fractions and mixed numbers
- how to use measurement tools, read a table, use a map, and fill in timecards
- how to use a calculator for operations with fractions
- how to fill in fractions on a standard grid

LESSON 5

Fraction Basics

Facts About Fractions

$\frac{1}{8}$ ← numerator, ← denominator

A **fraction** is part of a whole. When a whole is broken into parts, the parts are fractions of the whole. A fraction is written with one number over another number. The top number is called the **numerator.** The bottom number is called the **denominator.**

Proper Fractions

Example A pizza is cut into 8 equal pieces. After dinner, only 1 piece is left. So $\frac{1}{8}$ of the pizza is left, and $\frac{7}{8}$ of the pizza is gone.

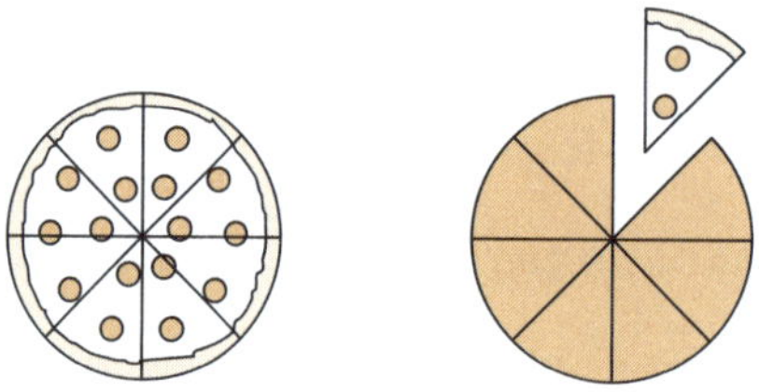

$\frac{1}{8}$ $\frac{\text{number of pieces left}}{\text{number of equal parts in whole pizza}}$

$\frac{7}{8}$ $\frac{\text{number of pieces eaten}}{\text{number of equal parts in whole pizza}}$

In the example above, the fractions $\frac{1}{8}$ and $\frac{7}{8}$ show part of one whole thing — the pizza. A fraction can also describe part of a group.

TIP
The denominator tells how many equal parts the whole is divided into. The numerator tells how many of those parts you are working with.
$\frac{\text{1 part out of total}}{\text{8 total parts}}$

Example Margo bought 4 apples. She ate 3 apples. She ate $\frac{3}{4}$ of the apples. She saved $\frac{1}{4}$ of the apples.

$\frac{3}{4}$ $\frac{\text{number of apples eaten}}{\text{number of apples in group}}$

Fractions like $\frac{1}{8}$, $\frac{7}{8}$, $\frac{1}{4}$, and $\frac{3}{4}$ are **proper fractions.** In a proper fraction, the numerator (top number) is less than the denominator (bottom number). A proper fraction has a value less than 1.

Example This rectangle has been divided into 8 parts. Three parts are shaded. So $\frac{3}{8}$ of the rectangle is shaded. In the proper fraction, $\frac{3}{8}$, the numerator, 3, is less than the denominator, 8. The value of $\frac{3}{8}$ is less than 1 because the shaded part is less than the whole rectangle.

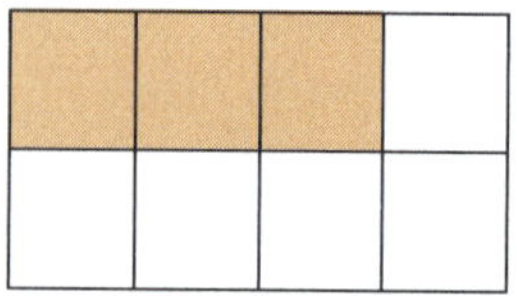

PRACTICE

A. **Write the fraction that names the shaded part. The first one is done for you.**

1. 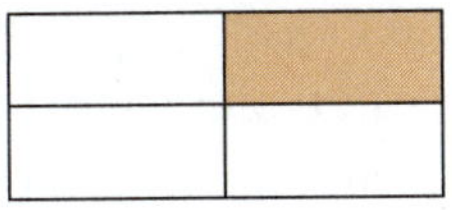 $\frac{1}{4}$

2.  ______

3. 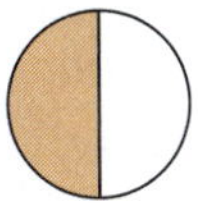 ______

4. 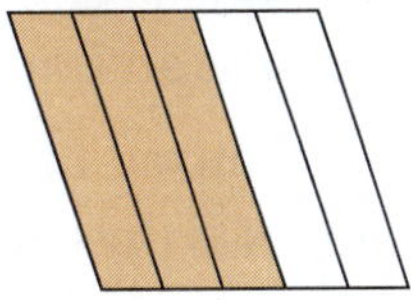 ______

5. 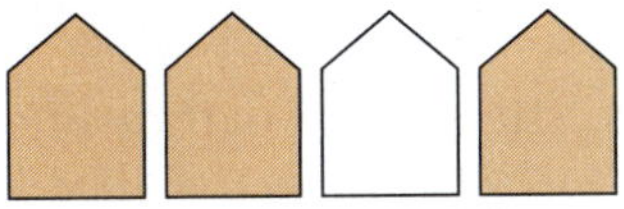 ______

6. 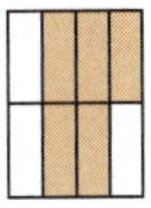 ______

B. **Divide and shade the figure to represent the given fraction. The first one is started for you.**

7. $\frac{4}{5}$ 

8. $\frac{3}{4}$ 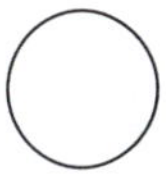

9. $\frac{2}{3}$ 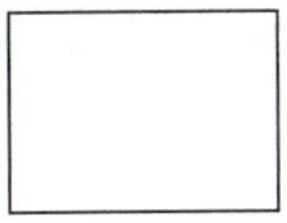

10. $\frac{5}{6}$ 

11. $\frac{7}{8}$ 

12. $\frac{1}{5}$ 

C. **Solve each problem.**

13. Mr. Martinez averages $14 per hour as a waiter, including tips. His hourly pay is $5 per hour. What fraction of his earnings comes from his hourly pay?

14. Harry's Restaurant employs 25 people. Eight employees are bus boys. What fraction of the employees are bus boys?

15. Cliff needs $24 for a new movie video. So far he has saved $17. What fraction of the price has Cliff saved so far?

16. Donna's basketball team won 8 out of the last 15 games. What fraction of the games did her team win?

Check your answers on page 275.

Improper Fractions and Mixed Numbers

There are two other kinds of fractions in addition to proper fractions. They are improper fractions and mixed numbers.

The value of an **improper fraction** is equal to or greater than 1. The numerator is greater than or equal to the denominator. Fractions such as $\frac{4}{4}$, $\frac{4}{3}$, $\frac{3}{2}$ are examples of improper fractions. Note that when the whole number 1 is written as an improper fraction, the numerator equals the denominator.

A **mixed number** is another way to show a value greater than 1. A mixed number is the sum of a whole number and a proper fraction. Fractions such as $1\frac{1}{2}$, $2\frac{1}{6}$, and $3\frac{3}{4}$ are examples of mixed numbers.

Improper fractions can be written as mixed numbers. Each of the 4 squares below has been divided into 4 parts. The shaded part can be called $\frac{13}{4}$ (an improper fraction) or $3\frac{1}{4}$ (a mixed number).

$= \frac{13}{4} = 3\frac{1}{4}$

Changing an Improper Fraction to a Mixed Number

Example Write $\frac{9}{5}$ as a mixed number.

Step 1 Divide the numerator by the denominator.

$$5\overline{)9}\quad \text{quotient } 1,\quad -5,\quad 4$$

Step 2 In this case there is a remainder, 4. Write the remainder over the original denominator. When there is no remainder, the improper fraction equals a whole number.

$$\frac{9}{5} = 1\frac{4}{5}$$

> **TIP**
> You can also change a whole number to an improper fraction. Write the whole number as the numerator with 1 as the denominator.
> $4 = \frac{4}{1}$

Changing a Mixed Number to an Improper Fraction

A mixed number can always be written as an improper fraction because both name a value greater than 1.

Example Write $2\frac{3}{8}$ as an improper fraction.

Step 1 Multiply the whole number part of the mixed number by the denominator of the fraction. Put this product over the original denominator.

$$2\frac{3}{8} = \frac{2 \times 8}{8} + \frac{3}{8}$$

Step 2 Add the numerator of the fraction to the result.

$$= \frac{16}{8} + \frac{3}{8}$$

Step 3 Write the total over the original denominator.

$$= \frac{19}{8}$$

PRACTICE

A. Write the improper fraction and mixed number that name the shaded part. The first one is done for you.

1. 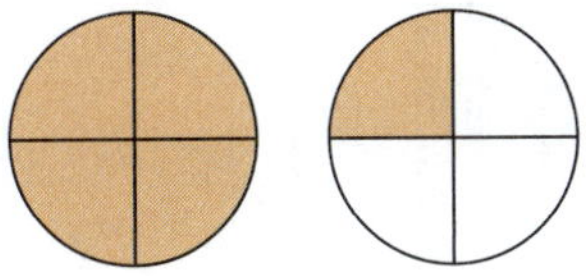 $\frac{5}{4}$ ______ $1\frac{1}{4}$ ______

2. 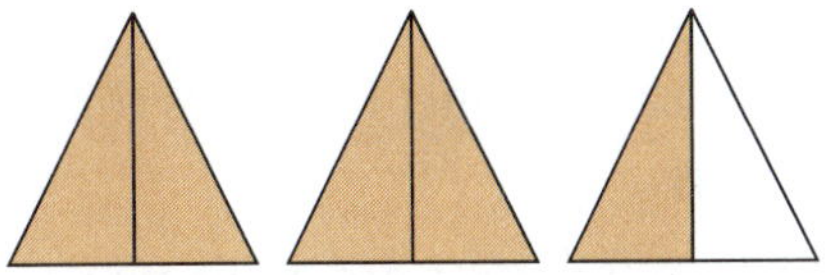 ______ ______

3. 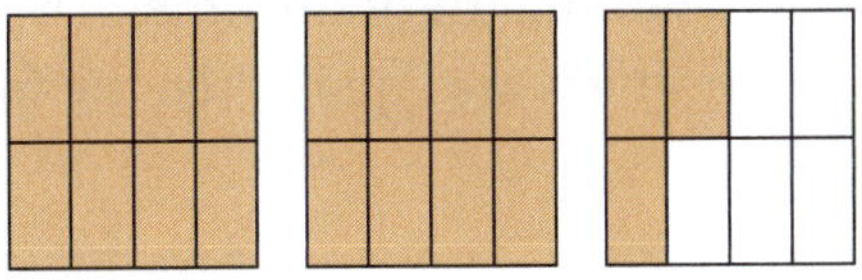 ______ ______

B. Change each improper fraction to a whole or mixed number. Show your work. The first one is done for you.

4. $\frac{7}{4} = 4\overline{)7}$ with quotient 1; -4; remainder 3; $= 1\frac{3}{4}$

5. $\frac{18}{6}$

6. $\frac{14}{5}$

7. $\frac{10}{3}$

8. $\frac{20}{4}$

9. $\frac{15}{8}$

C. Change each whole or mixed number to an improper fraction. Show your work. The first one is started for you.

10. $1\frac{5}{6}$

$$\frac{1 \times 6}{6} + \frac{5}{6} = \frac{6 + 5}{6} =$$

11. $2\frac{1}{2}$

12. 7

13. $3\frac{4}{5}$

14. 12

15. $10\frac{3}{4}$

D. Solve each problem.

16. A recipe calls for $3\frac{1}{2}$ cups of flour. Write an improper fraction to show the total amount in $3\frac{1}{2}$ cups.

17. Stephen worked 6 hours on Monday and $\frac{1}{3}$ hour Monday night at home. Write a mixed number that describes the number of hours Stephen worked on Monday.

Check your answers on pages 275–276.

Equivalent Fractions

Different fractions can represent the same value. The shaded fraction bars shown below are equivalent.

$\frac{1}{2}$ is equal to $\frac{4}{8}$

$\frac{1}{2}$ and $\frac{4}{8}$ are different names for the same value. Fractions that name the same value are called **equivalent fractions.**

> **TIP**
> The fraction $\frac{1}{2}$ names the same value as $\frac{2}{4}$ and $\frac{4}{8}$. Therefore, $\frac{1}{2}$ is equivalent to $\frac{2}{4}$ and $\frac{4}{8}$.

You can **cross-multiply** to find if two fractions are equivalent. Multiply the numerator of each fraction by the denominator of the other. The results are called **cross-products.** If the cross-products are equal, then the fractions are equivalent.

Example Are $\frac{6}{8}$ and $\frac{3}{4}$ equivalent fractions?

Step 1 Cross-multiply. $\frac{6}{8} \times \frac{3}{4}$

$6 \times 4 = 24$
$8 \times 3 = 24$

Step 2 Look at the results. If the results are the same, then the fractions are equivalent.

$\frac{6}{8}$ and $\frac{3}{4}$ are equivalent fractions because 24 and 24 are the same.

PRACTICE

Determine if the fractions are equivalent. Show your work. The first one is done for you.

1. $\frac{2}{4}$ $\frac{6}{12}$

   $\frac{2}{4} \times \frac{6}{12}$ $2 \times 12 = 24$
   $4 \times 6 = 24$

   $\frac{2}{4}$ and $\frac{6}{12}$ are equivalent.

2. $\frac{2}{3}$ $\frac{12}{18}$
3. $\frac{1}{2}$ $\frac{5}{10}$
4. $\frac{3}{6}$ $\frac{5}{12}$
5. $\frac{4}{6}$ $\frac{8}{12}$
6. $\frac{4}{7}$ $\frac{8}{14}$
7. $\frac{5}{6}$ $\frac{25}{30}$
8. $\frac{3}{5}$ $\frac{6}{15}$
9. $\frac{6}{10}$ $\frac{24}{40}$
10. $\frac{3}{5}$ $\frac{20}{25}$
11. $\frac{1}{2}$ $\frac{4}{8}$
12. $\frac{2}{3}$ $\frac{6}{9}$
13. $\frac{7}{8}$ $\frac{21}{24}$
14. $\frac{3}{8}$ $\frac{6}{16}$
15. $\frac{8}{9}$ $\frac{16}{18}$
16. $\frac{1}{4}$ $\frac{8}{16}$

 Check your answers on page 276.

TIP

You may want to look for the largest number to divide into the numerator and denominator. In the example, you can divide both the top and bottom numbers by 12. (This is the same as dividing by 6 in step 1 and dividing by 2 in step 2.)

$$\frac{12 \div 12}{24 \div 12} = \frac{1}{2}$$

Reducing a Fraction to Lowest Terms

The numerator and denominator of a fraction are sometimes called **terms.** Thus, **reducing** a fraction to **lower terms** means finding an equivalent fraction with a smaller numerator and denominator.

To reduce a fraction, divide the numerator and denominator by the same number. The fraction has been reduced to its **lowest terms** when no number except 1 will divide evenly into both the numerator and denominator.

Example Reduce $\frac{12}{24}$ to lowest terms.

Step 1 Choose a number that will divide into both. Try 6. $\frac{12 \div 6}{24 \div 6} = \frac{2}{4}$

Step 2 You may need to divide more than once to reduce a fraction to lowest terms. Use 2. $\frac{2 \div 2}{4 \div 2} = \frac{1}{2}$

✓ Check your answer by cross-multiplying.

$\frac{12}{24} \times \frac{1}{2}$ $\quad 12 \times 2 = 24$ $\quad 24 \times 1 = 24$

Since the cross-products are the same, the fractions are equivalent.

PRACTICE

Reduce each fraction to lowest terms. Check each answer by cross-multiplying. The first one is done for you.

1. $\frac{3}{6} = \frac{3 \div 3}{6 \div 3} = \frac{1}{2}$

   $\frac{3}{6} \times \frac{1}{2}$ $\quad 3 \times 2 = 6$ $\quad 6 \times 1 = 6$

2. $\frac{8}{12}$
3. $\frac{4}{10}$
4. $\frac{6}{8}$
5. $\frac{20}{25}$
6. $\frac{12}{36}$
7. $\frac{6}{9}$
8. $\frac{8}{16}$
9. $\frac{16}{20}$
10. $\frac{10}{14}$
11. $\frac{5}{15}$
12. $\frac{20}{24}$
13. $\frac{18}{27}$
14. $\frac{18}{24}$
15. $\frac{22}{33}$
16. $\frac{21}{28}$

Check your answers on page 276.

Raising a Fraction to Higher Terms

Raising a fraction to higher terms means finding an equivalent fraction with a greater numerator and denominator.

To raise a fraction to higher terms, multiply the numerator and the denominator by the same number.

TIP
To find the number that should be multiplied by 4 to get 20, you can divide 4 into 20.
$20 \div 4 = 5$

Example Raise $\frac{1}{4}$ to an equivalent fraction with a denominator of 20.

$$\frac{1}{4} = \frac{?}{20}$$

Step 1 Think, "What number should be multiplied by 4 to get 20?" The number 5 is multiplied by 4 to get 20.

$$4 \times ? = 20$$

Step 2 Multiply both the numerator and the denominator by 5.

$$\frac{1}{4} = \frac{1 \times 5}{4 \times 5} = \frac{5}{20}$$

✔ Check your work by cross-multiplying.

$$\frac{1}{4} \times \frac{5}{20} \qquad 1 \times 20 = 20 \qquad 4 \times 5 = 20$$

Since the cross-products are the same, the fractions $\frac{1}{4}$ and $\frac{5}{20}$ are equivalent.

PRACTICE

Find an equivalent fraction with the given denominator. Show your work. Check each answer by cross-multiplying. The first one is done for you.

1. $\frac{3}{4} = \frac{?}{8}$

   $\frac{3 \times 2}{4 \times 2} = \frac{6}{8}$

   $\frac{3}{4} \times \frac{6}{8}$ $3 \times 8 = 24$ $4 \times 6 = 24$

2. $\frac{2}{3} = \frac{?}{12}$
3. $\frac{1}{6} = \frac{?}{18}$
4. $\frac{7}{10} = \frac{?}{20}$
5. $\frac{5}{8} = \frac{?}{24}$
6. $\frac{2}{5} = \frac{?}{30}$
7. $\frac{4}{9} = \frac{?}{18}$
8. $\frac{1}{4} = \frac{?}{12}$
9. $\frac{7}{8} = \frac{?}{32}$
10. $\frac{3}{5} = \frac{?}{25}$
11. $\frac{7}{12} = \frac{?}{36}$
12. $\frac{3}{10} = \frac{?}{40}$
13. $\frac{4}{9} = \frac{?}{45}$
14. $\frac{9}{16} = \frac{?}{32}$
15. $\frac{3}{7} = \frac{?}{56}$
16. $\frac{18}{25} = \frac{?}{100}$

 Check your answers on pages 276–277.

Comparing Fractions

TIP
Recall the following:
> means *greater than*
< means *less than*
= means *equals*

Recall the symbols >, <, and = that were used to compare whole numbers. You wrote these symbols between two numbers to show how the value of the numbers compare. You can also use these symbols for comparing fractions.

When comparing two fractions with the same denominator, the fraction with the greater numerator is the greater fraction.

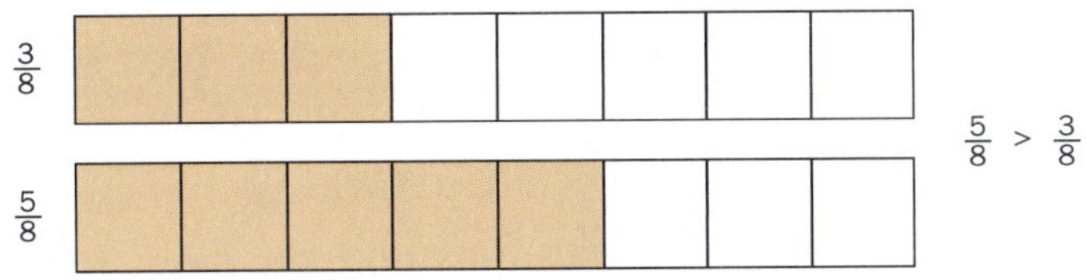

To compare two fractions with different denominators, cross-multiply and compare the results:

Example Compare $\frac{3}{4}$ and $\frac{2}{3}$.

Step 1 Cross-multiply.

$\frac{3}{4} \times \frac{2}{3}$ $3 \times 3 = 9$ numerator of first fraction
$2 \times 4 = 8$ numerator of second fraction

Step 2 Look at the results. Since 9 is greater than 8, then the first fraction is greater than the second fraction.

So $\frac{3}{4} > \frac{2}{3}$.

PRACTICE

Compare each pair of fractions. Write >, <, or =. The first one is done for you.

1. $\frac{4}{5}$ __>__ $\frac{5}{7}$

   $4 \times 7 = 28$; $5 \times 5 = 25$

   Since $28 > 25$, $\frac{4}{5} > \frac{5}{7}$

2. $\frac{3}{4}$ ____ $\frac{3}{8}$
3. $\frac{9}{16}$ ____ $\frac{12}{16}$
4. $\frac{2}{4}$ ____ $\frac{2}{3}$
5. $\frac{2}{3}$ ____ $\frac{3}{5}$
6. $\frac{8}{10}$ ____ $\frac{4}{5}$
7. $\frac{7}{8}$ ____ $\frac{3}{4}$
8. $\frac{5}{9}$ ____ $\frac{8}{9}$
9. $\frac{11}{12}$ ____ $\frac{8}{12}$
10. $\frac{5}{6}$ ____ $\frac{7}{8}$
11. $\frac{3}{8}$ ____ $\frac{4}{5}$
12. $\frac{3}{4}$ ____ $\frac{2}{3}$
13. $\frac{1}{3}$ ____ $\frac{3}{9}$
14. $\frac{15}{21}$ ____ $\frac{13}{21}$
15. $\frac{15}{24}$ ____ $\frac{18}{24}$
16. $\frac{3}{4}$ ____ $\frac{18}{24}$

Check your answers on page 277.

WORKPLACE MATH

Reading Measurement Tools

Example Rosa's chicken soup recipe calls for 10 ounces of chicken broth. Below is a measuring cup used for liquid measures.

Which of the following is the number of cups equal to the 10 ounces Rosa needs?

(1) 1 cup

(2) $1\frac{1}{4}$ cups

(3) $1\frac{1}{3}$ cups

(4) $1\frac{1}{2}$ cups

(5) $1\frac{3}{4}$ cups

Answer **(2)** is correct. First, find 10 ounces on the measuring cup. Find the same corresponding line for cups. **10 ounces = $1\frac{1}{4}$ cups,** so Rosa needs $1\frac{1}{4}$ cups of chicken broth.

Example Mr. Asano is a framer. A customer wants $2\frac{1}{2}$ inch matting around a photograph. Mr. Asano used his ruler to find the exact width of the matte.

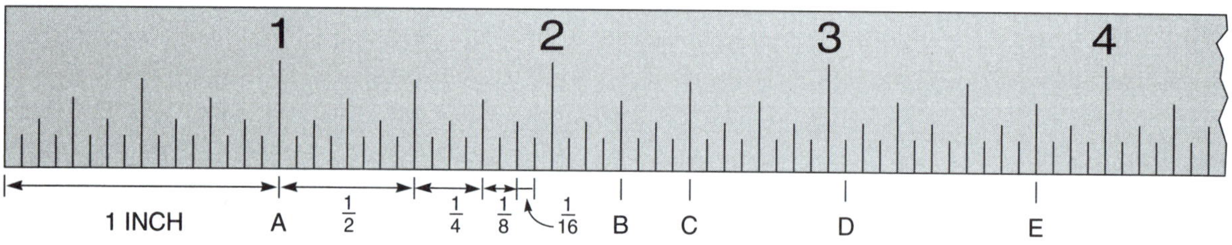

In the diagram above, which letter represents $2\frac{1}{2}$ inches?

(1) A
(2) B
(3) C
(4) D
(5) E

Answer **(3)** is correct. The $\frac{1}{2}$-inch mark is midway between two one-inch marks. The mark for $2\frac{1}{2}$ is in the middle of the 2-inch and 3-inch marks. The **$2\frac{1}{2}$-inch** mark is at letter **C**.

PRACTICE

Use the measuring cup or ruler from page 80 to answer questions 1 through 5. Circle the best answer for each question.

1. Mr. Asano measured a postcard that had a width of $3\frac{1}{16}$ inches. Which letter is at the $3\frac{1}{16}$-inch mark on Mr. Asano's ruler?

   (1) A
   (2) B
   (3) C
   (4) D
   (5) E

2. Andrea is making chicken-rice soup for her dinner party. The recipe calls for $\frac{3}{4}$ cup chicken broth. Which expression is correct about the fraction $\frac{3}{4}$?

   (1) $\frac{3}{4}$ is less than $\frac{1}{3}$
   (2) $\frac{3}{4}$ is greater than $\frac{1}{2}$
   (3) $\frac{3}{4}$ is less than $\frac{2}{3}$
   (4) $\frac{3}{4}$ is greater than 1
   (5) $\frac{3}{4}$ is greater than $1\frac{1}{4}$

3. Jaime wants to buy new tiling for his bathroom. The new tile has sides of $2\frac{1}{4}$ inches. Which expression is correct about the mixed number $2\frac{1}{4}$?

   (1) $2\frac{1}{4}$ is less than $2\frac{1}{8}$
   (2) $2\frac{1}{4}$ is equal to $2\frac{2}{8}$
   (3) $2\frac{1}{4}$ is less than $1\frac{3}{4}$
   (4) $2\frac{1}{4}$ is greater than $2\frac{3}{8}$
   (5) $2\frac{1}{4}$ is greater than $2\frac{1}{2}$

4. Rosa's measuring cup does not show the $\frac{1}{8}$ mark. Which statement best describes where a $\frac{1}{8}$ mark would go?

   (1) below the $\frac{1}{4}$ mark
   (2) between the $\frac{1}{4}$ and $\frac{1}{3}$ marks
   (3) between the $\frac{1}{3}$ and $\frac{1}{2}$ marks
   (4) between the $\frac{1}{2}$ and $\frac{2}{3}$ marks
   (5) between $\frac{3}{4}$ and 1 marks

5. When measuring length in inches, which of the following is true of $3\frac{1}{4}$?

   (1) $3\frac{1}{4}$ is less than 3
   (2) $3\frac{1}{4}$ is greater than $3\frac{1}{2}$
   (3) $3\frac{1}{4}$ is less than $3\frac{1}{16}$
   (4) $3\frac{1}{4}$ is greater than $3\frac{1}{8}$
   (5) $3\frac{1}{4}$ is greater than $3\frac{3}{4}$

6. Dean bought bananas, pears, and grapes at the local grocery store. The grapes weighed $1\frac{1}{2}$ pounds and the pears weighed $2\frac{1}{4}$ pounds. What is the total weight of the fruit Dean bought?

   (1) $1\frac{1}{2}$ pounds
   (2) $2\frac{1}{4}$ pounds
   (3) $3\frac{3}{4}$ pounds
   (4) 4 pounds
   (5) Not enough information is given.

7. **Reason** In a sentence or two, explain how you know $1\frac{1}{2} = 1\frac{2}{4}$. Include a drawing with your explanation.

Check your answers on page 277.

LESSON 5

Lesson 5 Review

Write the proper or improper fraction that names the shaded part.

1. 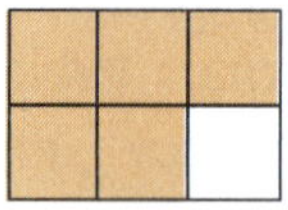 ____

2.  ____

3. 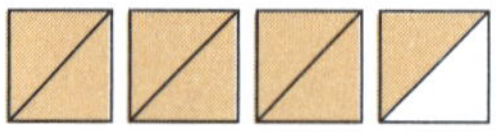 ____

4.  ____

Write the mixed number that names the shaded part.

5. 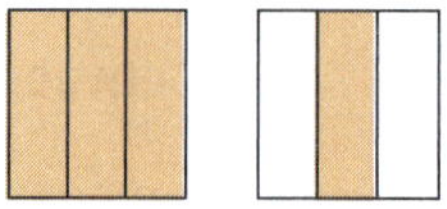 ____

6. 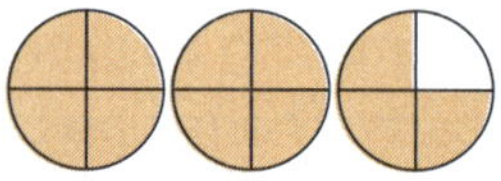 ____

Change each fraction to a whole or mixed number.

7. $\frac{11}{6}$ 8. $\frac{25}{5}$ 9. $\frac{19}{8}$ 10. $\frac{32}{7}$

Change each whole or mixed number to an improper fraction.

11. 7 12. $1\frac{7}{8}$ 13. $4\frac{3}{5}$ 14. $6\frac{2}{3}$

Reduce each fraction to lowest terms.

15. $\frac{5}{10}$ 16. $\frac{9}{12}$ 17. $\frac{16}{24}$ 18. $\frac{24}{32}$

Find an equivalent fraction with the given denominator.

19. $\frac{1}{2} = \frac{?}{18}$ 20. $\frac{5}{6} = \frac{?}{24}$ 21. $\frac{3}{4} = \frac{?}{32}$ 22. $\frac{4}{7} = \frac{?}{28}$

Compare each pair of fractions. Write >, <, or =.

23. $\frac{4}{5}$ ____ $\frac{3}{4}$ 24. $\frac{7}{5}$ ____ $\frac{5}{6}$

25. $\frac{4}{8}$ ____ $\frac{12}{24}$ 26. $\frac{8}{9}$ ____ $\frac{9}{12}$

Solve each problem.

27. There are 40 children who come to the day-care center where Ramona works. Of these, 13 come only for the after-school program. What fraction of the children come only after school?

28. Elmer baked 36 muffins for a bake sale. His children ate 5 muffins. What fraction of the muffins did his children eat?

Circle the best answer for each question.

29. Juan completed $\frac{6}{8}$ of his workday before taking a break. In lowest terms, how much of his workday did Juan complete?

(1) $\frac{2}{2}$

(2) $\frac{3}{2}$

(3) $\frac{3}{4}$

(4) $\frac{4}{6}$

(5) $\frac{12}{16}$

30. Val is making a tile collage to sell at a craft show. There are four big squares. Each big square of tile has four small squares. Fifteen of the small squares have a design. Write a mixed number that describes the number of big squares that have a design.

(1) $1\frac{5}{4}$

(2) 3

(3) $3\frac{3}{4}$

(4) 4

(5) $4\frac{1}{4}$

31. Betty lives 9 miles from work. Which improper fraction describes this distance?

(1) $\frac{9}{1}$

(2) $\frac{1}{9}$

(3) $\frac{9}{9}$

(4) $\frac{10}{9}$

(5) $\frac{9}{10}$

32. Lupe needs 6 ounces of water for a cake recipe. Use the measuring cup on page 80 to find how many cups of water Lupe needs.

(1) $\frac{1}{3}$

(2) $\frac{1}{2}$

(3) $\frac{2}{3}$

(4) $\frac{3}{4}$

(5) 1

33. On Friday, Stock 1 dropped $\frac{3}{4}$ point and Stock 2 dropped $\frac{5}{8}$ point. Based on this information, which statement is true?

(1) Stock 2 dropped more.

(2) Stock 1 dropped more.

(3) The stocks dropped the same amount.

(4) The stocks dropped at the same time.

(5) Both stocks sold at the same price.

34. There are 32 members in an investment club. Eight of the members are senior citizens. In lowest terms, what fraction of the club are senior citizens?

(1) $\frac{1}{2}$

(2) $\frac{1}{4}$

(3) $\frac{1}{8}$

(4) $\frac{1}{16}$

(5) $\frac{1}{32}$

Check your answers on page 278.

LESSON 6

Adding and Subtracting Fractions

Finding Common Denominators

Like fractions are fractions that have the same, or a common, denominator. The fractions $\frac{1}{4}$ and $\frac{3}{4}$ have a **common denominator** 4. The fractions $\frac{1}{6}$ and $\frac{2}{3}$ have different denominators, so they are **unlike fractions.**

The Lowest (or Least) Common Denominator

To add or subtract unlike fractions, you need to find a common denominator. The smallest number that each denominator will divide into evenly is called the **lowest common denominator.**

TIP

The lowest common multiple of two prime numbers is the product of the two numbers. Given $\frac{3}{5}$ and $\frac{2}{7}$, 5 and 7 are prime numbers. Multiply 5 by 7. 35 is the lowest common multiple.

$\frac{3}{5}$ becomes $\frac{21}{35}$

$\frac{2}{7}$ becomes $\frac{10}{35}$

Example Find the lowest common denominator for $\frac{3}{4}$ and $\frac{5}{6}$.

Step 1 Think of multiples of 4 (numbers that 4 divides into evenly). 4, 8, 12, 16

Step 2 Think of multiples of 6. 6, 12, 18

Step 3 Choose the first multiple that both 4 and 6 have in common as the lowest common denominator. 4, 8, (12), 16
6, (12), 18

The lowest multiple that both 4 and 6 will divide into evenly is 12. So the least common denominator for $\frac{3}{4}$ and $\frac{5}{6}$ is 12; $\frac{3}{4}$ becomes $\frac{9}{12}$ and $\frac{5}{6}$ becomes $\frac{10}{12}$.

$$\frac{3}{4} = \frac{3 \times 3}{4 \times 3} = \frac{9}{12} \qquad \frac{5}{6} = \frac{5 \times 2}{6 \times 2} = \frac{10}{12}$$

Example Find the least common denominator for $\frac{2}{3}$, $\frac{5}{8}$, and $\frac{15}{24}$.

Step 1 Think of multiples of 3, 8, and 24.

3: 3, 6, 9, 12, 15, 18, 21, (24), 27
8: 8, 16, (24), 32
24: (24), 48

Step 2 Choose the first multiple 3, 8, and 24 have in common. Use 24.

Step 3 Convert each fraction to an equivalent fraction with the common denominator.

$$\frac{2}{3} = \frac{2 \times 8}{3 \times 8} = \frac{16}{24} \qquad \frac{5}{8} = \frac{5 \times 3}{8 \times 3} = \frac{15}{24} \qquad \frac{15}{24}$$

PRACTICE

A. **Find the lowest common multiple for each set of denominators. Do not add or subtract the fractions. The first one is done for you.**

1. $\frac{1}{2}$ and $\frac{4}{6}$ 6

   multiples of 2: 2, 4, (6), 8

   multiples of 6: (6), 12, 18

2. $\frac{3}{4}$ and $\frac{4}{9}$
3. $\frac{1}{2}$ and $\frac{4}{9}$
4. $\frac{1}{3}$ and $\frac{2}{7}$
5. $\frac{1}{4}$ and $\frac{7}{12}$
6. $\frac{2}{3}$ and $\frac{4}{5}$
7. $\frac{1}{2}, \frac{2}{5},$ and $\frac{9}{10}$
8. $\frac{2}{3}, \frac{3}{4},$ and $\frac{5}{8}$
9. $\frac{1}{4}, \frac{2}{5},$ and $\frac{1}{10}$

B. **Find the lowest common denominator for each pair of fractions. Then change the fractions to have common denominators. The first one is started for you.**

10. $\frac{3}{5}$ and $\frac{5}{6}$

    multiples of 5: 5, 10, 15, 20, 25, (30)

    multiples of 6: 6, 12, 18, 24, (30)

    $\frac{3}{5} = \frac{3 \times 6}{5 \times 6} = \frac{18}{30}$

    $\frac{5}{6} =$

11. $\frac{1}{5}$ and $\frac{6}{7}$
12. $\frac{1}{2}$ and $\frac{1}{5}$
13. $\frac{4}{5}$ and $\frac{7}{8}$
14. $\frac{3}{4}$ and $\frac{4}{16}$
15. $\frac{2}{3}$ and $\frac{6}{8}$
16. $\frac{4}{6}$ and $\frac{6}{9}$
17. $\frac{2}{5}$ and $\frac{7}{8}$
18. $\frac{2}{3}$ and $\frac{9}{15}$
19. $\frac{3}{4}$ and $\frac{11}{14}$
20. $\frac{10}{12}$ and $\frac{3}{16}$
21. $\frac{5}{6}$ and $\frac{7}{8}$

C. **Change each set of fractions to have the lowest common denominator. The first one is started for you.**

22. $\frac{5}{8}, \frac{1}{4},$ and $\frac{1}{2}$

    multiples of 8: (8), 16

    multiples of 4: 4, (8), 12

    multiples of 2: 2, 4, 6, (8)

    $\frac{1}{4} = \frac{?}{8}$

23. $\frac{2}{3}, \frac{3}{5},$ and $\frac{1}{2}$
24. $\frac{5}{9}, \frac{1}{3}$ and $\frac{5}{6}$

Check your answers on page 279.

Adding Fractions and Mixed Numbers

To add like fractions, add only the numerators. Write the sum over the common denominator. Reduce your answer to lowest terms. If the answer is an improper fraction, change it to a mixed number.

Example Add: $\frac{5}{8} + \frac{7}{8}$

Step 1 Add the numerators. $5 + 7 = 12$

Step 2 Write 12 over the common denominator 8. $\frac{12}{8}$

Step 3 Change $\frac{12}{8}$ to a mixed number; reduce. $\frac{12}{8} = 1\frac{4}{8} = 1\frac{1}{2}$

So $\frac{5}{8} + \frac{7}{8} = \frac{12}{8} = 1\frac{4}{8} = 1\frac{1}{2}$.

To add unlike fractions, change the unlike fractions to like fractions.

Example Add: $\frac{1}{3} + \frac{3}{4}$

Step 1 Use multiples of 3 and 4 to find a common denominator. Think, "What number will both 3 and 4 divide into evenly?" The lowest common denominator is 12.

3: 3, 6, 9, (12), 15
4: 4, 8, (12), 16

Step 2 Change $\frac{1}{3}$ to a like fraction with a denominator of 12. $\frac{1 \times 4}{3 \times 4} = \frac{4}{12}$

Step 3 Change $\frac{3}{4}$ to a like fraction with a denominator of 12. $\frac{3 \times 3}{4 \times 3} = \frac{9}{12}$

Step 4 Add. Change $\frac{13}{12}$ to a mixed number. $\frac{4}{12} + \frac{9}{12} = \frac{13}{12} = 1\frac{1}{12}$

Adding Mixed Numbers

When you add mixed numbers, add the fractions first, changing unlike fractions to like fractions. Add the whole numbers. If the sum of the fractions is an improper fraction, change it to a whole or mixed number. Add any whole numbers from the fraction total to the whole number total. Reduce your answer to lowest terms.

Example Add: $2\frac{5}{6} + 3\frac{1}{2}$

Step 1 Find a common denominator and change to like fractions.

Step 2 Add the fractions and then add the whole numbers.

$$\begin{array}{r} 2\frac{5}{6} \\ +\,3\frac{1}{2} \\ \hline \end{array} \qquad \begin{array}{r} 2\frac{5}{6} \\ +3\frac{3}{6} \\ \hline =\,5\frac{8}{6} \end{array}$$

Step 3 Change the improper fraction ($\frac{8}{6}$) to a mixed number ($1\frac{2}{6}$). Add the whole number from the fraction total to the whole number total ($5 + 1\frac{2}{6}$). Reduce your answer to lowest terms.

$5\frac{8}{6} = 5 + 1\frac{2}{6}$
$= 6\frac{2}{6} = 6\frac{1}{3}$

TIP
To add a mixed number and a whole number, add the whole numbers. Write the fraction next to the whole number total.
$2\frac{5}{6} + 3 = 5\frac{5}{6}$

PRACTICE

A. **Add the fractions below. Reduce your answers to lowest terms. The first one is done for you.**

1. $\frac{2}{9} + \frac{5}{9} = \frac{7}{9}$

2. $\frac{7}{10} + \frac{9}{10} =$

3. $\frac{1}{4} + \frac{3}{5} =$

4. $\frac{5}{6} + \frac{2}{3} =$

5. $\frac{3}{8} + \frac{2}{3} =$

6. $\frac{2}{7} + \frac{4}{5} =$

B. **Add the mixed numbers below. Reduce your answers to lowest terms. The first one is started for you.**

7. $2\frac{1}{5} + 3\frac{2}{5}$ = $\frac{3}{5}$

8. $4\frac{5}{12} + 2\frac{7}{12}$

9. $1\frac{1}{4} + 3\frac{2}{3}$

10. $5\frac{2}{3} + 2\frac{2}{5}$

11. $6\frac{5}{8} + \frac{1}{6}$

12. $3\frac{2}{3} + 2\frac{3}{4}$

C. **Solve each problem. Show your work. Reduce your answers to lowest terms.**

13. Angela hiked $4\frac{3}{8}$ miles one day and $2\frac{5}{16}$ miles the next day. How many total miles did Angela hike?

14. Keith works at Whigam Woods. He needs to saw two pieces of wood for a customer. One piece of wood is $12\frac{1}{4}$ feet long, the other is $9\frac{3}{16}$ feet long. How long are the two pieces of wood altogether, in feet?

15. One weekend Raul worked $5\frac{3}{4}$ hours on Saturday and $3\frac{2}{3}$ hours on Sunday. How many hours did Raul work that weekend?

16. To make banana nut bread, Jeanne needs $1\frac{3}{4}$ cups flour, $1\frac{1}{4}$ cups bran, $\frac{1}{2}$ cup nuts, and $\frac{2}{3}$ cup sugar. How many cups do these dry ingredients total?

Check your answers on pages 279–280.

Subtracting Fractions and Mixed Numbers

To subtract like fractions, subtract only the numerators. Write the difference over the common denominator.

Example Subtract: $\frac{7}{8} - \frac{1}{8}$

Step 1 Subtract the numerators. $7 - 1 = 6$

Step 2 Write 6 over the common denominator 8. $\frac{6}{8}$

Step 3 Reduce $\frac{6}{8}$ to lowest terms. $\frac{6 \div 2}{8 \div 2} = \frac{3}{4}$

To subtract unlike fractions, change the unlike fractions to like fractions.

Example Subtract: $\frac{4}{5} - \frac{3}{10}$

Step 1 Use multiples to find a common denominator. Think, "What number will both 5 and 10 divide into evenly?" 10 is the lowest common denominator.

5: 5, (10), 15
10: (10), 20, 30

Step 2 Change $\frac{4}{5}$ to a like fraction with a denominator of 10. $\frac{4 \times 2}{5 \times 2} = \frac{8}{10}$

Step 3 Subtract and reduce the answer to lowest terms. $\frac{8}{10} - \frac{3}{10} = \frac{5}{10} = \frac{1}{2}$

TIP
Any fraction with the same number in both the numerator and denominator has a value of 1. So, $\frac{3}{3}$, $\frac{5}{5}$, and $\frac{100}{100}$ each has a value of 1.

To subtract a fraction or mixed number from a whole number, regroup 1 from the whole number and rewrite it as a fraction.

Example Subtract: $5 - 2\frac{3}{8}$

Step 1 Regroup 1 from the whole number 5 and rewrite it as a fraction with a common denominator of 8.

Step 2 Subtract. The answer, $2\frac{5}{8}$, is in lowest terms.

$$\begin{array}{rcr} 5 & = & 4\frac{8}{8} \\ -2\frac{3}{8} & = & -2\frac{3}{8} \\ \hline & & 2\frac{5}{8} \end{array}$$

Subtracting Mixed Numbers

When you subtract mixed numbers, sometimes the fraction being subtracted is larger than the fraction it is being subtracted from.

Example Subtract: $3\frac{1}{4} - 1\frac{5}{6}$

Step 1 Find a common denominator and write equal fractions.

Step 2 Regroup 1 $(\frac{12}{12})$ from the whole number column. Add it to $\frac{3}{12}$ in the fraction column. $(\frac{12}{12} + \frac{3}{12} = \frac{15}{12})$

Step 3 Subtract. The answer, $1\frac{5}{12}$, is in lowest terms.

$$\begin{array}{rcrcr} 3\frac{1}{4} & = & 3\frac{3}{12} & = & 2\frac{15}{12} \\ -1\frac{5}{6} & = & -1\frac{10}{12} & = & -1\frac{10}{12} \\ \hline & & & & 1\frac{5}{12} \end{array}$$

PRACTICE

A. **Subtract the fractions below. Reduce your answers to lowest terms. The first one is done for you.**

1. $\frac{11}{12} - \frac{5}{12} = \frac{6}{12} = \frac{1}{2}$

2. $\frac{5}{6} - \frac{3}{8} =$

3. $\frac{9}{16} - \frac{1}{4} =$

4. $6 - 2\frac{2}{3} =$

5. $\frac{7}{8} - \frac{2}{3} =$

6. $\frac{4}{5} - \frac{3}{4} =$

B. **Subtract the mixed numbers below. Reduce your answers to lowest terms. The first one is started for you.**

7. $4\frac{7}{8} - 1\frac{3}{8} = 3\frac{4}{8} =$

8. $7\frac{3}{4} - 2\frac{2}{3}$

9. $8 - 5\frac{2}{9}$

10. $6\frac{1}{6} - 3\frac{5}{6}$

11. $9\frac{1}{4} - 8\frac{5}{8}$

12. $7\frac{2}{5} - 1\frac{2}{3}$

C. **Solve each problem. Show your work. Reduce your answers to lowest terms.**

13. Chin swims $9\frac{3}{4}$ miles a week. So far this week he has swum $3\frac{1}{4}$ miles. How many more miles does Chin need to swim this week?

14. At the beginning of the week, the cook at Pasta Unlimited had 35 pounds of pasta. On Monday he used $8\frac{9}{16}$ pounds. How many pounds of pasta are left?

15. Elena mailed two packages at the post office. One package weighed $4\frac{3}{4}$ pounds. The other package weighed $2\frac{1}{2}$ pounds. How many more pounds did the heavier package weigh?

16. Yuki needs a shelf. She cut a board $4\frac{2}{3}$ feet long from a board that is $8\frac{1}{2}$ feet long. How many feet of board are left?

Check your answers on pages 280–281.

DATA ANALYSIS

Reading a Table

Number information, or **data,** can be organized in tables. The table is built with **rows** and **columns.** The rows go across and the columns go up and down. The rows and columns have **labels.** You can find the information you need where the row you want meets the column you need.

Example David, a tailor, is sewing bridesmaid dresses for Jan and Sue. He will use the table at right to find how much fabric to buy. David needs to make dress style C in size 10 for Jan and in size 14 for Sue.

Yardage Requirements

Sizes	10	12	14
Dress A	$2\frac{7}{8}$	3	$3\frac{1}{8}$
Dress B	3	$3\frac{1}{8}$	$3\frac{1}{4}$
Dress C	$3\frac{1}{4}$	$3\frac{3}{8}$	$3\frac{5}{8}$

Which is the correct expression to find the total number of yards of fabric he will need?

(1) $10 - 3\frac{1}{4}$

(2) $3\frac{1}{4} + 3\frac{5}{8}$

(3) $3\frac{1}{4} - 3\frac{5}{8}$

(4) $10 + 3\frac{1}{4}$

(5) $3\frac{5}{8} - 3\frac{1}{4}$

Answer **(2)** is correct. David uses the table to find the yardage he needs: the row for Style C meets the column for size 10 at $3\frac{1}{4}$ and for size 14 at $3\frac{5}{8}$.

Example David may use dress style B instead. David knows he needs $6\frac{7}{8}$ yards for the dresses in style C and $6\frac{1}{4}$ yards for the dresses in style B. Which is the correct expression to find how many more yards of fabric he will need for dresses in style C?

(1) $6\frac{7}{8} \div 6\frac{1}{4}$

(2) $6\frac{7}{8} \times 6\frac{1}{4}$

(3) $6\frac{7}{8} + 6\frac{1}{4}$

(4) $6\frac{7}{8} - 6\frac{1}{4}$

(5) $6 - 6 - \frac{7}{8} - \frac{1}{4}$

Answer **(4)** is correct. To find how many more yards of fabric he will need for dresses in style C, David needs to subtract. $\mathbf{6\frac{7}{8} - 6\frac{1}{4}}$.

PRACTICE

Circle the best answer for each question.

Questions 1–5 refer to the table below.

Yardage Requirements
Child's Jumper and Blouse

Sizes	4	5	6
Jumper	$1\frac{3}{4}$	$1\frac{7}{8}$	2
Blouse A	$1\frac{1}{8}$	$1\frac{1}{4}$	$1\frac{3}{8}$
Blouse B	$1\frac{1}{8}$	$1\frac{1}{8}$	$1\frac{1}{4}$
Facing, Blouse B	$\frac{3}{4}$	$\frac{3}{4}$	$\frac{7}{8}$

1. Charnelle decides to make matching jumpers for her two daughters. One daughter wears size 4 and the other wears size 6. Which is the correct expression to find how many yards of fabric she needs?

(1) $1\frac{3}{4} + 1\frac{7}{8}$

(2) $1\frac{3}{4} - 2$

(3) $1\frac{3}{4} + 2$

(4) $2 - 1\frac{1}{3}$

(5) Not enough information is given.

2. Misha decides to make blouse B and a jumper for her daughter in size 5. Which is the correct expression to find how many yards of fabric and facing she needs?

(1) $1\frac{3}{4} + 1\frac{1}{8} + \frac{3}{4}$

(2) $1\frac{7}{8} + 1\frac{1}{8} + \frac{3}{4}$

(3) $1\frac{7}{8} - 1\frac{1}{8} + \frac{3}{4}$

(4) $1\frac{7}{8} + 1\frac{1}{4} - \frac{3}{4}$

(5) $1\frac{7}{8} + 1\frac{1}{4} + \frac{7}{8}$

3. Russ has a fabric remnant that is 5 yards long. The remnant will be used to make a jumper in size 4. Which is the correct expression to find how many yards of fabric will be left?

(1) $5 - 1\frac{1}{8}$

(2) $5 + 1\frac{3}{4}$

(3) $5 - 1\frac{7}{8}$

(4) $5 + \frac{3}{4}$

(5) $5 - 1\frac{3}{4}$

4. A tailor is making a jumper in size 4 and blouse A in size 6. Which is the correct expression to find the total number of yards of fabric he needs?

(1) $1\frac{3}{4} + 1\frac{3}{8}$

(2) $1\frac{3}{4} - 1\frac{3}{8}$

(3) $1\frac{3}{4} + 1\frac{1}{4}$

(4) $1\frac{7}{8} + 1\frac{3}{8}$

(5) $1\frac{3}{4} + 1\frac{1}{8}$

5. Yvonne has a piece of fabric that is $1\frac{1}{2}$ yards long. Choose the largest size pattern for which Yvonne would have enough fabric.

(1) blouse A, size 4

(2) blouse A, size 6

(3) jumper, size 6

(4) jumper, size 4

(5) jumper, size 5

Check your answers on page 281.

MEASUREMENT AND GEOMETRY

Using a Map

Example Kim drives a delivery truck for a laundry service. The map shows her daily route. Kim begins at point A, drives east to point B, then north to point C. Next she goes west and south through points D, E, and F. Finally, Kim drives south to point A.

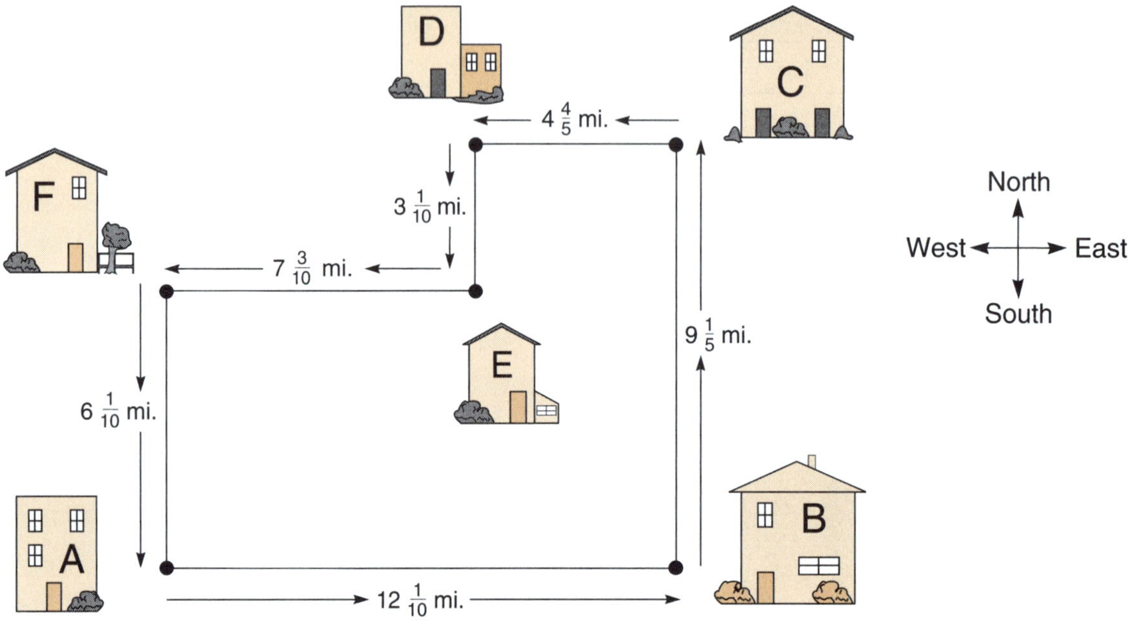

Kim's boss, Miguel, wants her to be at point D on the map by noon. How many miles will Kim drive by noon?

(1) $16\frac{1}{10}$

(2) $24\frac{11}{10}$

(3) $26\frac{1}{10}$

(4) $26\frac{6}{10}$

(5) $26\frac{6}{20}$

Answer **(3)** is correct. Kim covers three distances to get from point A to point D by noon: $12\frac{1}{10}$ miles from A to B, $9\frac{1}{5}$ miles from B to C, $4\frac{4}{5}$ miles from C to D.

$$12\frac{1}{10} + 9\frac{1}{5} + 4\frac{4}{5} = 12\frac{1}{10} + 9\frac{2}{10} + 4\frac{8}{10} = 25\frac{11}{10} = 25 + 1\frac{1}{10} = 26\frac{1}{10} \textbf{ miles}$$

PRACTICE

Circle the best answer for each question.

<u>Questions 1 through 6</u> refer to the following information.

José restocks soda machines for a vending company. The map below shows the route José follows each day.

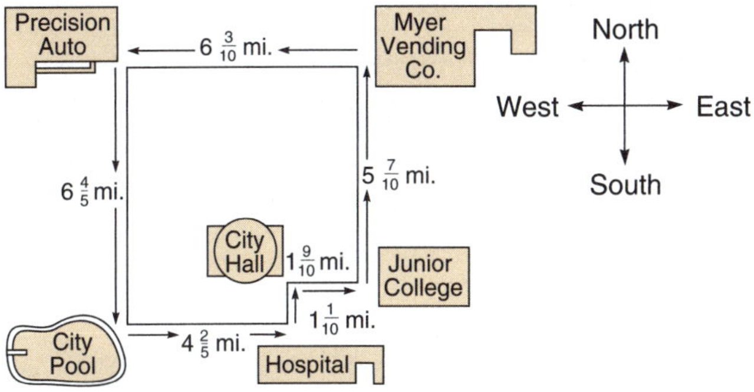

1. During the morning, José restocks machines for all the customers from Myer Vending Co. to the hospital. How many miles does he drive during the morning?

   (1) $13\frac{1}{10}$

   (2) $16\frac{1}{2}$

   (3) $16\frac{4}{5}$

   (4) $16\frac{9}{10}$

   (5) $17\frac{1}{2}$

2. One-Stop Gas and Food is $3\frac{3}{10}$ miles past the city pool on the way to the hospital. How many miles does José drive to get from One-Stop Gas and Food to the hospital?

   (1) $7\frac{7}{10}$

   (2) $7\frac{1}{5}$

   (3) $1\frac{1}{5}$

   (4) $1\frac{1}{10}$

   (5) $1\frac{1}{15}$

3. During the afternoon, José drives from the hospital to Myer Vending Co. How many miles does he drive during the afternoon?

   (1) $8\frac{7}{10}$

   (2) $8\frac{1}{10}$

   (3) $7\frac{7}{10}$

   (4) $7\frac{17}{30}$

   (5) 3

4. How many miles closer to Myer Vending Co. is the junior college than Precision Auto?

   (1) $1\frac{3}{5}$

   (2) $1\frac{2}{5}$

   (3) $1\frac{1}{10}$

   (4) $\frac{3}{5}$

   (5) $\frac{2}{5}$

5. José lives near Myer Vending Co. How many miles does José drive from home to work at Myer Vending Co.?

   (1) $1\frac{1}{10}$

   (2) $4\frac{2}{5}$

   (3) $5\frac{7}{10}$

   (4) $6\frac{3}{10}$

   (5) Not enough information is given.

6. **Connect** Explain how you would estimate the total number of miles of José's route.

Check your answers on page 281.

WORKPLACE MATH

Filling in Timecards

Kareem is an office clerk in a grocery store. His job is to complete the timecards for the store's employees. At the end of the day, Kareem records the time employees worked to the nearest quarter of an hour.

- 15 minutes $= \frac{1}{4}$ hour, because $\frac{15}{60} = \frac{1}{4}$
- 30 minutes $= \frac{1}{2}$ hour, because $\frac{30}{60} = \frac{1}{2}$
- 45 minutes $= \frac{3}{4}$ hour, because $\frac{45}{60} = \frac{3}{4}$

NAME: Jan Lopez
WEEK OF: 3/4 – 3/11

IN	MON.	3/4	8:15 A.M.
OUT	MON.	3/4	12:36 P.M.
IN	MON.	3/4	1:28 P.M.
OUT	MON.	3/4	5:17 P.M.
IN	TUES.	3/5	8:05 A.M.

Example First, Kareem finds the number of hours Jan worked Monday morning. To the nearest quarter of an hour, how many hours did Jan work in the morning?

(1) $3\frac{1}{4}$

(2) 4

(3) $4\frac{1}{6}$

(4) $4\frac{1}{4}$

(5) $5\frac{1}{4}$

Answer **(4)** is correct. Jan started at 8:15 A.M. Kareem thinks of 8:15 as $8\frac{1}{4}$. Jan went to lunch at 12:36 P.M. Since 36 is closest to 30 minutes, Kareem thinks of 12:36 as $12\frac{1}{2}$. Then he subtracts the smaller mixed number from the larger one.

$\mathbf{12\frac{1}{2} - 8\frac{1}{4} = 12\frac{2}{4} - 8\frac{1}{4} = 4\frac{1}{4}}$, so Jan worked about $4\frac{1}{4}$ **hours** in the morning.

Example Next, Kareem finds about how many hours Jan worked after lunch on Monday. To the nearest quarter of an hour, how many hours did Jan work after lunch?

(1) $4\frac{3}{4}$

(2) $4\frac{1}{2}$

(3) 4

(4) $3\frac{3}{4}$

(5) $3\frac{1}{2}$

Answer **(4)** is correct. Jan came back from lunch at 1:28 P.M. Kareem changes 1:28 to $1\frac{1}{2}$. Jan left work at 5:17 P.M. Since 17 is closest to 15, Kareen changes 5:17 to $5\frac{1}{4}$. Then he subtracts the smaller mixed number from the larger one.

$\mathbf{5\frac{1}{4} - 1\frac{1}{2} = 5\frac{1}{4} - 1\frac{2}{4} = 4\frac{5}{4} - 1\frac{2}{4} = 3\frac{3}{4}}$ **hours,** so Jan worked $3\frac{3}{4}$ **hours** after lunch.

PRACTICE

Circle the best answer for each question.

1. Kareem knows Jan worked $4\frac{1}{4}$ hours in the morning and $3\frac{3}{4}$ hours in the afternoon. What was the total number of hours Jan worked during the day?
   (1) 7
   (2) $7\frac{3}{4}$
   (3) 8
   (4) $8\frac{1}{4}$
   (5) $8\frac{1}{2}$

Questions 2 through 4 refer to the timecard at right.

NAME:	Paul Wagner		
WEEK OF:			3/12 – 3/19
IN	TUES.	3/13	8:50 A.M.
OUT	TUES.	3/13	12:32 P.M.
IN	TUES.	3/13	1:24 P.M.
OUT	TUES.	3/13	5:10 P.M.

2. About how many hours did Paul work in the morning?
   (1) $3\frac{1}{4}$
   (2) $3\frac{1}{2}$
   (3) $3\frac{3}{4}$
   (4) $4\frac{1}{4}$
   (5) $5\frac{1}{4}$

3. About how many hours did Paul work in the afternoon?
   (1) $3\frac{3}{4}$
   (2) 4
   (3) $4\frac{1}{2}$
   (4) $4\frac{3}{4}$
   (5) $6\frac{3}{4}$

4. How many total hours did Paul work on Tuesday?
   (1) $7\frac{1}{2}$
   (2) 8
   (3) $8\frac{1}{4}$
   (4) 9
   (5) $9\frac{1}{2}$

5. On Wednesday Paul arrived to work at 8:25 A.M. Later he took a 30-minute lunch and then worked the rest of the afternoon. How many hours did Paul work on Wednesday morning?
   (1) $\frac{1}{2}$ hour
   (2) $3\frac{1}{2}$ hours
   (3) 4 hours
   (4) 8 hours
   (5) Not enough information is given.

6. Maria worked $38\frac{1}{4}$ hours one week and $40\frac{3}{4}$ hours the following week. How many more hours did Maria work the second week than the first?
   (1) 2
   (2) $2\frac{1}{4}$
   (3) $2\frac{1}{2}$
   (4) $2\frac{3}{4}$
   (5) Not enough information is given.

7. **Reason** Explain how you would change the following minutes to fractional parts of an hour. What fraction did you get for each?
   20 minutes
   40 minutes
   50 minutes

Check your answers on pages 281–282.

Adding and Subtracting Fractions on the Calculator

Many scientific calculators have a fraction key. The CASIO *fx-260SOLAR* calculator has the fraction key, (a b/c). This key allows you to enter a fraction or a mixed number without first converting to a decimal. Also, the answer on the display can show a whole number, fraction, mixed number, or decimal.

As with whole numbers, the (+) and (−) keys are used to add and subtract. Once you know how to use the fraction key, adding and subtracting fractions and mixed numbers will be easier.

The CASIO *fx-260SOLAR* Calculator

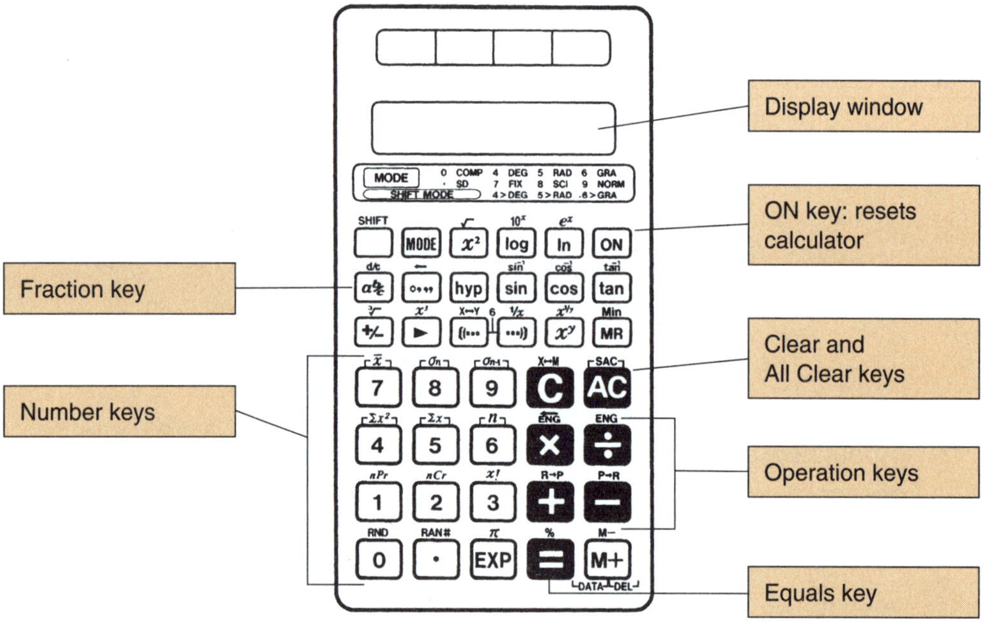

Adding Fractions and Mixed Numbers

Use the (a b/c) key and the (+) key to add fractions and mixed numbers.

Example Sammi worked $3\frac{1}{4}$ hours in the morning and $4\frac{1}{2}$ hours in the afternoon. How many hours did she work in all?

Step 1 Use the fraction key to enter $3\frac{1}{4}$.

Enter 3 (a b/c) 1 (a b/c) 4 The display will show [3 ⌟ 1 ⌟ 4.]

Step 2 Now use the addition key and the fraction key to add $4\frac{1}{2}$ to $3\frac{1}{4}$.

Enter (+) 4 (a b/c) 1 (a b/c) 2 The display will show [4 ⌟ 1 ⌟ 2.]

Step 3 To find the total of the two amounts, press (=).

The display will show [7 ⌟ 3 ⌟ 4.]. Sammi worked $7\frac{3}{4}$ **hours** in all.

Subtracting Fractions and Mixed Numbers

Use the (a b/c) key and the (−) key to subtract fractions and mixed numbers.

Example Neal worked $8\frac{1}{2}$ hours on Monday and $6\frac{3}{4}$ hours on Tuesday. How many more hours did he work on Monday than on Tuesday?

Step 1 Use the fraction key to enter $8\frac{1}{2}$.

Enter 8 (a b/c) 1 (a b/c) 2 The display will show [8⌟1⌟2.]

Step 2 Now use the subtraction key and the fraction key to subtract $6\frac{3}{4}$ from $8\frac{1}{2}$.

Enter (−) 6 (a b/c) 3 (a b/c) 4 The display will show [6⌟3⌟4.]

Step 3 To find the difference, press (=).

The display will show [1⌟3⌟4.]. Neal worked **$1\frac{3}{4}$ hours** more on Monday.

PRACTICE

A. Use your calculator to find the answer.

1. $\frac{3}{4} + \frac{5}{16} =$ ______
2. $\frac{11}{15} - \frac{2}{3} =$ ______
3. $\frac{7}{8} + \frac{2}{5} =$ ______
4. $\frac{34}{45} - \frac{5}{9} =$ ______
5. $\frac{23}{40} + \frac{3}{5} =$ ______
6. $2\frac{1}{4} + 1\frac{7}{8} =$ ______
7. $5\frac{1}{5} - 2\frac{3}{10} =$ ______
8. $7\frac{1}{3} + 2\frac{1}{8} =$ ______
9. $8\frac{3}{10} - 5\frac{8}{15} =$ ______
10. $20\frac{4}{5} - 12\frac{7}{8} =$ ______

B. Read each problem carefully. Use your calculator to solve the problems.

11. Alfredo has a board that measures $8\frac{1}{4}$ feet. How much will he have left after cutting off $4\frac{1}{8}$ feet?

______

12. Marcia needs $12\frac{1}{2}$ feet of gutter for one side of a shed and $8\frac{5}{16}$ feet of gutter for another side. How many feet of gutter does she need?

______

13. Andy needs $2\frac{3}{4}$ cups of chocolate chips and $1\frac{1}{2}$ cups of peanut butter chips for a cookie recipe. How many cups of chips are needed in all?

______

14. Carol worked $38\frac{1}{3}$ hours last week and $42\frac{1}{2}$ hours this week. How many more hours did she work this week than last week?

______

Check your answers on page 282.

LESSON 6

Lesson 6 Review
Lesson 6 Review

Lesson 6 Review

Add or subtract. Reduce your answers to lowest terms.

1. $\frac{15}{16} - \frac{9}{16} =$

2. $\frac{1}{5} + \frac{7}{10} =$

3. $8 - 2\frac{1}{6} =$

4. $\frac{5}{6} + \frac{3}{8} =$

5. $\begin{array}{r} 3\frac{7}{9} \\ +4\frac{2}{9} \\ \hline \end{array}$

6. $\begin{array}{r} 4\frac{9}{10} \\ -2\frac{3}{10} \\ \hline \end{array}$

7. $\begin{array}{r} 2\frac{5}{8} \\ +2\frac{3}{4} \\ \hline \end{array}$

Use your calculator to solve these problems.

8. $9\frac{1}{5} - 8\frac{3}{5} =$ ______

9. $3 - \frac{7}{8} =$ ______

10. $6\frac{1}{3} + 2\frac{3}{4} =$ ______

Questions 11 through 13 refer to the table at right.

Yardage Requirements
Prom Dress

Sizes	8	10	12
Dress A	$5\frac{3}{4}$	$5\frac{7}{8}$	$6\frac{3}{8}$
Dress B			
Main Color	$5\frac{1}{8}$	$5\frac{1}{4}$	$5\frac{5}{8}$
Contrasting Color	$1\frac{3}{8}$	$1\frac{3}{8}$	$1\frac{1}{2}$
Lace A trim	$2\frac{1}{2}$	$2\frac{5}{8}$	$2\frac{3}{4}$
Lace B trim	$2\frac{1}{8}$	$2\frac{1}{8}$	$2\frac{1}{4}$

11. Jenny would like to make dress B in size 12. How many yards of fabric will she need?
    (1) $3\frac{1}{4}$
    (2) $6\frac{5}{8}$
    (3) $7\frac{1}{8}$
    (4) $8\frac{1}{2}$
    (5) $9\frac{1}{8}$

12. Caryn is trying to decide whether she wants to make dress A or dress B. How many more yards of lace are needed if she chooses dress A in size 8 rather than dress B in size 8?
    (1) $\frac{1}{8}$
    (2) $\frac{1}{6}$
    (3) $\frac{3}{8}$
    (4) $\frac{5}{8}$
    (5) $4\frac{5}{8}$

13. Amanda is making two style A prom dresses in sizes 8 and 10. What is the total amount of lace she will need for the two dresses?
    (1) $2\frac{1}{4}$
    (2) $2\frac{3}{4}$
    (3) $5\frac{1}{8}$
    (4) $10\frac{3}{8}$
    (5) $11\frac{5}{8}$

Solve each problem.

Questions 14 through 17 refer to the following information.

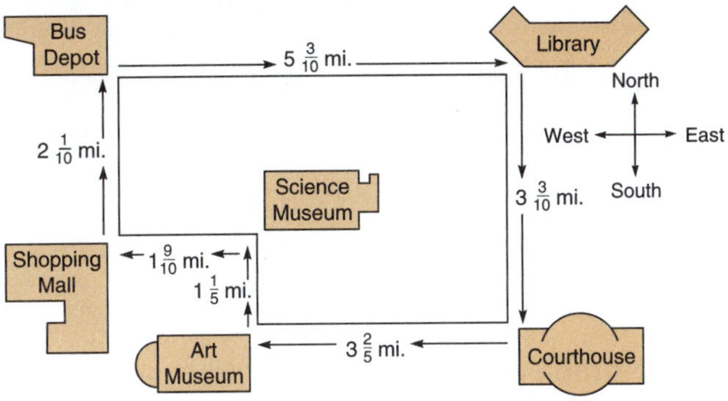

Sundra is a city bus driver. The map above shows the route she drives starting at the bus depot.

14. The busiest part of Sundra's route is between the courthouse and the shopping mall. About how many miles is this part of the route?

15. Exactly how many miles is the part of Sundra's route between the courthouse and the shopping mall?

16. Sundra then returns to the bus depot. What is the total distance from the courthouse to the bus depot?

17. Sundra's entire route is $17\frac{1}{5}$ miles long. Estimate how many miles she has left to drive after she has reached the library.

Questions 18 through 20 refer to the timecard below.

NAME:		Joy Chan	
WEEK OF:		4/4 – 4/11	
IN	MON.	4/4	8:24 A.M.
OUT	MON.	4/4	12:13 P.M.
IN	MON.	4/4	12:58 P.M.
OUT	MON.	4/4	5:43 P.M.

18. Rounded to the nearest quarter of an hour, how many hours did Joy work in the morning?

19. How many hours did Joy work in the afternoon?

20. How many total hours did Joy work on Monday?

Check your answers on pages 282–284.

Lesson 7

Multiplying and Dividing Fractions

Multiplying and Dividing Fractions

Multiplying Fractions and Mixed Numbers

Multiplying Fractions

You do not need to find a common denominator to multiply and divide fractions. To multiply one fraction by another, multiply the numerators. Then multiply the denominators.

Example Multiply: $\frac{2}{3} \times \frac{3}{8}$

Step 1 Multiply the numerators. $\frac{2}{3} \times \frac{3}{8} = \frac{6}{?}$

Step 2 Multiply the denominators. $\frac{2}{3} \times \frac{3}{8} = \frac{6}{24}$

Step 3 Reduce the answer to lowest terms. $\frac{6}{24} = \frac{6 \div 6}{24 \div 6} = \frac{1}{4}$

A shortcut called **canceling** will make your work easier. Let's work the same problem again using the shortcut.

TIP

Canceling is a way to simplify a problem with fractions before you multiply.

$\frac{\overset{1}{1}}{\underset{1}{\cancel{2}}} \times \frac{\overset{2}{\cancel{4}}}{5} = \frac{2}{5}$

Example Multiply: $\frac{2}{3} \times \frac{3}{8}$

Step 1 Look at the numerator of one fraction and the denominator of the other fraction. Think of a number that will evenly divide into both. Both 2 and 8 can be divided by 2. Divide both by 2. Cross out the 2 and write 1 to show 2 ÷ 2 = 1. Cross out the 8 and write 4 to show 8 ÷ 2 = 4. $\frac{\overset{1}{\cancel{2}}}{3} \times \frac{3}{\underset{4}{\cancel{8}}}$

Step 2 Look at the numerator of the second fraction and the denominator of the other fraction and repeat the process. Since both numbers are 3, you can divide each by 3. $\frac{\overset{1}{\cancel{2}}}{\underset{1}{\cancel{3}}} \times \frac{\overset{1}{\cancel{3}}}{\underset{4}{\cancel{8}}} = \frac{1}{4}$

Step 3 Multiply the new numerators. Multiply the new denominators.

Multiplying Mixed Numbers

You can also multiply fractions by mixed numbers and whole numbers.

Example Multiply: $4\frac{1}{2} \times 2\frac{1}{3}$

Step 1 Change the mixed numbers to improper fractions.

$$4\frac{1}{2} \times 2\frac{1}{3} = \frac{9}{2} \times \frac{7}{3}$$

Step 2 Use canceling to reduce fractions.

Step 3 Multiply. If your answer is an improper fraction, simplify it by changing the improper fraction to a mixed whole number.

$$\frac{\overset{3}{\cancel{9}}}{2} \times \frac{7}{\underset{1}{\cancel{3}}} = \frac{21}{2} = 10\frac{1}{2}$$

PRACTICE

A. **Multiply the fractions. Use canceling before you multiply, or reduce your answers to lowest terms. The first one is done for you.**

1. $\frac{\overset{1}{\cancel{2}}}{3} \times \frac{1}{\underset{2}{\cancel{4}}} = \frac{1}{6}$
2. $\frac{3}{8} \times \frac{4}{5} =$
3. $\frac{3}{7} \times \frac{2}{9} =$

B. **Multiply the fractions by whole numbers or mixed numbers. Use canceling before you multiply, or reduce your answers to lowest terms. The first one is started for you.**

4. $\frac{3}{8} \times 6 = \frac{3}{8} \times \frac{6}{1} =$
5. $5 \times 2\frac{3}{10} =$
6. $\frac{5}{8} \times 2\frac{2}{5} =$
7. $3\frac{1}{2} \times \frac{8}{9} =$
8. $2\frac{1}{3} \times 1\frac{1}{2} =$
9. $1\frac{4}{5} \times 6\frac{2}{3} =$

C. **Solve each problem. Show your work. Use canceling before you multiply, or reduce your answers to lowest terms.**

10. On Saturday 24 people bought cars at Best Car Sales. Of the 24 customers, $\frac{3}{4}$ bought new cars. How many people bought new cars?

11. Donna can walk at the rate of $3\frac{1}{2}$ miles per hour. If she keeps up this pace, how far can she walk in $1\frac{1}{2}$ hours?

12. The Sugar Bowl ordered $15\frac{3}{4}$ pounds of almonds. The nut supplier sent only $\frac{2}{3}$ of the order. How many pounds of almonds did the supplier send?

13. Tony can pick $4\frac{1}{4}$ pints of berries per hour. How many pints can he pick in 6 hours?

Check your answers on page 284.

Dividing Fractions and Mixed Numbers

Dividing Fractions

Dividing fractions is the same as multiplying fractions—with one important difference. You must **invert** the fraction you are dividing by. When you invert the fraction you are dividing by, also change the division sign to a multiplication sign. **Note:** Do not cancel before inverting.

TIP

Invert means to turn over or to switch the numerator and the denominator. For example:

$\frac{2}{3}$ becomes $\frac{3}{2}$

$\frac{1}{4}$ becomes $\frac{4}{1} = 4$

$2 = \frac{2}{1}$ becomes $\frac{1}{2}$

Example Divide: $\frac{3}{8} \div \frac{1}{2}$

Step 1 Invert the fraction you are dividing by. Change the ÷ sign to ×.

$$\frac{3}{8} \div \frac{1}{2} = \frac{3}{8} \times \frac{2}{1}$$

Step 2 Use canceling to reduce both fractions. Multiply the fractions.

$$\frac{3}{\cancel{8}_4} \div \frac{\cancel{2}^1}{1} = \frac{3}{4}$$

Dividing Mixed Numbers

To divide by a mixed number or a whole number, change it to an improper fraction before you invert.

Example Divide: $6 \div 2\frac{2}{5}$

$$6 \div 2\frac{2}{5}$$

Step 1 Change the whole number and mixed number to improper fractions.

$$6 = \frac{6}{1}$$

$$2\frac{2}{5} = \frac{2 \times 5}{5} + \frac{2}{5} = \frac{12}{5}$$

$$6 \div 2\frac{2}{5} = \frac{6}{1} \div \frac{12}{5}$$

Step 2 Invert the fraction you are dividing by, and change the ÷ sign to ×.

$$\frac{6}{1} \times \frac{5}{12}$$

Step 3 Use canceling to reduce. Multiply.

$$\frac{\cancel{6}^1}{1} \times \frac{5}{\cancel{12}_2} = \frac{5}{2}$$

Step 4 Reduce your answer to lowest terms.

$$\frac{5}{2} = 2\frac{1}{2}$$

Example Divide: $5\frac{1}{3} \div 4\frac{1}{2}$

$$5\frac{1}{3} \div 4\frac{1}{2}$$

Step 1 Change the mixed numbers to improper fractions. **Note:** Do not cancel before inverting.

$$5\frac{1}{3} = \frac{5 \times 3}{3} + \frac{1}{3} = \frac{16}{3}$$

$$4\frac{1}{2} = \frac{4 \times 2}{2} + \frac{1}{2} = \frac{9}{2}$$

$$5\frac{1}{3} \div 4\frac{1}{2} = \frac{16}{3} \div \frac{9}{2}$$

Step 2 Invert the fraction you are dividing by, and change the ÷ sign to ×.

$$\frac{16}{3} \times \frac{2}{9}$$

Step 3 Multiply. (Note that there is no canceling to be done.)

$$\frac{16}{3} \times \frac{2}{9} = \frac{32}{27}$$

Step 4 Reduce your answer to lowest terms.

$$\frac{32}{27} = 1\frac{5}{27}$$

PRACTICE

A. **Divide the fractions. Use canceling whenever possible. Reduce your answers to lowest terms. The first one is done for you.**

1. $\frac{4}{5} \div \frac{1}{3} = \frac{4}{5} \times \frac{3}{1}$
   $= \frac{12}{5} = 2\frac{2}{5}$

2. $\frac{5}{6} \div \frac{5}{8} =$

3. $\frac{4}{15} \div \frac{4}{5} =$

4. $\frac{2}{3} \div \frac{5}{6} =$

B. **Divide by whole numbers or mixed numbers. Use canceling whenever possible. Reduce your answers to lowest terms. The first one is started for you.**

5. $\frac{7}{12} \div 3 = \frac{7}{12} \times \frac{3}{1} =$
   $\frac{7}{12} \times \frac{1}{3} =$

6. $2 \div 1\frac{1}{7} =$

7. $\frac{5}{8} \div 1\frac{5}{6} =$

8. $3\frac{3}{4} \div \frac{3}{10} =$

9. $1\frac{2}{5} \div 2\frac{1}{4} =$

10. $4\frac{1}{4} \div 1\frac{1}{2} =$

C. **Solve each problem. Show your work. Use canceling whenever possible. Reduce your answers to lowest terms.**

11. The 15 acres of vacant land across the street from Dominic's house are being developed as building lots. Each lot is to be $\frac{5}{8}$ acre. How many building lots will there be?

12. Sharla bought $5\frac{1}{2}$ pounds of ground beef. Before freezing the ground beef, she plans to make it into hamburger patties that weigh $\frac{1}{4}$ pound each. How many patties can Sharla make?

13. A hiking trail in a national park is $6\frac{1}{4}$ miles long. Roger averages $2\frac{1}{2}$ miles per hour. How many hours will it take him to hike the trail?

14. Gina wants to have music playing continuously in her store during business hours. The store is open $12\frac{1}{2}$ hours every day. Each music tape lasts $1\frac{1}{4}$ hours. How many tapes does she need each day?

Check your answers on page 284.

PROBLEM SOLVING

Make a Diagram

Sometimes making a diagram may help you understand and solve a problem.

Example Tien works for a company that makes and repairs furniture. She cuts boards into lengths needed to make sofas. Tien has an order form that tells her to cut four pieces of lumber measuring $16\frac{1}{8}$ inches each. Tien knows that each cut she makes will waste $\frac{1}{8}$ inch of the board. Tien makes a diagram of the situation.

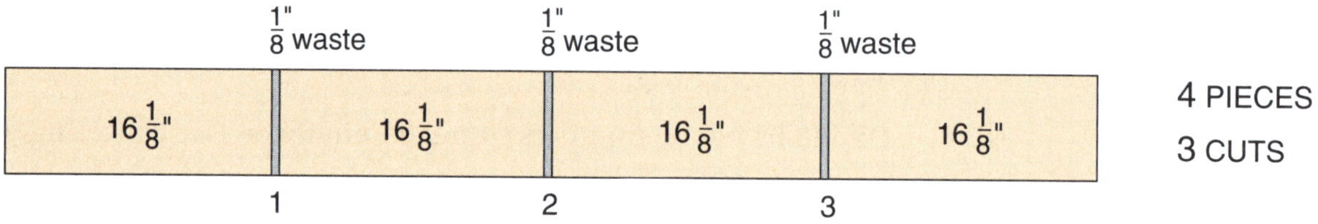

Look at the diagram above. Which expression finds the total amount of waste that will result from cutting the board?

(1) $3 \times \frac{1}{8}$

(2) $3 \div \frac{1}{4}$

(3) $3 \times 16\frac{1}{8}$

(4) $4 \times \frac{1}{8}$

(5) $4 \times 16\frac{1}{8}$

Answer (1) is correct. Tien will make three cuts; each cut will waste $\frac{1}{8}$ inch of wood. $\mathbf{3 \times \frac{1}{8} = \frac{3}{1} \times \frac{1}{8} = \frac{3}{8}}$ **inch,** so $\frac{3}{8}$ **inch** will be wasted in the cutting.

Example Tien wants to use 1 board for this order. Think of the number of pieces needed, the length of the pieces, and the amount of wood that will be wasted. In inches, what is the minimum length of board she could use?

(1) 48

(2) $64\frac{1}{2}$

(3) $64\frac{7}{8}$

(4) $64\frac{6}{8}$

(5) 65

Answer (3) is correct. Tien needs $64\frac{1}{2}$ inches for the four pieces of wood ($16\frac{1}{8} \times 4$) and $\frac{3}{8}$ inch for the waste. $64\frac{1}{2} + \frac{3}{8} = 64\frac{4}{8} + \frac{3}{8} =$ **$64\frac{7}{8}$ inches,** so Tien needs a board with a minimum length of $64\frac{7}{8}$ inches.

PRACTICE

Circle the best answer for each question.

Questions 1 through 3 refer to the following information.

Tien needs to cut five shelves for a cabinet. Each shelf is to be $1\frac{2}{3}$ feet long. She would like to cut the shelves from one piece of board and not have any wood left over. Each cut will waste $\frac{3}{16}$ inch of wood.

1. Not including the waste, how many inches of wood are needed for five shelves?
   (1) 180
   (2) 100
   (3) 96
   (4) 60
   (5) 8

2. How many cuts are needed to make five shelves from one board? Make a diagram to support your answer.

   (1) 1
   (2) 2
   (3) 3
   (4) 4
   (5) 5

3. What is the total amount of waste, in inches, Tien will have from cutting the shelves?
   (1) $\frac{1}{4}$
   (2) $\frac{3}{8}$
   (3) $\frac{9}{16}$
   (4) $\frac{5}{8}$
   (5) $\frac{3}{4}$

Questions 4 through 6 refer to the following information.

Miriam works at the same company Tien does. She needs to cut four boards, each $6\frac{7}{8}$-feet long. Miriam would like to cut the four boards from a single, longer board.

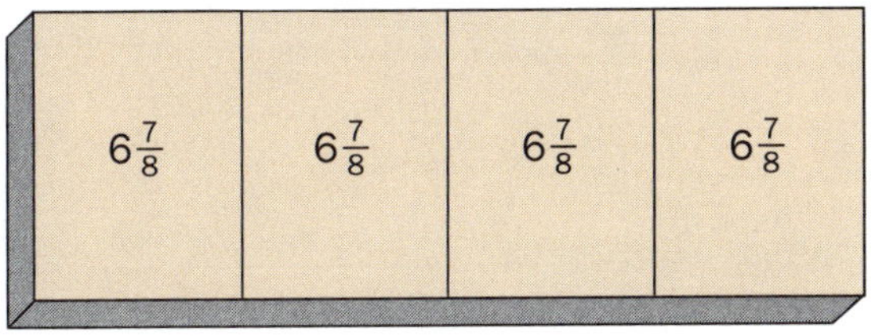

4. Not including the waste, what is the total length, in feet, of the four boards Miriam needs to cut?
   (1) $6\frac{7}{8}$
   (2) 24
   (3) $24\frac{7}{8}$
   (4) $27\frac{1}{2}$
   (5) 52

5. If a total of $2\frac{1}{4}$ inches will be wasted by the cuts made to the longer board, how many total inches will be wasted with each cut?
   (1) $\frac{2}{3}$
   (2) $\frac{3}{4}$
   (3) $2\frac{1}{4}$
   (4) $2\frac{3}{4}$
   (5) $6\frac{3}{4}$

6. **Communicate** Draw a diagram that shows how many cuts you would make to a board to get six smaller, but equal, boards.

Check your answers on pages 284–285.

Multiplying and Dividing Fractions and Mixed Numbers on the Calculator

On the CASIO *fx*-260, you can use the fraction key [a b/c] to multiply and divide fractions and mixed numbers. As with whole numbers, the [×] and [÷] keys are used to multiply and divide.

The CASIO *fx-260SOLAR* Calculator

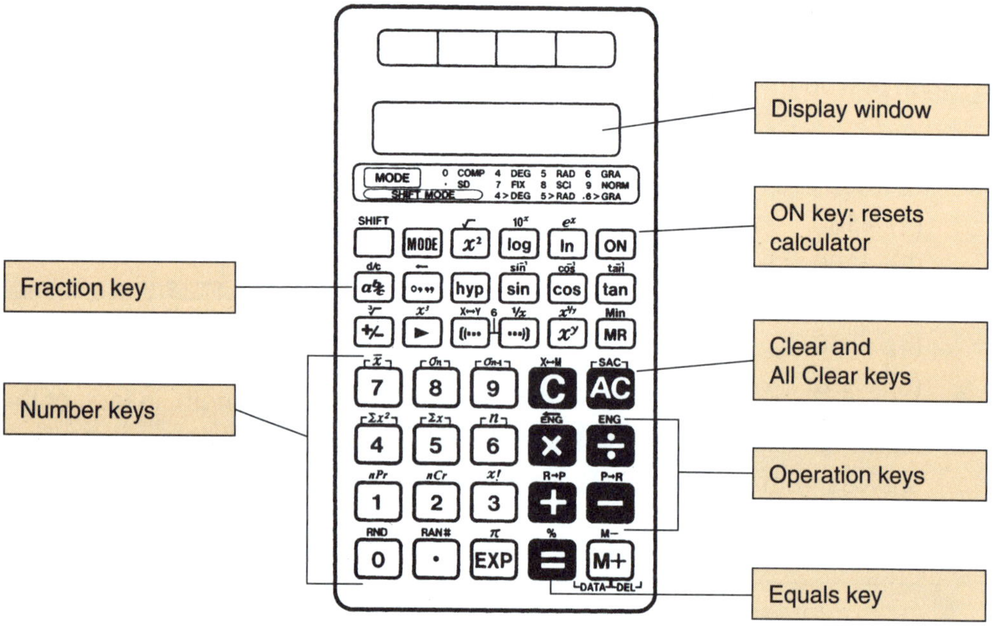

Use the [a b/c] key and the [×] key to multiply fractions and mixed numbers.

Example Monica bought 5 packages of peanuts. Each package contained $2\frac{1}{4}$ ounces. How many ounces of peanuts did Monica buy in all?

Step 1 Use the fraction key to enter $5 \times 2\frac{1}{4}$.

Enter 5 [×] 2 [a b/c] 1 [a b/c] 4

The display will show the last number entered. [2⌟1⌟4.]

Step 2 To find the product of the two amounts, press [=].

The display will show [11⌟1⌟4.]. Monica bought **$11\frac{1}{4}$ ounces** of peanuts.

Example If a recipe calls for $\frac{3}{4}$ teaspoon salt, how much salt should Andre use if he triples the recipe?

Step 1 Enter 3 [×] 3 [a b/c] 4 The display shows the last amount entered. [3⌟4.]

Step 2 Press [=]. The display will show [2⌟1⌟4.]. Andre should use **$2\frac{1}{4}$ teaspoons** of salt.

Dividing Fractions and Mixed Numbers

You can use the [a b/c] key and the [÷] key to divide fractions and mixed numbers.

Example Bert bought a large package of chicken wings that he wants to divide into smaller bags to freeze. How many $1\frac{1}{2}$-pound bags can he make from a $7\frac{1}{2}$-pound package?

Step 1 Use the fraction key to enter $7\frac{1}{2} \div 1\frac{1}{2}$.

Enter 7 [a b/c] 1 [a b/c] 2 [÷] 1 [a b/c] 1 [a b/c] 2

The display will show the last number entered. [1⌟1⌟2.]

Step 2 Now press [=]. The display will show the final result. Bert can fill 5 of the $1\frac{1}{2}$-pound bags.

PRACTICE

A. Use your calculator to find the answers.

1. $5 \times \frac{7}{10} =$ ______
2. $\frac{4}{5} \times \frac{8}{9} =$ ______
3. $2\frac{1}{4} \div 4 =$ ______
4. $9 \div \frac{3}{5} =$ ______
5. $10\frac{1}{2} \times \frac{1}{8} =$ ______
6. $2\frac{1}{5} \div 1\frac{1}{10} =$ ______
7. $8\frac{3}{8} \times 40 =$ ______
8. $3\frac{1}{3} \div \frac{1}{3} =$ ______
9. $20\frac{1}{4} \times \frac{1}{5} =$ ______
10. $35 \div 1\frac{3}{4} =$ ______

B. Use your calculator to solve these problems.

11. Robert has a bag that contains 20 cups of flour. How many loaves of bread can he make if each loaf uses $2\frac{1}{2}$ cups of flour?

    ______

12. Carla uses about $3\frac{3}{4}$ feet of ribbon for every package she wraps. How many feet of ribbon will she need to wrap 15 packages?

    ______

13. Mike's dog gets $2\frac{1}{3}$ cups of dry food at mealtime. Feeding his dog once a day, how many cups of dog food will Mike use in 14 days?

    ______

14. Emily made $4\frac{1}{2}$ pounds of fudge. If she wants to evenly divide it among 6 friends, how much fudge will each friend get?

    ______

Check your answers on page 285.

Filling In Fractions on a Standard Grid

Some mathematics test questions require you to fill in your answer as a fraction or an improper fraction on a standard grid.

To correctly fill in a fraction or mixed number, use the information below.

- You can begin your answer in any column as long as it fits on the grid.
- Write the answer including the fraction bar in the boxes on the top row.
- Fill in the correct circle below each filled-in box. Fill in only one circle in each column.
- Leave any unused column blank.

Example Renee used $\frac{1}{4}$ cup of water and $\frac{1}{2}$ cup of milk for a recipe. How many cups of liquid did she use?

Step 1 Add to find the total amount of liquid used.
$\frac{1}{4} + \frac{1}{2} = \frac{1}{4} + \frac{2}{4} = \frac{3}{4}$ cup

Step 2 Write the fraction in the top row of the grid. Be sure to write the fraction bar in its own column.

Step 3 Fill in the matching circle under each filled-in column. Leave any unused columns blank.

Entering Mixed Numbers on a Standard Grid

To fill in a mixed number on the standard grid, you must first change it to an improper fraction or a decimal.

Example Cameron travels a total of $9\frac{1}{4}$ miles to and from work. How many miles does he travel one way?

Step 1 To find half of the distance, multiply by $\frac{1}{2}$. This is the same as dividing by 2.
$9\frac{1}{4} \times \frac{1}{2} = \frac{37}{4} \times \frac{1}{2} = \frac{37}{8}$

Note: Do not change the answer to a mixed number.

Step 2 Write the improper fraction in the top row of the grid. Be sure to write the fraction bar in its own column.

Step 3 Fill in the matching circle under each filled-in column. Leave any unused columns blank.

PRACTICE

Solve each problem. Then fill in your answer on the grid.

1. Sally plans to build a house on two lots that are side by side. One lot is $1\frac{2}{3}$ acres and the other is $1\frac{3}{4}$ acres. What is the total acreage of the two lots?

   Mark your answer in the circles in the grid.

2. A piece of wire $5\frac{1}{2}$ feet long is cut into 4 equal pieces. How many feet is each piece?

   Mark your answer in the circles in the grid.

3. If Marcus jogs $2\frac{1}{2}$ miles every day, how many miles will he jog in 5 days?

   Mark your answer in the circles in the grid.

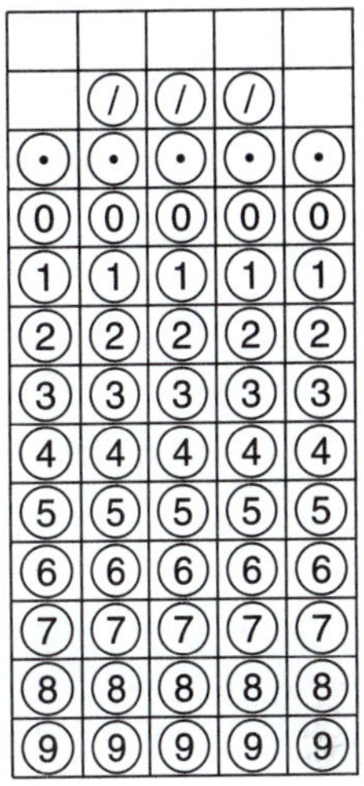

4. Pat worked construction for $7\frac{3}{4}$ hours on Monday and $9\frac{1}{4}$ hours on Friday. How many hours longer did Pat work on Friday than on Monday?

   Mark your answer in the circles in the grid.

Check your answers on page 285.

LESSON 7

Lesson 7 Review
Lesson 7 Review

Lesson 7 Review

Multiply or divide. Reduce your answers to lowest terms.

1. $\frac{4}{5} \times \frac{1}{3} =$

2. $\frac{4}{9} \div \frac{1}{6} =$

3. $\frac{5}{8} \div \frac{3}{5} =$

4. $\frac{2}{7} \times \frac{5}{12} =$

5. $7 \div \frac{7}{8} =$

6. $6 \times \frac{3}{4} =$

7. $\frac{4}{15} \times \frac{9}{10} =$

8. $\frac{2}{5} \div \frac{4}{9} =$

9. $4 \times 2\frac{5}{6} =$

10. $5 \div 1\frac{1}{9} =$

11. $2\frac{3}{4} \times \frac{2}{3} =$

12. $4\frac{1}{2} \div 1\frac{1}{5} =$

13. $3\frac{1}{5} \div 2\frac{2}{5} =$

14. $2\frac{5}{12} \times 1\frac{5}{7} =$

Use your calculator to solve these problems.

15. $\frac{7}{12} \times 3\frac{3}{5} =$ ____________

16. $\frac{5}{6} \div 2\frac{1}{2} =$ ____________

17. $2\frac{6}{7} \times 1\frac{5}{16} =$ ____________

18. $3\frac{1}{3} \div 4\frac{1}{6} =$ ____________

Circle the best answer for each question.

19. Richard cooks at a restaurant. He is making meatballs, and the recipe calls for $\frac{3}{4}$ cup of bread crumbs. He plans to double the recipe. Which is the correct expression to find the number of cups of bread crumbs he should use?

(1) $\frac{3}{4} \times \frac{1}{2}$
(2) $\frac{3}{4} \times 1$
(3) $\frac{3}{4} \times 2$
(4) $\frac{3}{4} \div 2$
(5) $2 \div \frac{3}{4}$

20. Carla's recipe for potato salad serves 18 people and calls for $1\frac{7}{8}$ cups of mayonnaise. Carla wants to make enough salad to serve 6 people. How many cups of mayonnaise should she use?

(1) $\frac{5}{16}$
(2) $\frac{5}{8}$
(3) $\frac{15}{16}$
(4) $3\frac{3}{4}$
(5) $5\frac{5}{8}$

21. John wants seven shelves for his bookcase. Each shelf is to be $2\frac{1}{2}$ feet long. From one long board, how many cuts are needed to make the seven shelves? Make a diagram to support your answer.

(1) 3
(2) 4
(3) 5
(4) 6
(5) 7

Solve each problem. Then fill in the answer in the grid.

22. Delia is a secretary at Auto Insurance Agency. She works $7\frac{1}{2}$ hours a day. Delia answers the telephone $\frac{1}{3}$ of each day. How many hours does Delia spend answering the telephone each day?

Mark your answer in the circles in the grid.

23. Maggie makes braided rugs to sell at craft shows. She needs to cut strips of fabric $1\frac{1}{8}$ inches wide for the rugs. How many strips can she cut from a piece of fabric 45 inches wide?

Mark your answer in the circles in the grid.

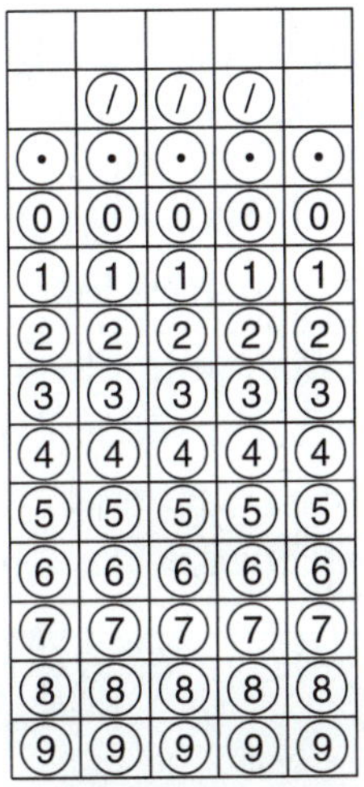

Check your answers on page 286.

Math at Work

Construction: Carpenter

Carpenters build many different types of structures. They measure, cut, and fit together the materials and pieces used to make a building's framework—the walls, floors, roofs, and spaces inside. Carpenters also work on the finishing touches inside buildings. They hang doors or cabinets, put on molding and baseboards, and build bookcases. Carpenters may also work on renovations and additions to existing buildings.

Carpenters must have excellent math and measuring skills. They apply these skills when they read blueprints, and measure and cut the materials used in the building's construction. Blueprints are line drawings that show the exact position and measurements of each part of the building. Blueprints frequently include fractional measurements that the carpenter must be able to interpret and use.

Look at the chart showing some careers in construction.

- Do any of the careers interest you? If so, which ones?
- What information would you need to find out more about those careers? On a separate piece of paper, write some questions that you would like answered. You can find out more information about those careers in the *Occupational Outlook Handbook* at your local library or online.

Some Careers in Construction

Carpet Installer
removes any existing carpet and lays new padding and carpeting

Electrician
installs, repairs, and rewires buildings for electricity

Painter
prepares walls for painting; selects color and applies paint to surface

Tilesetter
prepares floor for new tile; lays tiles in specific pattern

Carpenters often purchase the materials to be used on the job. Use the diagram below to answer the questions that follow.

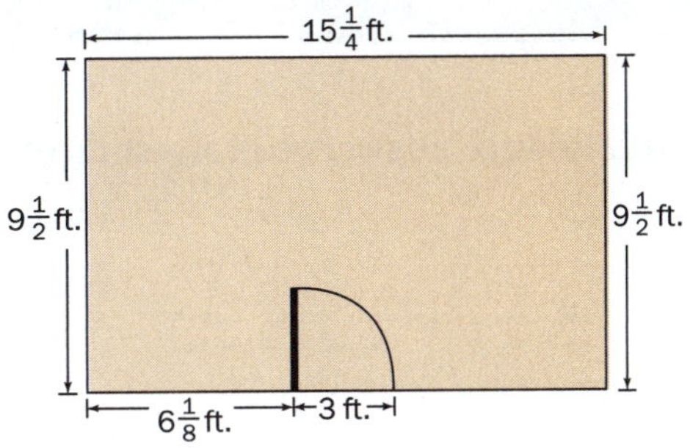

Jake has been hired to put baseboards and ceiling molding around the bedroom in the diagram above.

Definitions you will need:

Baseboard: decorative wood strips along the perimeter of a room. Strips do not go across doorways, but they do go under windows.

Ceiling molding: decorative wood strips placed along the ceiling of a room. Strips do run on ceiling area over doorways and windows.

1. What is the measurement of the wall to the right of the doorway?

   (1) $15\frac{1}{4}$ feet

   (2) $9\frac{1}{8}$ feet

   (3) $6\frac{1}{8}$ feet

   (4) 3 feet

   (5) Not enough information is given.

2. How many feet of baseboard does Jake need for this room?

   (1) $49\frac{1}{2}$

   (2) $46\frac{1}{2}$

   (3) $30\frac{1}{2}$

   (4) 19

   (5) Not enough information is given.

3. What is the total length of ceiling molding and baseboard that Jake needs for this room? Explain how you got your answer.

Check your answers on page 287.

Unit 2 Review Fractions

Solve. Show your work. Reduce answers to lowest terms.

1. $\frac{5}{6} + \frac{3}{4} =$

2. $\frac{9}{10} - \frac{2}{3} =$

3. $\frac{2}{5} + \frac{3}{10} + \frac{1}{4} =$

4. $\begin{array}{r} 4\frac{7}{8} \\ +2\frac{1}{3} \\ \hline \end{array}$

5. $\begin{array}{r} 3\frac{2}{5} \\ -1\frac{3}{4} \\ \hline \end{array}$

6. $\frac{7}{12} \div \frac{5}{8} =$

7. $\frac{15}{16} \times \frac{8}{9} =$

8. $3\frac{1}{3} \times 2\frac{1}{2} =$

9. $3\frac{3}{8} \div 3\frac{3}{5} =$

Use your calculator to find each answer.

10. $\frac{1}{4} + \frac{2}{3} =$ ______

11. $2\frac{3}{4} - 1\frac{1}{3} =$ ______

12. $11 \times 5\frac{1}{16} =$ ______

13. $18 \div 2\frac{1}{4} =$ ______

14. $19\frac{3}{8} - 13\frac{5}{12} =$ ______

15. $\frac{3}{5} + \frac{2}{9} + \frac{1}{3} =$ ______

Solve. Show your work. Reduce answers to lowest terms.

16. The Rockets lost 3 of the 20 games played this season. What fraction of the games did the Rockets lose?

17. Raul works at a garden supply store. The store receives flower seeds in 30-ounce boxes. Raul repackages the seeds in $\frac{3}{4}$-ounce packets. How many packets can Raul fill from each box?

18. Keiko needs a piece of wood that is $55\frac{1}{2}$ inches long to trim a window. How many feet of wood does she need?

19. Bob bought a jumbo roll of wrapping paper. The paper on the roll is $8\frac{1}{3}$ yards long. How many feet of paper are on the roll?

Question 20 refers to the following map.

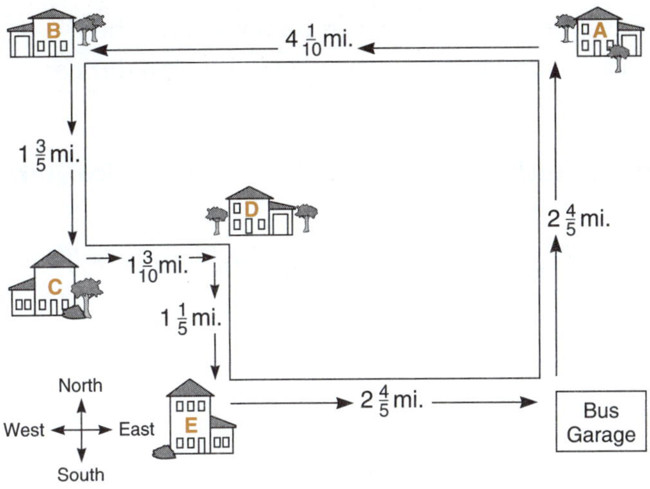

20. Henry completes his bus route at point C. How many miles he must drive to get from the end of his route to the bus garage?

(1) $2\frac{4}{5}$

(2) $4\frac{4}{5}$

(3) $5\frac{3}{10}$

(4) $9\frac{4}{5}$

(5) $13\frac{4}{5}$

21. A recipe that serves 8 people calls for $2\frac{1}{3}$ cups of milk. Clara wants to serve 16 people. Which is the correct expression to find the number of cups of milk she should use?

(1) $2\frac{1}{3} \times \frac{1}{2}$

(2) $2\frac{1}{3} \times 2$

(3) $2\frac{1}{3} \div 2$

(4) $2\frac{1}{3} \times 16$

(5) $16 \div 2\frac{1}{3}$

Questions 22 and 23 refer to the following table.

Yardage Requirements: Skirt and Blouse

Sizes	10	12	14
Blouse A	$2\frac{3}{4}$	$2\frac{7}{8}$	3
Blouse B	$2\frac{7}{8}$	3	$3\frac{1}{4}$
Skirt A, B	$1\frac{1}{4}$	$1\frac{1}{4}$	$1\frac{5}{8}$

22. Which is the correct expression to find how many more yards of fabric Pearl needs to make blouse B in size 14 than skirt B in size 14?

(1) $3\frac{1}{4} + 1\frac{5}{8}$

(2) $3\frac{1}{4} + 1\frac{1}{4}$

(3) $3\frac{1}{4} - 1\frac{1}{4}$

(4) $3\frac{1}{4} - 1\frac{5}{8}$

(5) $3 - 1\frac{5}{8}$

23. Gloria is making a skirt and blouse A in size 12. How many yards of fabric does she need?

Mark your answer in the circles in the grid.

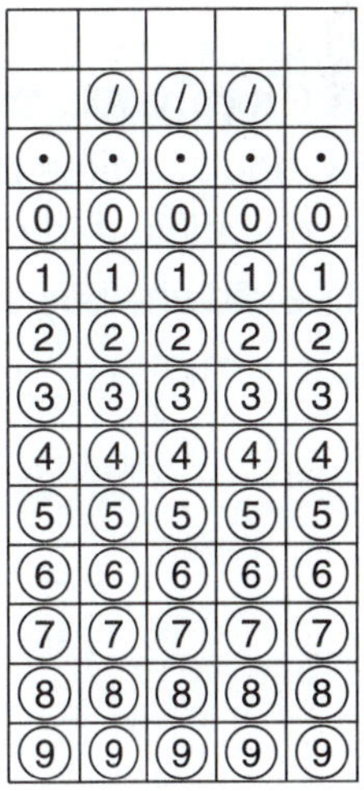

Math Extension

Find 10 different examples showing fractions used in everyday life. Start with recipes and food labels. You might also look at instructions for home-improvement projects and sewing patterns. Find out how fractions are used in lumber yards and in sports.

Check your answers on pages 287–288.

Mini-Test • Unit 2

Directions: This is a 15-minute practice test. After 15 minutes, mark the last number you finished. Then complete the test and check your answers. If most of your answers were correct but you did not finish, try to work faster next time.

Part I Directions: Choose the one best answer to each question. You MAY use your calculator.

1. Angela made 30 cups of snack mix. If one serving is $1\frac{1}{4}$ cups of snack mix, how many servings does Angela have?

   (1) $7\frac{1}{2}$

   (2) 18

   (3) 24

   (4) $28\frac{3}{4}$

   (5) $37\frac{1}{2}$

2. Lana earns $8 per hour working at the Cornerstone Bakery. How much does she earn in $6\frac{1}{2}$ hours?

   (1) $65
   (2) $52
   (3) $48
   (4) $32
   (5) $14

3. At Frank's Grocery, Allison bought $1\frac{2}{3}$ pounds of regular coffee beans and $1\frac{3}{4}$ pounds of decaffeinated coffee beans. How much coffee did she buy altogether?

   (1) $2\frac{1}{2}$

   (2) $2\frac{7}{12}$

   (3) $2\frac{5}{7}$

   (4) $3\frac{5}{12}$

   (5) $4\frac{3}{7}$

Question 4 refers to the triangular herb garden shown below.

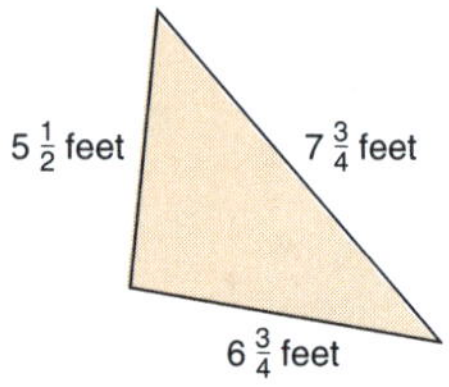

4. What is the total distance around the herb garden in feet?

   (1) $12\frac{1}{4}$

   (2) $14\frac{1}{2}$

   (3) 18

   (4) 20

   (5) Not enough information is given.

5. A welder needs to replace $36\frac{3}{8}$ inches of tubing. If she has $39\frac{5}{16}$ inches of new tubing, how many inches will she need to cut off?

   Mark your answer in the circles in the grid.

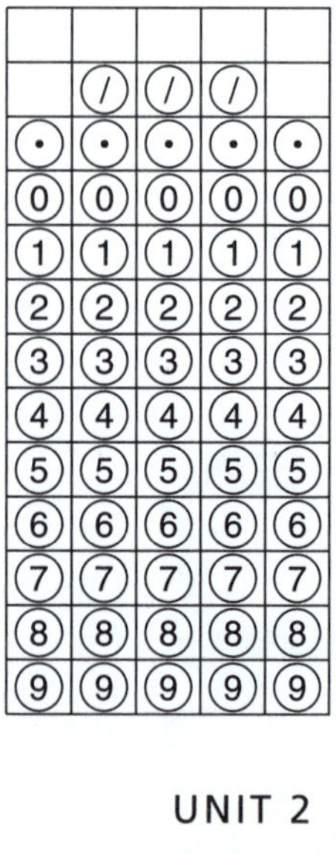

Part II **Directions**: Choose the **one best answer** to each question. You MAY NOT use your calculator.

6. A concrete wall measuring $32\frac{1}{2}$ feet by $5\frac{1}{4}$ feet needs to be painted. Which of the following expressions could be used to find how many square feet of wall needs painting?

   (1) $32\frac{1}{2} + 5\frac{1}{4}$

   (2) $32\frac{1}{2} - 5\frac{1}{4}$

   (3) $2 \times 32\frac{1}{2} + 2 \times 5\frac{1}{4}$

   (4) $32\frac{1}{2} \div 5\frac{1}{4}$

   (5) $32\frac{1}{2} \times 5\frac{1}{4}$

Questions 7 and 8 refer to the following graph.

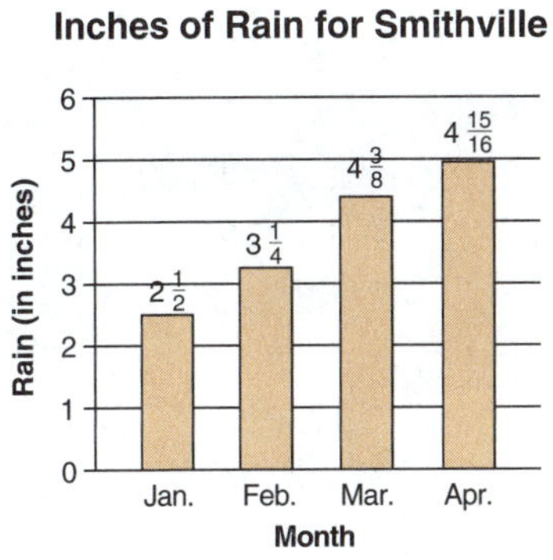

7. How many more inches of rain did Smithville get in March than in February?

   (1) $\frac{3}{8}$

   (2) $\frac{9}{16}$

   (3) $1\frac{1}{8}$

   (4) $1\frac{1}{2}$

   (5) $1\frac{3}{4}$

8. What was the total amount of rainfall for January and February?

   (1) $5\frac{1}{2}$

   (2) $5\frac{3}{4}$

   (3) $5\frac{7}{8}$

   (4) $6\frac{1}{8}$

   (5) $6\frac{1}{4}$

9. Jane froze 24 bags of green beans. If each bag holds $1\frac{1}{2}$ pounds, how many pounds of green beans did she freeze?

   (1) 16

   (2) $17\frac{1}{2}$

   (3) 24

   (4) $25\frac{1}{2}$

   (5) 36

10. Which of the following expressions can be used to find how many $\frac{1}{4}$-pound boxes can be filled using $2\frac{1}{2}$ pounds of nails?

    (1) $2\frac{1}{2} + \frac{1}{4}$

    (2) $2\frac{1}{2} - \frac{1}{4}$

    (3) $2\frac{1}{2} \times \frac{1}{4}$

    (4) $2\frac{1}{2} \div \frac{1}{4}$

    (5) $2\frac{1}{2} \div 4$

11. Out of the $7\frac{1}{2}$ hours Josh worked on Monday, he spent $\frac{1}{3}$ of his time making labels. How many hours did Josh spend making labels?

    Mark your answer in the circles in the grid.

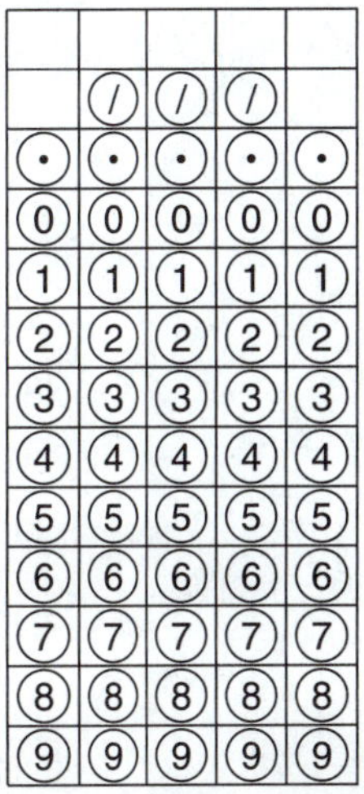

Check your answers on pages 288–289.

Unit 3

Decimals

Decimals are one way of showing parts of a whole. Money is probably the most common use of decimals. Many things that you purchase are weighed and measured using decimals. Gasoline is sold in decimal amounts. Foods are often weighed on scales that show weights in decimal form. In the news, decimals are used to report stock prices and some sport statistics.

Write the price of something you often buy at the store. ______________

Write the normal body temperature for a person. ____________________

Write the dial number of your favorite radio station. _________________

Thinking About Decimals

You may not realize how often you talk, read, or think about numbers as you go about your daily life. Think about your recent activities.

Check the box for each activity you have done recently.

- ☐ Did you write a check to pay a bill?
- ☐ Did you read a shelf label at a store to find a price?
- ☐ Did you calculate how much your next paycheck should be?
- ☐ Did you estimate the cost of an item?
- ☐ Did you put gasoline in a car?
- ☐ Did you buy produce or any kind of food that is sold by weight?

Write some other activities where you used decimals.

____

____

____

Previewing the Unit

In this unit, you will learn:

- what decimals mean in various situations
- how to perform basic operations with decimals
- how to use decimals in metric measurements
- how to add, subtract, multiply, and divide decimals on a calculator

LESSON 8

Decimal Basics

Money and Decimals

Example Suppose you stop to put gas in your car at the local gas station. You look at the pump and find you owe $9.75 for 8.2 gallons of gasoline.

A decimal, like a fraction, shows part of a whole number. In fact, decimals and fractions can represent the same amount.

Decimals are fractions that use the place value system.

Our money system is based on decimals. Dollars are represented by whole numbers. Cents represent 100 parts of a dollar and are written as decimals.

Which is the correct statement to describe the 7 in $9.75?

(1) 7 cents
(2) 7 dollars
(3) 70 cents
(4) 70 dollars
(5) 700 cents

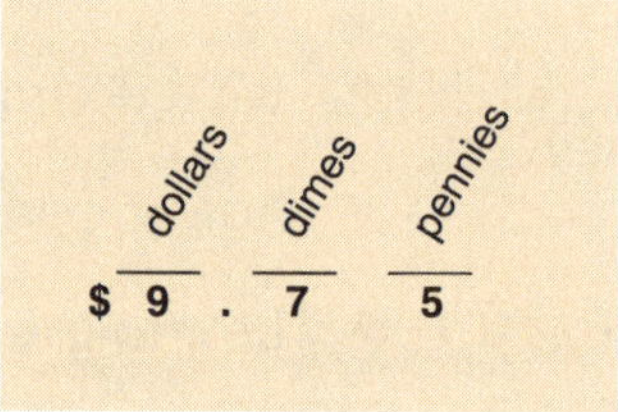

Answer **(3)** is correct. The 75 means 75 cents or 75 out of 100 parts of a dollar. The 7 has a value of **70/100** of a dollar, or **70 cents.** This also means 7 dimes since a dime is 10 out of 100 parts of a dollar.

Example The chart below shows the decimal place values when the decimal is not a money amount.

Which is the value of the 5 in 9.75?

(1) five tenths
(2) five hundredths
(3) five thousandths
(4) five ten thousandths
(5) five hundred thousandths

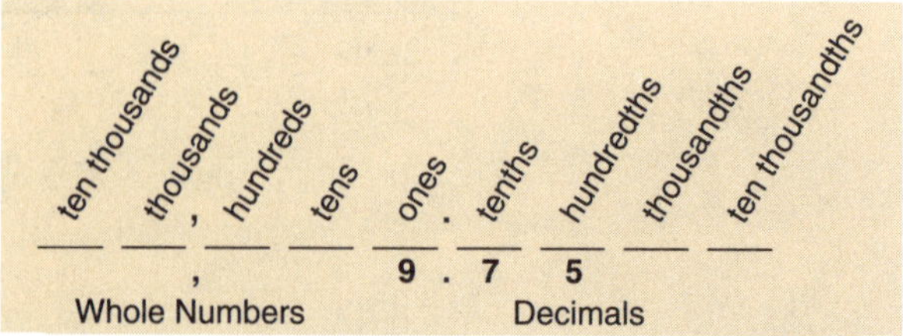

Answer **(2)** is correct. Look at the place value chart. The 5 is two places to the right of the decimal point. It has a value of **five hundredths.**

PRACTICE

Circle the best answer for each question.

1. What is the value of 6 in $6.54?

(1) 6 cents
(2) 6 dimes
(3) 6 dollars
(4) 60 dollars
(5) 600 dollars

2. What is the value of 4 in $1.45?

(1) 4 cents
(2) 4 dollars
(3) 40 cents
(4) 45 cents
(5) 400 cents

3. What is the value of 2 in $28.31?

(1) 2 dollars
(2) 20 cents
(3) 20 dollars
(4) 200 cents
(5) 200 dollars

4. Ramon wrote an amount of money on a check. How should Ramon write out this amount?

(1) 49 dollars and 5 cents
(2) 49 dollars and 50 cents
(3) 49 dollars and 55 cents
(4) 49 dollars and 550 cents
(5) Not enough information is given.

Questions 5 through 7 refer to the place value chart on page 120.

5. What is the value of 6 in the decimal 14.1736?

(1) six tenths
(2) six hundredths
(3) six thousandths
(4) six ten thousandths
(5) six hundred thousandths

6. What is the value of 2 in the decimal 183.0625?

(1) two tenths
(2) two hundredths
(3) two thousandths
(4) two ten thousandths
(5) two hundred thousandths

7. What is the value of 1 in the decimal 364.1507?

(1) one tenth
(2) one hundredth
(3) one thousandth
(4) one ten thousandth
(5) one hundred thousandth

8. **Communicate** Explain how the number of zeros in multiples of ten (10, 100, 1,000, etc.) can help you remember decimal place names. For example:

> Ten (10) has one zero, so one place after the decimal point (.1) is the tenths place.

Explain the next three places to the right of the decimal point (.01, .001, .0001).

Check your answers on page 289.

Reading and Writing Decimals

Place value helps you read and write decimals. In a group of decimal digits, the place value of the last digit is the value of the digits. For example, 0.45 is read **"forty-five hundredths"** because 5 (the last digit) is in the hundredths place.

Example Read the decimal 3.82.

Step 1 State the whole number. 3.82 "Three

Step 2 State *and* for the decimal point. and

Step 3 State the decimal part and the place-value name for the last decimal digit. 3.**82** eighty-two hundredths"

Example Write *five and twelve thousandths* using digits.

Step 1 Write the whole number. 5

Step 2 Write a decimal point for the word *and*. 5.

Step 3 Think of how many decimal places you will need. Write 12 so that the last digit is in the thousandths place. Since there is no digit in the tenths place, fill it with a placeholder zero. 5.012

TIP
A single zero before (to the left of) the decimal point means the decimal has no whole number part. 0.135 is less than one

When the decimal has no whole number part, you just state the decimal name. 0.35 is "thirty-five hundredths."

PRACTICE

Write each decimal in words. The first one is done for you.

1. 4.17 four and seventeen hundredths
2. 0.256 ____________
3. 2.09 ____________
4. 6.805 ____________

Write each decimal using digits. The first one is started for you.

5. Twenty-four and three hundred fifty-six thousandths 24____________
6. Three and seventy-eight hundredths ____________
7. Four hundred ninety-one thousandths ____________
8. Two hundred sixty-seven and three tenths ____________
9. Fourteen and thirty-six thousandths ____________
10. Five and seven thousand eighty-four ten thousandths ____________

 Check your answers on page 289.

Relating Decimals and Fractions

You can change decimals to fractions.

Example Change 0.25 to a fraction in lowest terms.

Step 1 For the numerator of the fraction, write the original number without its decimal point. $\frac{25}{}$

Step 2 The denominator is the place value of the decimal. The place value of 0.25 is hundredths. $\frac{25}{100}$

Step 3 Reduce the fraction to lowest terms. $\frac{25}{100} = \frac{25 \div 25}{100 \div 25} = \frac{1}{4}$

TIP
Remember the zeros in the decimal 0.0064 have no value; they are only placeholders.

Example Change 0.0064 to a fraction in lowest terms.

Step 1 Write the numerator of the fraction. $\frac{64}{}$

Step 2 Write the place value of the decimal. The place value of 0.0064 is ten thousandths. $\frac{64}{10,000}$

Step 3 Reduce the fraction to lowest terms. $\frac{64}{10,000} = \frac{64 \div 16}{10,000 \div 16} = \frac{4}{625}$

If the decimal is a mixed decimal, that is, has a whole number part, the whole number is placed in front of the fraction. $1.25 = 1\frac{1}{4}$

PRACTICE

Change each decimal to a fraction in lowest terms. The first one is done for you.

1. 0.5
   $0.5 = \frac{5}{10} = \frac{1}{2}$
2. 0.45
3. 0.15
4. 0.64
5. 0.125
6. 1.8
7. 0.10
8. 0.005
9. 0.32
10. 0.375
11. 2.319
12. 0.004
13. 3.028
14. 0.85
15. 0.625
16. 4.0024

Check your answers on page 289.

Comparing and Rounding Decimals

Comparing Decimals

Would you rather pay $0.79 or $0.99 per pound for apples? You would pick $0.79 per pound because it is the smaller amount. To answer the question, you compared two decimals and found the smaller one.

TIP
You can add zeros after the last digit in a decimal without changing the value of the decimal.
.500 is the same as .5

Example Compare 0.325 and 0.5

Step 1 Line up the decimal points.
0.325
0.5

Step 2 If the two decimals have a different number of places, add zeros to the end of the decimal with the fewer digits so that the two decimals have the same number of digits.
0.325
0.5**00**

Step 3 Compare as you would with whole numbers. Since 325 is less than 500, 0.325 is less than 0.500.
0.<u>3</u>25
0.<u>5</u>00

0.325 < 0.5
or **0.5 > 0.325**

Rounding Decimals

You can follow the same steps to round decimals as you did with whole numbers.

TIP
If the number to the right of the underlined digit is less than 5, do not change the underlined digit; if it is 5 or more, add 1 to the underlined digit.

Example Round 5.1628 to the nearest thousandth.

Step 1 Underline the digit in the place you want to round the number. (2 is in the thousandths place.)
5.16<u>2</u>8

Step 2 Look at the digit to the right of the underlined digit.
5.16<u>2</u>8

Step 3 Decide if the underlined digit needs to change. Since 8 is greater than 5, add 1 to the underlined digit.
5.16<u>3</u> _

Step 4 Drop the digit to the right of the underlined digit.
5.163

PRACTICE

A. Compare each pair of numbers. Write $>$, $<$, or $=$. The first one is done for you.

1. 0.27 __$<$__ 0.72	2. 0.43 ______ 0.09	3. 0.73 ______ 0.542
4. 8.058 ______ 8.58	5. 2.58 ______ 2.580	6. 53.005 ______ 52.008
7. 0.863 ______ 0.9	8. 2.001 ______ 2.01	9. 1.32 ______ 1.319
10. 5 ______ 5.00	11. 0.95 ______ 0.954	12. 0.005 ______ 0.05
13. 0.43 ______ 0.34	14. 0.54 ______ 0.054	15. 5.2 ______ 5.200

B. Round each decimal to the given place value. The first one is done for you.

16. Round 6.3782 to the nearest hundredth. __6.38__
17. Round 8.276 to the nearest whole number. ______
18. Round 46.3518 to the nearest tenth. ______
19. Round 71.0483 to the nearest thousandth. ______
20. Round 56.2819 to the nearest hundredth. ______
21. Round 0.5462 to the nearest tenth. ______
22. Round 101.667 to the nearest whole number. ______
23. Round 25.0235 to the nearest thousandth. ______

C. Solve each problem.

24. Last season Adrian had a batting average of 0.215 and Angel had a batting average of 0.213. Who had the higher batting average?

25. The population of Sumner City is 95.7 million. What is the population of Sumner City to the nearest million?

26. A box of Brand A cereal costs $2.83. A box of Brand B cereal costs $2.79. Both boxes weigh the same. Which cereal costs less?

27. Saban stocks the shelves at a local health food store. The store buys spices in large quantities, then sells them in smaller quantities. Saban has written down the weights of the newly packaged garlic powder. They are

0.75 oz., 0.57 oz., 0.69 oz., and 0.5 oz.

If Saban wants to record the weights of the garlic powder from lightest to heaviest, in what order would these amounts be listed?

Check your answers on pages 289–290.

MEASUREMENT AND GEOMETRY

Using Metric Measurement

Example Charlie is building fences around rectangular flower beds that are in front of an office building. The measurements on the landscaping plans are in meters. The plan for one of the flower beds is shown at the right.

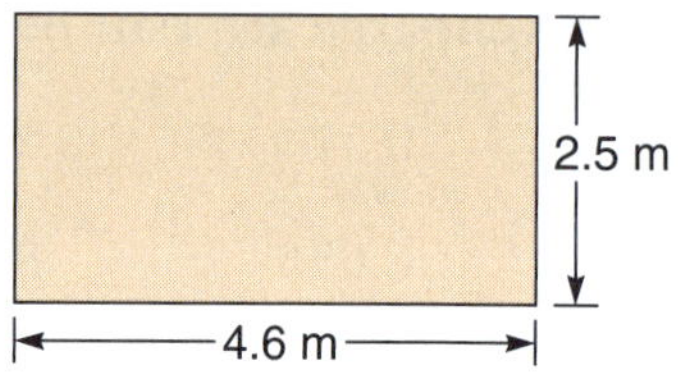

A meter (m) is a measure of length in the metric system. A meter is a little longer than a yard. (1 yard = 36 inches; 1 meter = 39.37 inches)

Parts of a meter are expressed as decimals.

Charlie sees that the flower bed is a rectangle that measures 4.6 meters long and 2.5 meters wide. Which is the correct expression to find the perimeter of the flower bed in meters? Remember, perimeter is the distance around a figure.

(1) 4.6 + 2.5
(2) 4.6 + 2.5 + 2.5 + 2.5
(3) 4.6 + 4.6 + 2.5 + 2.5
(4) 4.6 + 4.6 + 4.6 + 4.6
(5) 4.6 + 4.6 + 2.5 + 2.5 + 2.5

4.6
4.6
2.5
+2.5
14.2

Answer (3) is correct. Charlie needs to find the total of all the sides, so he adds all four lengths. **4.6 + 4.6 + 2.5 + 2.5 = 14.2 meters,** so Charlie needs **14.2 meters** of fencing.

Example Charlie decides to leave an 0.8-meter opening in the center of one of the 4.6-meter sides. He wants to find how much fencing he will need for the side with the opening. Which is the correct expression to find how much fencing Charlie will need for this side?

(1) 4.6 + 0.8
(2) 4.6 − 0.8
(3) 0.8 − 4.6
(4) 4.6 + 4.6 + 0.8 + 0.8
(5) 4.6 + 4.6 + 4.6 − 0.8

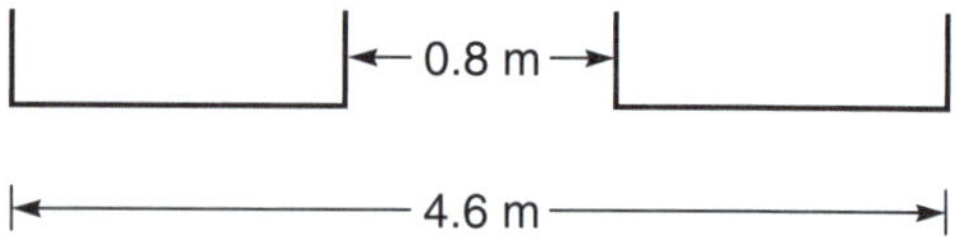

3 16
4.6
−0.8
3.8

Answer (2) is correct. Charlie needs to find the difference between the length of the side and the opening, so he subtracts. **4.6 − 0.8 = 3.8 meters,** so Charlie needs **3.8 meters** of fencing for the side.

PRACTICE

Circle the best answer for each question.

Questions 1 and 2 refer to the flower bed shown below.

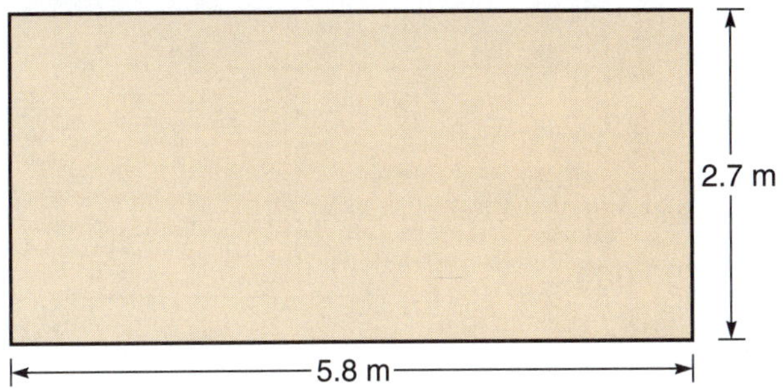

1. Shauna works for a landscaping service. She is putting edging around the flower bed shown here. Which is the correct expression to find the perimeter of the flower bed?

   (1) 5.8 + 2.7
   (2) 5.8 + 2.7 + 2.7
   (3) 5.8 + 5.8 + 2.7
   (4) 5.8 + 5.8 + 2.7 + 2.7
   (5) 5.8 + 5.8 + 5.8 + 5.8

2. Shauna decides to leave a 0.7-meter opening in the center of one of the 2.7-meter sides. Which is the correct expression to find how many meters of edging Shauna will need for this side?

   (1) 2.7 + 0.7
   (2) 2.7 − 0.7
   (3) 2.7 + 2.7 + 0.7 + 0.7
   (4) 5.8 + 5.8 + 2.7 + 2.7
   (5) 5.8 + 5.8 + 2.7 − 2.7

3. Luis wants to put a fence around a rectangular swimming pool that is 9.6 meters long. How many meters of fencing will he need?

   (1) 9.6
   (2) 19.2
   (3) 28.8
   (4) 38.4
   (5) Not enough information is given.

Questions 4 and 5 refer to the playground shown below.

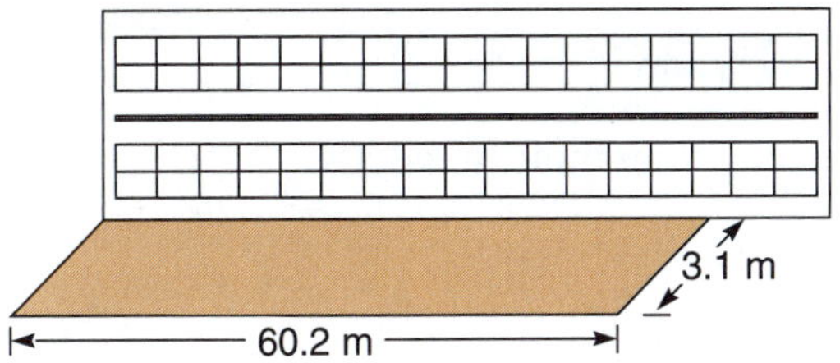

4. One side of a rectangular playground is next to the school building. The principal wants to rope off the other three sides. Which is the correct expression to find how many meters of rope are needed?

   (1) 60.2 + 3.1
   (2) 60.2 − 3.1
   (3) 60.2 + 3.1 + 3.1
   (4) 60.2 + 60.2 + 3.1
   (5) 60.2 + 60.2 + 3.1 + 3.1

5. Stan may want to paint a yellow strip (a place where teachers will stand) that is 15.8 meters long in the center of the 60.2-meter side. Which is the correct expression to find how many meters of the 60.2-meter side will not have a yellow strip?

   (1) 15.8 + 60.2
   (2) 15.8 − 60.2
   (3) 60.2 − 15.8
   (4) 60.2 + 3.1 + 3.1 − 15.8
   (5) Not enough information is given.

6. **Extend** Measure the length and width of the top of your desk in meters. Write these measurements to the nearest tenth and to the nearest hundredth of a meter.

Check your answers on page 290.

LESSON 8

Lesson 8 Review

Write the place value of each underlined digit.

1. $\$3.2\underline{7}$ ______________________
2. $14.078\underline{2}$ ______________________
3. $79.58\underline{6}$ ______________________
4. $\$0.\underline{5}1$ ______________________

Write each decimal in words.

5. 0.82 ___________________________________
6. 3.513 ___________________________________
7. 18.3406 ______________________________
8. 790.004 ______________________________

Write each decimal using digits.

9. two and forty-three hundredths ______________________
10. two thousand three hundred forty-three ten thousandths ____________
11. thirty-seven and six hundred twenty-one thousandths ____________
12. four hundred twenty-nine and nine hundredths ____________

Change each decimal to a fraction or mixed number in lowest terms.

13. 0.05
14. 0.28
15. 0.75
16. 0.025
17. 0.96
18. 3.205
19. 5.0008
20. 9.35

Compare each pair of numbers. Write >, <, or =.

21. 0.759 ______ 0.795
22. 0.326 ______ 0.54
23. 0.6 ______ 0.60
24. 32.574 ______ 32.547

Round each decimal to the given place value.

25. Round 3.196 to the nearest hundredth. ______________________
26. Round 6.453 to the nearest tenth. ______________________
27. Round 0.724 to the nearest whole number. ______________________
28. Round 42.9358 to the nearest thousandth. ______________________

Circle the best answer for each question.

29. Julio's doctor prescribed sixteen thousandths of a gram of a medicine for him. How do you write *sixteen thousandths* in digits?

(1) 0.0016
(2) 0.016
(3) 0.16
(4) 1,600
(5) 16,000

30. Lucy's vegetable garden is rectangular and measures 5.2 meters long by 3.4 meters wide. Which is the correct expression to find the perimeter of the garden?

(1) 5.2 + 3.4
(2) 5.2 + 3.4 + 3.4
(3) 5.2 + 5.2 + 3.4
(4) 5.2 + 5.2 + 3.4 + 3.4
(5) 5.2 + 5.2 + 5.2 + 3.4

31. Calvin had 4.2 meters of fabric. He used 2.7 meters of fabric to recover a chair. Which expression shows how many meters of fabric Calvin has left?

(1) 4.2 − 2.7
(2) 4.2 + 2.7
(3) 4.2 ÷ 2.7
(4) 2.7 ÷ 4.2
(5) 4.2 × 2.7

32. Al ran a 100-meter race in 15.25 seconds; Ron finished in 15.06 seconds. Which statement is true about the race?

(1) Al finished before Ron.
(2) Both finished the race at the same time.
(3) Al's time was faster than Ron's time.
(4) Ron finished before Al.
(5) Ron's time was slower than Al's time.

33. Masami is a butcher. He packaged a piece of meat that weighed 3.059 pounds. Which words best describe the number of pounds of meat in that package?

(1) three and fifty-nine hundredths
(2) three and fifty-nine thousandths
(3) three and five hundredths
(4) three and nine thousandths
(5) three and five hundred nine thousandths

34. How does 24.6305 compare to 24.6350?

(1) 24.6305 > 24.6350
(2) 24.6350 = 24.6305
(3) 24.6350 < 24.6305
(4) 24.6350 > 24.6305
(5) 24.6305 = 24.6350

35. Last week Mary Lou paid $1.325 per gallon of gasoline. Rounded to the nearest cent, how much did she spend per gallon of gasoline?

(1) $1.00
(2) $1.30
(3) $1.32
(4) $1.33
(5) $1.40

36. Betty finished reading 0.4 of a new mystery. What fraction of the book has Betty read so far?

(1) $\frac{4}{5}$
(2) $\frac{2}{5}$
(3) $\frac{1}{25}$
(4) $\frac{1}{250}$
(5) Not enough information is given.

Check your answers on pages 290–291.

LESSON 9

Adding and Subtracting Decimals

Adding and Subtracting Decimals

Decimal Operations

Adding and subtracting decimals, including money, is like adding and subtracting whole numbers. The important step is to line up the decimal points because it keeps place values together.

Example Add: 4.6 + 14 + 3.38 + 0.125

Step 1 Rewrite the problem, lining up the decimal points. Put zeros to the right of the last decimal digits so that all numbers have the same number of digits.

```
   4.600
  14.000
   3.380
+  0.125
```

Step 2 Add, regrouping as needed. Put the decimal point in the answer directly under the decimal points in the problem.

```
  11 1
   4.600
  14.000
   3.380
+  0.125
  22.105
```

Example Subtract: 15.2 − 9.75

Step 1 Rewrite the problem, lining up the decimal points. Put a zero to the right of the last decimal digit so that both numbers have the same number of digits.

```
  15.20
−  9.75
```

Step 2 Subtract, regrouping as needed. Put the decimal point in the answer.

```
  15.20
−  9.75
   5.45
```

TIP

If necessary, add zeros so that all the numbers have the same number of digits. Remember, you can add zeros after the last digit in a decimal without changing the value.

4.6 = 4.600

Adding and Subtracting Decimals on a Calculator

To add or subtract decimals on a calculator, use the decimal point key at the appropriate place between the digits of the decimal.

Example 50.32 + 4.236 − 24.005

[AC] 50 [.] 32 [+] 4 [.] 236 [−] 24 [.] 005 [=] 30.551

For more practice adding and subtracting decimals with the calculator, see Calculator Handbook pages 246–247.

PRACTICE

A. **Add or subtract. The first one is done for you.**

1. $\begin{array}{r} {}^{1\ \ 1} \\ 3.726 \\ +4.915 \\ \hline 8.641 \end{array}$

2. $\begin{array}{r} 17.368 \\ -\ \ 8.415 \\ \hline \end{array}$

3. $\begin{array}{r} 24.28 \\ -13.7\ \ \\ \hline \end{array}$

4. $\begin{array}{r} 28.467 \\ +37.29\ \ \\ \hline \end{array}$

5. $\begin{array}{r} 12.05 \\ -\ \ 6.8\ \ \\ \hline \end{array}$

6. $\begin{array}{r} 3.7\ \ \\ 14.24 \\ +\ 2.3\ \ \\ \hline \end{array}$

7. $\begin{array}{r} 24.81\ \ \\ 35.7\ \ \ \ \\ +28.274 \\ \hline \end{array}$

8. $\begin{array}{r} 8.4\ \ \\ -0.31 \\ \hline \end{array}$

9. $\begin{array}{r} 14.8\ \ \ \ \\ -\ \ 6.753 \\ \hline \end{array}$

10. $\begin{array}{r} 12\ \ \ \ \ \\ +\ \ 5.88 \\ \hline \end{array}$

B. **Add or subtract. Rewrite the problems below. Line up the decimal points. The first one is started for you.**

11. 2.43 + 0.57 + 3.18 =

    $\begin{array}{r} 2.43 \\ 0.57 \\ +3.18 \\ \hline \end{array}$

12. 67.31 − 49.826 =

13. 57.43 − 29.5 =

14. 0.3 + 21.508 + 3.4 =

15. 4.26 + 5.1 + 8.39 =

16. 12.6 − 8.897 =

C. **Solve the problems below with a calculator. Refer to pages 246–247 in the Calculator Handbook for additional information.**

17. Phalla had a balance of $256.81 in her checking account. She wrote checks for $2.80, $34.33, and $62.18. She deposited $122.77 and $83.93. Find Phalla's new balance.

    ____

18. One weekend Roberto worked a few hours of overtime. He earned $43.50 on Saturday and $72.50 on Sunday. How much did Roberto earn that weekend?

    ____

Check your answers on pages 291–292.

PROBLEM SOLVING

Multi-Step Problems

The key to solving a problem with more than one step is to think through the problem and decide what to do in each step before you solve the problem.

Example Eva works at a flower shop. She records each sale on a sales slip.

Ms. Morris is buying two flower arrangements: one costs $19.95 and the other costs $52.75. Eva needs to find three numbers: *SUBTOTAL, TAX,* and *TOTAL.*

- **A subtotal is a total of part of a group of numbers; here, it is the sum of the cost of the two flower arrangements before tax.**
- **Tax is additional money people must pay based on the amount of the subtotal.**
- **The total is the complete cost of the order, including tax.**

Compton Floral
444 W. Orange Grove
Margate, FL 33063

Sm. Arrangement	$19	95
Large Arrangement	52	75
SUBTOTAL		
TAX		
TOTAL	$	

First, Eva finds the subtotal. Which is the correct expression to find the subtotal?

(1) $19.95 − $52.75
(2) $19.95 + $52.75
(3) $52.75 − $19.95
(4) $52.75 × $19.95
(5) $52.75 ÷ $19.95

Answer **(2)** is correct. Eva adds the cost of the two arrangements to get the subtotal. **$19.95 + $52.75 = $72.70,** so the subtotal is **$72.70.**

Example Next Eva needs to find the total. She knows the sales tax for this subtotal is $4.91. Which is the correct expression to find the total cost of this sale?

(1) subtotal − tax
(2) subtotal + tax
(3) tax − subtotal
(4) subtotal × tax
(5) subtotal ÷ tax

Answer **(2)** is correct. Eva adds the tax to the subtotal to find the total cost of this sale. **$72.70 + $4.91 = $77.61,** so the total cost of the sale is **$77.61.**

PRACTICE

Circle the best answer for each question.

Questions 1 through 3 refer to the following sales slip.

Frank's Coats
526 E. Pine
Chicago, IL 60611

Sm. Jacket	$39	99
Large Jacket	58	00
SUBTOTAL		
TAX		
TOTAL	$	

1. Mrs. Wilson wants to buy two winter jackets for her children. Which is the correct expression to find the subtotal of this sale before tax?
   (1) $39.99 + $58.00
   (2) $39.99 − $58.00
   (3) $58.00 ÷ $39.99
   (4) $58.00 − $39.99
   (5) $58.00 × $39.99

2. The tax for the items is $6.86. Which is the correct expression to find the total including tax of this sale?
   (1) subtotal − $6.86
   (2) $6.86 − subtotal
   (3) subtotal + $6.86
   (4) subtotal × $6.86
   (5) subtotal ÷ $6.86

3. Mrs. Wilson decides to put the coats on layaway and pays $45.00 to hold them. Which is the correct expression to find how much Mrs. Wilson still owes on the coats?
   (1) total − $45.00
   (2) total + $45.00
   (3) $45.00 − total
   (4) total × $45.00
   (5) total ÷ $45.00

4. A customer is buying a pair of jeans that costs $30.00, a sweatshirt that costs $24.00, and a T-shirt for $12.00. Which is the correct expression to find the subtotal of this sale?
   (1) $30.00 − $24.00 + $12.00
   (2) $30.00 + $24.00 + $12.00
   (3) $24.00 − $30.00 − $12.00
   (4) $30.00 × $24.00 × $12.00
   (5) $30.00 ÷ $24.00 + $12.00

5. Lucia is a sales clerk at a camping store. A customer bought 3 shirts for $18 each and 2 plastic water containers for $5.99 each. Which is the correct expression to find the total of this sale before tax?
   (1) (3 + $18) × (2 + $5.99)
   (2) (3 × $18) × (2 × $5.99)
   (3) (3 × $18) − (2 + $5.99)
   (4) (3 + $18) + (2 × $5.99)
   (5) (3 × $18) + (2 × $5.99)

6. A customer uses a discount coupon at a store. The store clerk figures the subtotal, discount, and tax. Which is the correct expression to find the amount the customer will pay?
   (1) subtotal + discount + tax
   (2) subtotal − discount + tax
   (3) subtotal − discount − tax
   (4) subtotal + discount − tax
   (5) subtotal − discount × tax

7. **Communicate** A customer buys three items: $16.00, $21.50, and $32.25. The sales tax is $5.23, and the customer gives $80.00 to a cashier. Explain in three steps without calculation how to find the correct change to give the customer.

Check your answers on page 292.

ALGEBRA TOPIC

Order of Operations

There are four **operations** to use with numbers. You can add, subtract, multiply, or divide. Numbers and operation symbols together make a **mathematical expression.** When an expression has more than one operation, the order in which you do the operations can change the answer. To solve an expression, always follow the order of operations:

Step 1 Do any operations in parentheses first.

Step 2 Multiply and divide working from left to right.

Step 3 Add and subtract working from left to right.

Example What is the value of the expression $80 − 3 × ($8.50 + $6.50)?

Step 1 Do the operation in parentheses. $80 − 3 × **($8.50 + $6.50)**
$80 − 3 × **($15)**

Step 2 Do the multiplication operation. $80 − **3 × $15**
$80 − **$45**

Step 3 Do the subtraction operation. $80 − $45 = $35

The value of the expression is **$35.**

You also need the order of operations to choose the correct expression for solving a word problem. First, think about what you would do to solve the problem. Then see which expression uses those operations in the correct order.

TIP
It may help to put into words what you would do to solve the problem. "Multiply to find the cost of the tires, add the oil change, and subtract the total from $120."

Example Alicia bought 2 tires for $45.89 each. She also got an oil change for $22.25. Alicia gave the clerk $120. Which is the correct expression to find the amount of her change?

(1) ($120 − 2 × $45.89) + $22.25
(2) ($120 − 2) × ($45.89 + $22.25)
(3) $120 − (2 × $45.89 + $22.25)
(4) $120 − 2 × $45.89 + $22.25
(5) $120 − (2 × $45.89) + $22.25

Answer **(3)** is correct. To find the amount of change, subtract the amount of the purchase from $120.

$120 − (2 × $45.89 + $22.25) = $120 − ($91.78 + $22.25)
= $120 − ($114.03) = $5.97.

Use the correct order of operations to show that the other four answer options give different incorrect answers.

PRACTICE

Circle the best answer for each question.

1. Lorene earns $14.50 an hour. She worked 38 hours last week and 40 hours this week. Which is the correct expression to find Lorene's earnings in the 2-week period?

   (1) $14.50 + (38 × 40)
   (2) ($14.50 × 40) + 38
   (3) ($14.50 × 38) + 40
   (4) $14.50 × (38 + 40)
   (5) $14.50 × 38 + 40

2. Glen has two part-time jobs. At one job he works 24 hours and earns $9.95 an hour. At the other job he works 18 hours and earns $11.75 an hour. Which expression finds how much Glen earns a week?

   (1) 24 × ($9.95 + 18) × $11.75
   (2) 24 × ($9.95 + $11.75) × 18
   (3) $9.95 × (24 + 18) × $11.75
   (4) (18 × $9.95) + (24 × $11.75)
   (5) (24 × $9.95) + (18 × $11.75)

<u>Question 3</u> refers to the following table.

Bowling Scores	
Player 1	205
Player 2	165
Player 3	104
Player 4	190

3. Aida scored 40 points above the mean (average) score of the other four bowlers. Which is the correct expression to find Aida's score?

   (1) 205 ÷ 4 + 40
   (2) (205 + 40) ÷ 4 + 40
   (3) (205 + 165 + 104 + 190) ÷ 4
   (4) (205 + 165 + 104 + 190 + 40) ÷ 4
   (5) (205 + 165 + 104 + 190) ÷ 4 + 40

4. Which expression describes the following: the sum of $6.50 and $9.25, multiplied by their difference?

   (1) ($6.50 + $9.25) + ($9.25 − $6.50)
   (2) ($6.50 + $9.25) × ($9.25 + $6.50)
   (3) ($6.50 − $9.25) × ($9.25 − $6.50)
   (4) ($6.50 + $9.25) + ($9.25 + $6.50)
   (5) ($6.50 + $9.25) × ($9.25 − $6.50)

5. A customer at Fabric Factory bought 3 yards of fabric for $11.00 a yard and 5.5 yards of fabric for $8.40 a yard. Which is the correct expression to find the total cost of the fabric?

   (1) (3 × $11.00) + (5.5 × $8.40)
   (2) (3 + $11.00) + (5.5 + $8.40)
   (3) (3 × $11.00) × (5.5 × $8.40)
   (4) (3 × $11.00) − (5.5 × $8.40)
   (5) Not enough information is given.

6. Miguel and Gina have $40 to spend on entertainment. They bought two movie tickets for $8.50 each, a large popcorn for $4.50, and two drinks for $3.50 each. Which is the correct expression to find out how much money they have left?

   (1) 2 × ($40 − $8.50 + $4.50 + 3.50)
   (2) $40 − 2 × ($8.50 + $4.50 + $3.50)
   (3) $40 − 2 × $8.50 − 2 × $3.50 − $4.50
   (4) $40 − 2 × $8.50 + 2 × $3.50 + $4.50
   (5) $40 − (2 × $8.50 + $3.50) − $4.50

7. **Reason** Use parentheses to rewrite and solve the expression 6 × 2 + 4 − 3 to give three different solutions.

Check your answers on page 292.

DATA ANALYSIS

Reading a Line Graph

Example Dennis Stone works for the Department of Water and Power for the city of Los Angeles. His boss, Lupe Navarro, asks him to find the levels of rainfall for Los Angeles this year.

Dennis used the **line graph** at right to answer questions about rainfall for Los Angeles. The graph shows the rainfall for the first 6 months of the year.

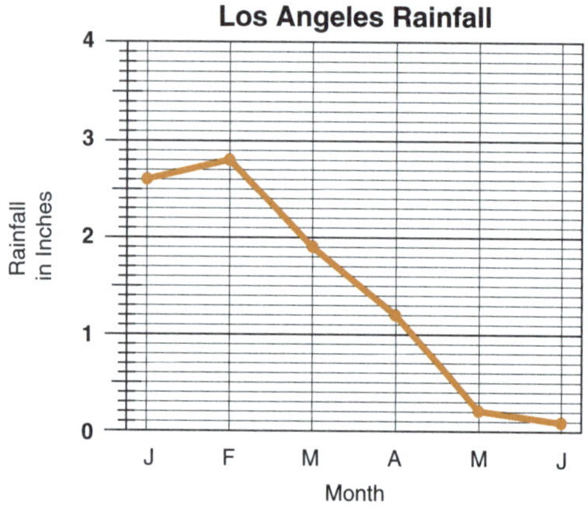

A line graph contains lines called **axes.** The **vertical axis** (up and down) is marked with a **scale.** For example, this scale represents the number of inches of rainfall in Los Angeles. The longer marks on the scale represent whole numbers. The smaller marks divide each inch into tenths. The **horizontal axis** (across) is marked with letters representing the first 6 months of the year.

Ms. Navarro wants Dennis to find how many inches of rainfall the city received in March of this year. Which is the correct answer?

(1) 0.2
(2) 1.2
(3) 1.9
(4) 2.1
(5) 2.9

Answer **(3)** is correct. Dennis finds the point for March above the first M on the horizontal axis. He reads across to the scale on the vertical axis. The point represents **1.9 inches** of rainfall.

Example Next, Dennis needs to find the difference in rainfall between February and March. Which is the correct expression to find how many more inches of rainfall the city received in February than in March?

(1) $2.8 + 1.9$
(2) $2.8 \div 1.9$
(3) 2.8×1.9
(4) $2.8 - 1.9$
(5) $1.9 \div 2.8$

Answer **(4)** is correct. The rainfall for February is 2.8 inches. Subtract to find the difference between 2.8 inches and 1.9 inches (rainfall for March). **$2.8 - 1.9 = 0.9$ inch,** so there was **0.9 inch** more rainfall in February than in March.

PRACTICE

Circle the best answer for each question.

Questions 1 through 3 refer to the following line graph.

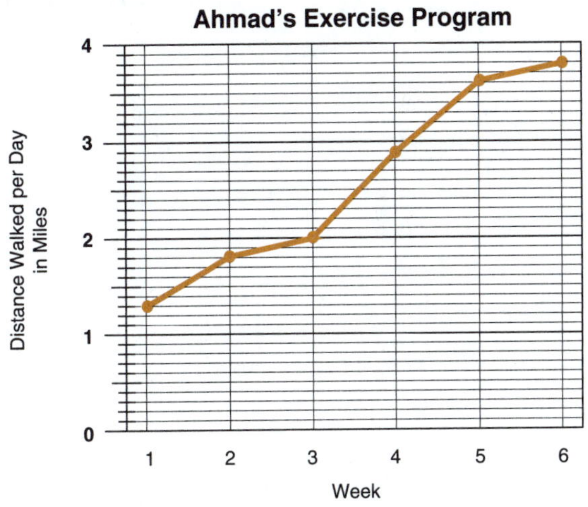

1. How many miles did Ahmad walk each day during week 4 of his exercise program?

   (1) 2.4
   (2) 2.9
   (3) 3.1
   (4) 3.9
   (5) 29

2. During which time period did Ahmad have the greatest increase in the distance he walked per day?

   (1) from week 1 to week 2
   (2) from week 2 to week 3
   (3) from week 3 to week 4
   (4) from week 4 to week 5
   (5) from week 5 to week 6

3. How many miles farther did Ahmad walk each day during week 6 of his program than during week 1?

   (1) 0.5
   (2) 2.5
   (3) 2.9
   (4) 3.0
   (5) 5.1

Questions 4 through 6 refer to the following line graph.

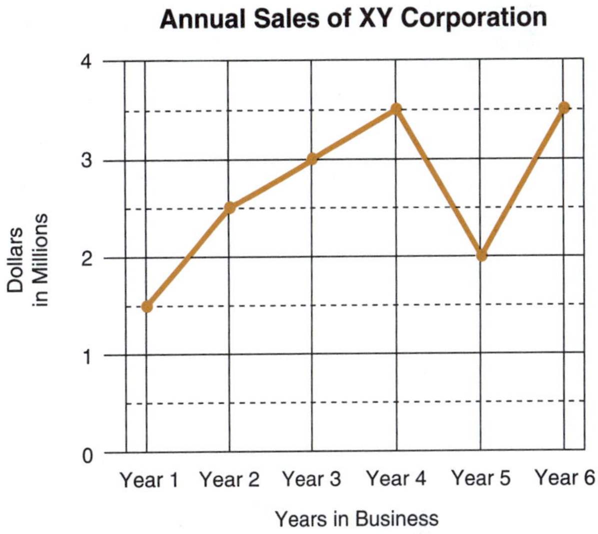

4. The XY Corporation experienced a decrease in sales in only one year. How much lower were the sales in that year than the sales of the year before?

   (1) $0.5 million
   (2) $1 million
   (3) $1.5 million
   (4) $2 million
   (5) Not enough information is given.

5. What were the total sales for Years 3 through 6?

   (1) $11 million
   (2) $12 million
   (3) $14 million
   (4) $15 million
   (5) $16 million

6. **Reason** Look at the graph above. Explain what steeper and flatter lines mean. (Hint: Think about increases and decreases.)

Check your answers on pages 292–293.

LESSON 9

Lesson 9 Review

Lesson 9 Review

Lesson 9 Review

Add or subtract.

1. $\begin{array}{r} 35.2 \\ +41.84 \\ \hline \end{array}$

2. $\begin{array}{r} 74.3 \\ -51.42 \\ \hline \end{array}$

3. $\begin{array}{r} 93.26 \\ -14.336 \\ \hline \end{array}$

4. $\begin{array}{r} 15.93 \\ +\ 4.895 \\ \hline \end{array}$

5. $\begin{array}{r} 8.45 \\ +15.316 \\ \hline \end{array}$

6. $\begin{array}{r} 4.26 \\ +3.947 \\ \hline \end{array}$

7. $\begin{array}{r} 7.389 \\ -0.49 \\ \hline \end{array}$

8. $\begin{array}{r} 5.9 \\ -4.78 \\ \hline \end{array}$

9. $\begin{array}{r} 3.81 \\ 12.463 \\ +\ 5.4 \\ \hline \end{array}$

10. $\begin{array}{r} 42.18 \\ 53.7 \\ +82.427 \\ \hline \end{array}$

11. 12.372 − 9.473 =

12. 7.37 − 5.893 =

13. 8.25 + 9.47 + 3.06 =

14. 15.43 + 2.8 + 3.816 =

15. 4.245 + 1.97 + 2.8 =

16. 3.8 − 1.924 =

17. 124.897 + 23.8764 = ____________

18. 4,234.256 − 2,897.9825 = ____________

Circle the best answer for each question.

19. Robyn had a gift certificate for $100. She bought a pair of jeans for $30.00 and a shirt for $24.00. Sales tax was $4.46. Which is the correct expression to find how much money Robyn had left?

(1) $100 − $30.00 − $24.00
(2) $100 − ($30.00 + $4.46)
(3) $100 − ($30.00 + $24.00 − $4.46)
(4) $100 − ($30.00 + $24.00 + $4.46)
(5) $100 + ($30.00 − $24.00 − $4.46)

20. What is the value for the expression 4 + (2 × 6) − (12 ÷ 2)?

(1) 2
(2) 6
(3) 10
(4) 18
(5) 30

Questions 21 through 24 refer to the following information.

George works in the electronics department of a large store. A customer is buying a portable tape player for $24.99 and a cassette tape for $4.99. Sales tax on this purchase is $2.10. The customer pays $40.00 in cash.

21. Estimate the subtotal before tax on this sale.

22. What is the exact subtotal before tax on this sale?

23. What is the total including tax on this sale?

24. How much change should George give his customer?

Circle the best answer for each question.

Questions 25 through 27 refer to the following graph.

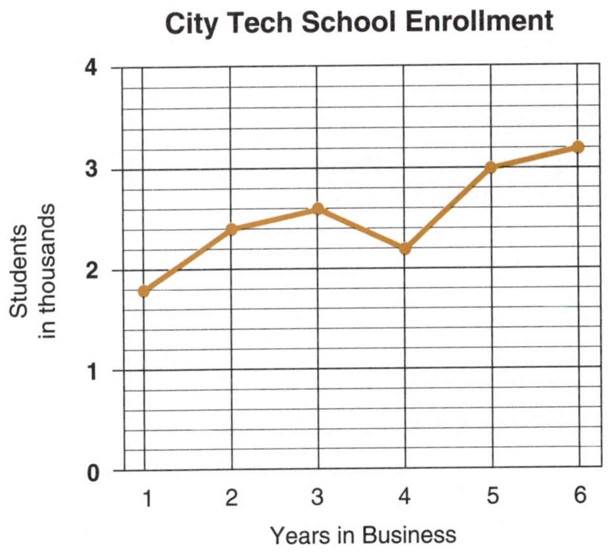

25. How many students were enrolled in Year 2?
(1) 1,800
(2) 2,200
(3) 2,400
(4) 3,000
(5) 4,200

26. In which year was there the greatest change in enrollment from the year before?
(1) Year 2
(2) Year 3
(3) Year 4
(4) Year 5
(5) Year 6

27. How many more students were enrolled in Year 6 than in Year 1?
(1) 200
(2) 1,200
(3) 1,400
(4) 2,400
(5) 5,000

Check your answers on pages 293–294.

LESSON 10

Multiplying and Dividing Decimals

Multiplying Decimals

Multiplying decimals is almost the same as multiplying whole numbers. There is only one difference. You have to place the decimal point in the answer.

> **TIP**
> Sometimes there are not enough digits in the answer. Then you need to use zeros as placeholders. Place the zeros to the *left* of the answer.
>
> | | |
> |---|---|
> | .014 | 3 places |
> | × 0.6 | +1 place |
> | 0.0084 | 4 places |

Example Multiply: 6.03 × 0.5

Step 1 Write the problem. In multiplication, you don't have to line up the decimal points. Line up the numerals on the right instead.

```
  6.03
×  0.5
```

Step 2 Multiply as you would with whole numbers.

```
  6.03
×  0.5
  3015
```

Step 3 Count the total number of decimal places in the problem. The number of decimal places is the number of digits to the right of the decimal points. There are two decimal places in 6.03 and one decimal place in 0.5. The answer will have 3 decimal places.

```
  6.03    ←   2 places
×  0.5    ←  +1 place
 3015,    ←   3 places
```

Step 4 Finally, place the decimal point in the answer. Start at the right and count the same number of decimal places from Step 3 to the left. Then write the decimal point.

```
  6.03    Count 3
×  0.5    places to
 3.015    the left.
```

PRACTICE

Multiply. The first one is done for you.

1. 13 × 0.4 = 5.2
2. 9.1 × 8
3. 12.3 × 0.5
4. 1.04 × 0.07
5. 0.75 × 0.5
6. 136 × 0.006
7. 128 × 0.2
8. 17.3 × 1.6
9. 0.42 × 0.03
10. 2.05 × 0.9
11. 5.27 × 3.6 =
12. 6.2 × 0.08 =
13. 28.32 × 5.4 =

MEASUREMENT AND GEOMETRY

Reading Scales

You may have noticed that when you shop for vegetables, fruits, cheese, or items from a deli counter, the clerk weighs the item on a **digital scale.** Below is a picture of one such scale.

TIP
A digital scale uses decimals to show the weight (in pounds), unit price (in dollars), and total price (in dollars).

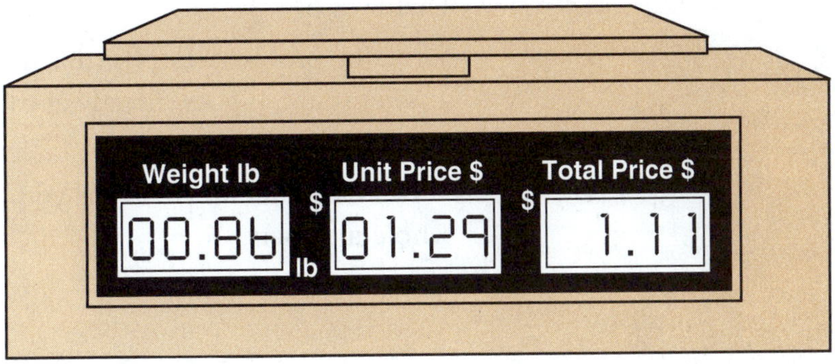

Example Pedro purchased two red peppers. Red peppers cost $1.29 per pound. How much did the red peppers weigh?

Step 1 Find the column on the scale that says, "Weight lb."

Step 2 Read the display. The display reads 00.86.

The red peppers weighed 0.86 pounds.

PRACTICE

Solve each problem. The first one is done for you.

1. What is the weight of $\frac{1}{5}$ pound of reggianno parmesan cheese on a digital scale?

   (Hint: To find the decimal value of a fraction, divide the numerator by the denominator.)

   00.20 pounds

2. Write an expression that would give you the total price of oranges, based on the display at right.

3. Fill in the display at the right to show the total cost of $2\frac{1}{4}$ pounds of ground round at $2.69 per pound.

4. Fill in the display at the right to show the total cost of $\frac{3}{4}$ pound of romaine lettuce at 69¢ per pound.

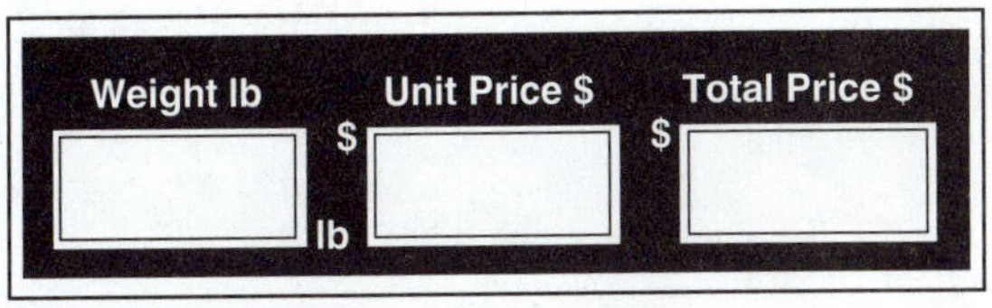

Check your answers on page 295.

Dividing Decimals

Dividing a decimal by a whole number is the same as dividing whole numbers. You just need to place the decimal point in the answer.

TIP

Sometimes you will need to put zeros at the end of a decimal to solve the problem.

For example, 1.2 ÷ 5.

```
  0.24
5)1.20
 - 10
   20
 - 20
    0
```

Example Divide: 8.6 ÷ 2

Step 1 Write the problem. Put the decimal point in the answer space. It goes directly above the decimal point in the problem.

```
2)8.6
```

Step 2 Solve the problem. Divide as you would with whole numbers.

```
  4.3
2)8.6
 -8
  06
 - 6
   0
```

Step 3 Check by multiplying.

```
  4.3
×   2
  8.6
```

To divide by a decimal, you need to change the decimal to a whole number by moving the decimal point as far to the right as you can. Then move the decimal point in the number you are dividing the same number of places to the right.

TIP

Sometimes you will need to use zeros as placeholders in the answer.

For example, 0.028 ÷ 4. Since 4 won't divide into 0 or 2, place zeros in the answer as needed.

```
  0.007
4)0.028
  - 28
     0
```

Example Divide: 2.032 ÷ 0.8

Step 1 Change 0.8 to a whole number. Move the decimal point one place to the right.

Step 2 Move the decimal point in 2.032 one place to the right.

```
0.8)2.0.32
```

Step 3 Place the decimal point in the answer space and divide.

```
   2.54
8)20.32
 -16
   4 3
  -4 0
     32
    -32
      0
```

Multiplying and Dividing Decimals on a Calculator

To multiply or divide decimals on a calculator, enter each decimal number including the decimal point, use the [×] or [÷] key between each number, then use the equals key to get the answer.

For more practice multiplying and dividing decimals with the calculator, see Calculator Handbook pages 248–249.

Example 6.8 × 5.3 ÷ 0.02

[AC] 6.8 [×] 5.3 [÷] 0.02 [=] 1802

PRACTICE

A. **Divide. The first problem is done for you.**

1. $6\overline{)2.82}$ = 0.47

   -24

   42

   -42

   0

2. $9\overline{)5.67}$

3. $7\overline{)41.3}$

4. $12\overline{)33.6}$

5. $13\overline{)110.5}$

6. $4\overline{)1.224}$

7. $21\overline{)0.882}$

8. $0.5\overline{)0.001}$

9. $0.08\overline{)0.0248}$

10. $0.12\overline{)1.0236}$

B. **Rewrite the problems below. Divide. The first one is started for you.**

11. $0.105 \div 6 =$

    $6\overline{)0.1050}$ = 0.01

    -6

    45

12. $0.51 \div 0.012 =$

13. $12.6 \div 3.6 =$

14. $0.015 \div 0.25 =$

15. $64 \div 0.004 =$

16. $26.2701 \div 5.1 =$

C. **Solve the problems below with a calculator. Refer to pages 248–249 in the Calculator Handbook for additional information.**

17. The Chou family paid $38.89 for electricity in August. The family pays $0.1045 for each kilowatt hour of electricity. To the nearest whole number, how many kilowatt hours of electricity did they use in August?

18. Last month Gabriel earned $2,541. He worked 38.5 hours each week. What was Gabriel's hourly pay? (Hint: Assume 4 weeks = 1 month)

Check your answers on pages 295–296.

MEASUREMENT AND GEOMETRY

Metric Conversion

Example Jane works in a factory that makes hair-care products. One of the products contains a fruit extract. The extract is weighed in **kilograms.** Jane fills plastic tubes with small amounts of the extract. These amounts are so small they are weighed in **grams.**

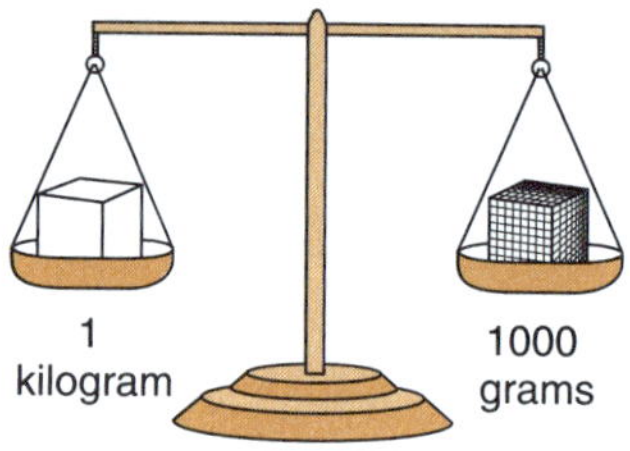

1 kilogram (kg) = 1,000 grams (g)

The numbers 10, 100, and 1,000 are called **powers of ten.** A simple way to multiply by a power of ten is to move the decimal point one place to the right for each zero in the power of ten.

This morning Jane received a shipment weighing 25.6 kilograms. Which is the correct expression to find how many grams the shipment weighs?

(1) 25.6×10
(2) 25.6×100
(3) $25.6 \div 100$
(4) $25.6 \times 1{,}000$
(5) $25.6 \times 10{,}000$

Answer **(4)** is correct. To convert 25.6 kilograms to grams, Jane needs to multiply by 1,000. Since there are three zeros in 1,000, she moves the decimal point three places to the right. **25.6 × 1,000 = 25,600**, so the shipment weighed **25,600 grams**.

TIP
To multiply by a power of ten, count the number of zeros and move the decimal the same number of places to the right.

$3.25 \times 10 = 32.5$

$3.25 \times 100 = 325.$

$3.25 \times 1{,}000 = 3{,}250.$

Example Jane's co-worker, Leon, prepares the liquid base for a hair-care product. The ingredients are measured in **milliliters.** After he finishes the job, he needs to record the amount in **liters** on a report form.

1 liter (L) = 1,000 milliliters (ml)

Today Leon prepares 45,000 milliliters of the liquid base. Which is the correct expression to find the amount in liters?

(1) $45{,}000 \div 10$
(2) $45{,}000 \div 100$
(3) $45{,}000 \div 1{,}000$
(4) $45{,}000 \times 1{,}000$
(5) $45{,}000 \times 10{,}000$

Answer **(3)** is correct. To convert 45,000 milliliters to liters, Leon needs to divide by 1,000. To divide by a power of ten, count the zeros. Move the decimal point the same number of places to the left. **45,000 ÷ 1,000 = 45**, so Leon has prepared **45 liters** of the liquid base.

TIP
To divide by a power of ten, count the number of zeros and move the decimal the same number of places to the left.

$325 \div 10 = 32.5$

$325 \div 100 = 3.25$

$325 \div 1{,}000 = 0.325$

PRACTICE

Circle the best answer for each question. Refer to the chart below if necessary.

Length	10 millimeters (mm) 100 centimeters (cm) 1,000 meters (m)	= 1 centimeter (cm) = 1 meter (m) = 1 kilometer (km)
Mass (Weight)	1,000 milligrams (mg) 1,000 grams (g)	= 1 gram (g) = 1 kilogram (kg)
Volume	1,000 milliliters (ml)	= 1 liter (L)

1. Ana's personal best in the long jump is 4.5 meters. How many centimeters did she jump?
   (1) 0.0045 cm
   (2) 0.045 cm
   (3) 0.45 cm
   (4) 4.5 cm
   (5) 450 cm

2. Rick ate a bowl of chili that had 5.5 grams of fat. How many milligrams of fat were in the chili?
   (1) 0.0055 mg
   (2) 0.55 mg
   (3) 550 mg
   (4) 5,500 mg
   (5) 55,000 mg

3. A square measures 0.4 meter on each side. What is the perimeter of the square in centimeters?
   (1) 1.6 cm
   (2) 160 cm
   (3) 1,600 cm
   (4) 16,000 cm
   (5) 160,000 cm

4. A tube contains 100 milliliters of lotion. How many liters of lotion does the tube contain?
   (1) 0.001 L
   (2) 0.01 L
   (3) 0.1 L
   (4) 1 L
   (5) 10 L

5. Monica ran a race in 4.5 seconds. How many meters long is the race?
   (1) 0.1 m
   (2) 1 m
   (3) 10 m
   (4) 100 m
   (5) Not enough information is given.

6. A can of turkey gravy weighs 298 grams. What is the weight of the can in milligrams?
   (1) 298,000 mg
   (2) 29,800 mg
   (3) 2,980 mg
   (4) 298 mg
   (5) Not enough information is given.

7. **Connect** Look at three labels on foods that you commonly eat. Change any amounts listed in grams to kilograms or amounts listed as liters to milliliters.

Check your answers on page 296.

WORKPLACE MATH

Filling Out Forms

Example Rafael is a stock supervisor at a large department store. He orders items as shelf supplies run low. He started to fill out the following order form.

Item Number	Description	Quantity	Unit Price	Amount
23F	Cotton washcloths green mist	12	$4.49	
24A	Cotton hand towels ivory	8	$7.99	
26A	Medium-support polyester-fill pillow	15		$375.00
28B	Cappuccino/espresso maker, Model 781	3	$49.99	
31G	Ceramic hand-painted mug from Italy	6	$9.00	
45C	Black tote bag	4	$135.00	
			Total	

Rafael needs to complete the order form. What is the total amount for the cotton washcloths?

(1) $4.61
(2) $44.90
(3) $53.88
(4) $63.92
(5) $150.00

Answer **(3)** is correct. Rafael multiplies the quantity (12) by the unit price ($4.49). **$4.49 × 12 = $53.88**, so the total amount for the washcloths is **$53.88**.

Example Rafael is ordering 15 medium-support polyester-fill pillows that cost a total of $375. What is the unit price for each pillow?

(1) $15.00
(2) $25.00
(3) $37.50
(4) $360.00
(5) Not enough information is given.

Answer **(2)** is correct. Rafael divides the total amount of the pillows ($375) by the quantity (15). **$375 ÷ 15 = $25**, so the unit price for each pillow is **$25**.

PRACTICE

Circle the best answer for each question. Question 1 through 8 refer to the order form on page 146.

1. What is the total amount for the cappuccino/espresso makers?

   (1) $49.99
   (2) $52.99
   (3) $99.98
   (4) $149.97
   (5) $199.96

2. The total amount for the black tote bags is $540. Which expression shows how Rafael arrived at that amount?

   (1) 6 × $9
   (2) 4 × $135
   (3) 5 + $135
   (4) 5 × $135
   (5) 5 ÷ $135

3. How many items did Rafael order all together?

   (1) 48
   (2) 45
   (3) 38
   (4) 30
   (5) 6

4. What is the total amount for the 12 cotton washcloths and the 8 cotton hand towels all together?

   (1) $12.48
   (2) $53.88
   (3) $63.92
   (4) $117.80
   (5) $142.80

5. Use your calculator to complete the amount column. What is the TOTAL for this order?

   (1) $231.47
   (2) $642.77
   (3) $686.79
   (4) $1,182.77
   (5) $1,236.77

6. The following day Rafael's boss asked him to order four blenders. What is the total amount for the four blenders?

   (1) $279.96
   (2) $500.00
   (3) $636.00
   (4) $716.00
   (5) Not enough information is given.

7. The following week Rafael needed to order 15 more hand-painted mugs. Including the first order and this order, what is the total amount for the mugs?

   (1) $54
   (2) $135
   (3) $189
   (4) $324
   (5) Not enough information is given.

8. **Reason** Suppose the black tote bags were on back order. This means they were not shipped with the rest of the order, and the store was not charged for them. Explain what the new total is, and how you arrived at it.

Check your answers on page 296.

WORKPLACE MATH

Ordering Supplies

When ordering supplies, you often order more than one of each item. For example, a supply order might include three boxes of pens or ten boxes of paper clips. The price of one item is called the **unit cost. Unit** means one item. **Total cost** is found by multiplying the unit cost by the number of units you want.

Total cost = Unit cost × Number of units

Example Victor orders the supplies for the county clerk's office. He needs to order 50 computer disks and 3 cases of computer paper. Each case contains 10 packets (reams) of paper.

Victor looks up the price of one case of computer paper in a catalog. The catalog states that each case costs $24.99. Which is the correct expression to find the total cost of the computer paper in Victor's order?

(1) $\$24.99 \times 10$
(2) $\$24.99 \times 3$
(3) $\$24.99 \times 3 \times 10$
(4) $\$24.99 \div 3 \times 10$
(5) $3 \times \$24.99 \div 10$

Answer **(2)** is correct. To find the total cost of the computer paper, Victor should multiply the unit cost ($24.99) by the number of units (3). **$24.99 × 3 = $74.97,** so the total cost of the 3 packages of computer paper is **$74.97.** (The fact that there are 10 reams of paper in each case is extra information.)

You can also find the unit cost of an item.

Unit cost = Total cost ÷ Number of units

Example Victor looks up the price of computer disks. He can buy a box of 50 disks for $21.99. Which is the correct expression to find the cost of one disk?

(1) $\$21.99 \times 50$
(2) $\$21.99 \times 1$
(3) $50 \div \$21.99$
(4) $\$21.99 \div 50$
(5) Not enough information is given.

Answer **(4)** is correct. To find the cost of each disk, Victor divides the total cost ($21.99) by the number of disks (50). Then he rounds his answer to the nearest cent. **$21.99 ÷ 50 = 0.4398,** so the cost of one computer disk is approximately **$0.44.**

PRACTICE

Circle the best answer for each question.

1. Erica orders supplies for her department. She needs to order six daily calendar refills ($0.99 each). Which is the correct expression to find the total cost of her order?

(1) $0.99 ÷ 6
(2) $0.99 × 0.6
(3) 6 ÷ $0.99
(4) $0.99 × 6
(5) 1 ÷ $0.99

2. The Music Mart is having a sale. You can buy five cassette tapes for $19.95. Fernando wants to buy one tape. Which is the correct expression to find the unit cost of one cassette tape?

(1) $19.95 ÷ 1
(2) $19.95 × 5
(3) $19.95 ÷ 5
(4) 5 ÷ $19.95
(5) $19.95 × 0.5

3. Rita is ordering supplies for her office. She can order six reams of color paper for $47.94. She only wants one ream of paper. Which is the correct expression to find the unit cost of one ream of paper?

(1) $47.94 ÷ 6
(2) 6 × $47.94
(3) 6 ÷ $47.94
(4) $47.94 × 0.6
(5) $47.94 ÷ 0.6

4. Kelvin is ordering school supplies for his students. Kelvin orders 25 colored pencils. The pencils cost $0.19 each. Which is the correct expression to find the total cost of the pencils?

(1) $0.19 × 0.25
(2) $0.19 ÷ 25
(3) $0.19 × 2.5
(4) 25 ÷ $0.19
(5) $0.19 × 25

5. Esther decides to stock up on paper towels for her day-care business. She wants to order 25 rolls of paper towels. Which is the correct expression to find the total cost of the paper towels?

(1) $0.59 × 25
(2) $0.59 ÷ 25
(3) 25 ÷ $0.59
(4) $0.59 × 2.5
(5) Not enough information is given.

6. Fantastic Lights is having a sale. You can buy a four-pack of frosted light bulbs for $1.69. Which is the correct expression to find the unit cost of one light bulb?

(1) $1.69 × 0.4
(2) 4 ÷ $1.69
(3) 4 × $1.69
(4) $1.69 ÷ 4
(5) Not enough information is given.

7. **Reason** Think about the equation Total cost = Unit cost × Number of units. Explain how to get the following equation: Unit cost = Total cost ÷ Number of units

Check your answers on pages 296–297.

Filling in Decimal Answers on a Standard Grid

You have already learned how to enter whole numbers and fractions on a grid.

Remember, your answer sheet may be scored by a machine. Here are some important points to keep in mind so that your answer will be scored correctly.

- Write your answer in the row of boxes at the top of the grid. This row is not scored by the machine, but you can use it as a guide to fill in the circles below.
- You can fill in only one circle in each column, so the decimal point must be placed in a column of its own.
- You can start your answer in any of the five columns as long as it fits entirely within the grid.
- Leave any unused columns blank.

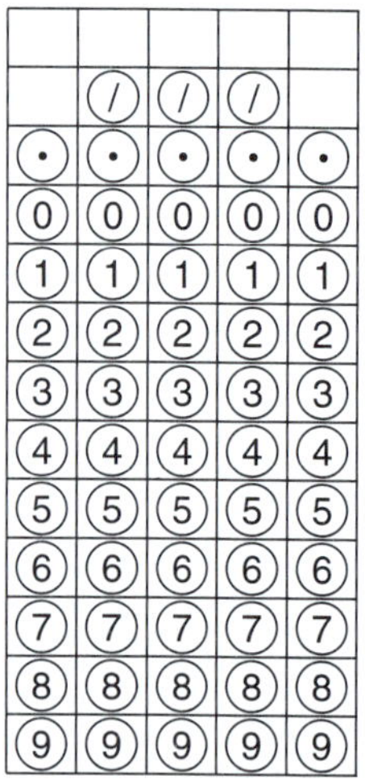

Example A rectangular painting is 2.9 meters long and 1.3 meters wide. What is the perimeter of the painting in meters?

Add the four sides to find the perimeter: 2.9 + 2.9 + 1.3 + 1.3 = **8.4 meters**

To fill in the grid, write the answer 8.4 in the row of boxes. Then fill in the correct circles on the grid. Since the answer can begin in any column, all the grids below would be scored as correct.

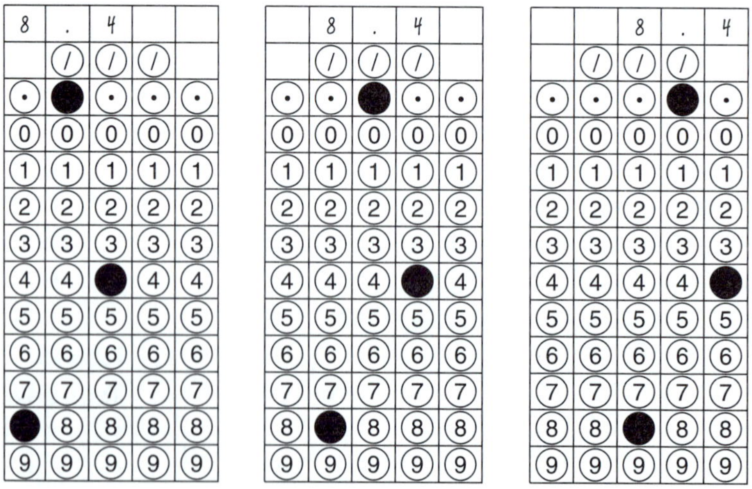

PRACTICE

Solve the following problems and record your answers in the grids provided.

1. Mike wants to start walking to and from work five days a week to get some exercise. He figures out that the walk one way is 1.25 miles. How many miles will Mike walk in a week going to and from work?

   Mark your answer in the circles in the grid.

2. A carton of canned pears weighs 177.6 ounces. If there are 12 cans in the carton, how many ounces does each can weigh?

   Mark your answer in the circles in the grid.

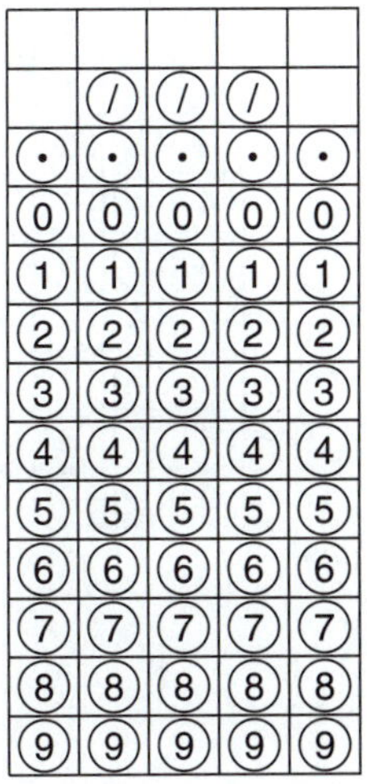

3. Pablo read on his gas bill that his family used an average of 1.45 units of natural gas per day in October. In August, his family averaged only 1.29 units per day. How many more units per day did the family use in October than in August?

   Mark your answer in the circles in the grid.

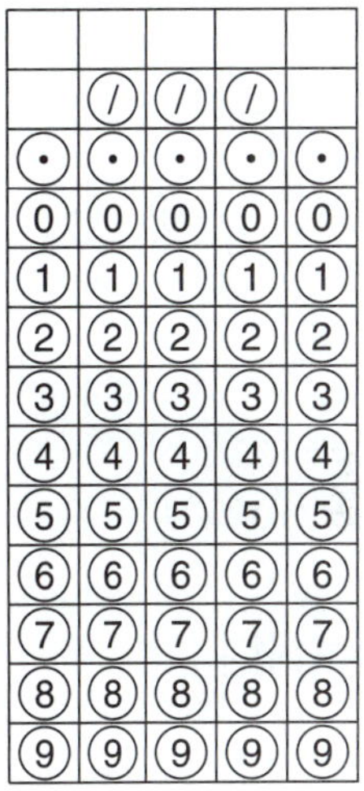

4. Each dose of cold medicine contains 0.035 gram of a decongestant. The bottle holds enough for 50 doses. How many grams of decongestant are in the bottle?

   Mark your answer in the circles in the grid.

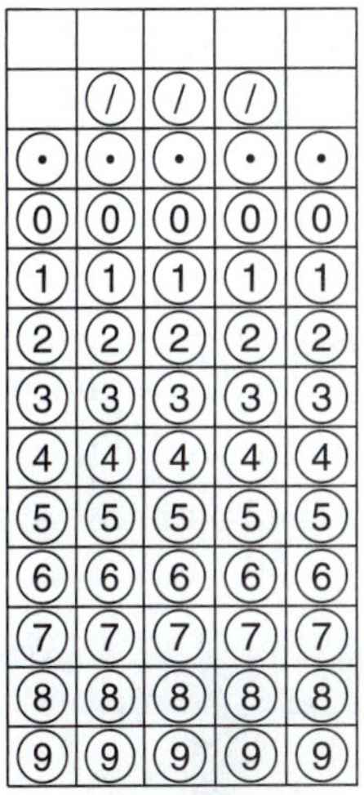

Check your answers on page 297.

10 LESSON

Lesson 10 Review
Lesson 10 Review

Lesson 10 Review

Multiply or divide.

1. $\begin{array}{r} 3.2 \\ \times\ \ 6 \\ \hline \end{array}$

2. $\begin{array}{r} 2.06 \\ \times 0.04 \\ \hline \end{array}$

3. $\begin{array}{r} 4.17 \\ \times\ \ 0.3 \\ \hline \end{array}$

4. $\begin{array}{r} 12.4 \\ \times\ \ 2.7 \\ \hline \end{array}$

5. $\begin{array}{r} 0.752 \\ \times\ \ 0.18 \\ \hline \end{array}$

6. $6\overline{)4.56}$

7. $4\overline{)3.3}$

8. $0.5\overline{)1.935}$

9. $0.08\overline{)5.42}$

10. $1.5\overline{)13.95}$

11. $0.56 \times 0.009 \div 0.02 =$ ______

12. $37.6 \div 0.16 \times 0.554 =$ ______

Solve each problem.

13. A can of tuna contains 12.5 grams of protein. How many milligrams of protein does the tuna contain?

14. Miguel works 5 days each week. He takes the same route to and from work each day. If he drives a total of 74.5 miles to and from work each week, how far does he drive to and from work each day?

15. Risa bought $1\frac{3}{4}$ pounds of ground beef for \$2.39 per pound. Fill in the display below to show the total cost of the ground beef.

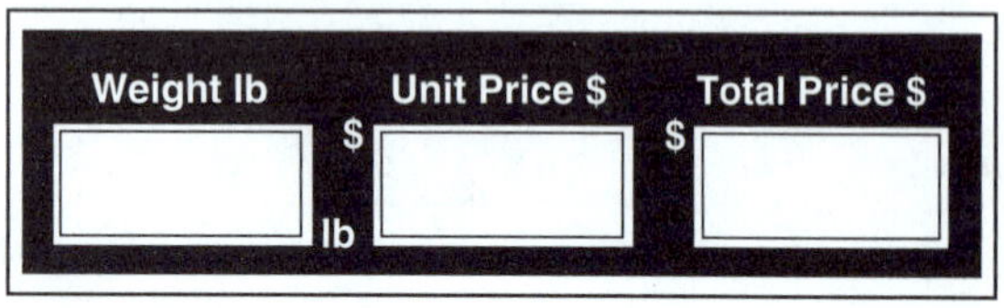

16. A steel cable is 16.734 meters long. If the cable is cut into three equal pieces, what will be the length, in meters, of each piece?

Mark your answer in the circles in the grid.

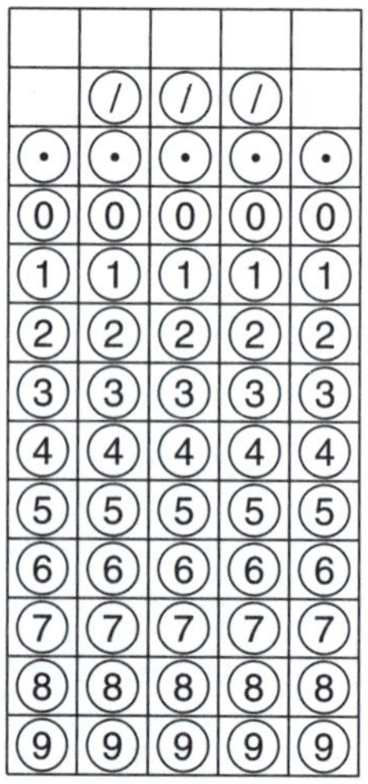

Circle the best answer for each question.

17. Anna bought a set of 4 tires for \$197.00. Which is the correct expression to find the cost of one tire?

(1) 4 × \$197.00
(2) \$197.00 ÷ 4
(3) \$197.00 × 0.4
(4) 4 ÷ \$197.00
(5) Not enough information is given.

18. Savemore Foods has bananas for \$0.60 per pound. Kai bought 4 pounds of bananas. How much will the bananas cost?

(1) \$0.15
(2) \$1.50
(3) \$2.40
(4) \$8.32
(5) \$11.20

19. Carolyn is a stock supervisor at a computer store. She filled out an order form with the following information: Item #1400S Scanner for PC, 4 at \$149.00 each; Item #74051 Anti-Static Cleaning Wipes, 7 at \$6.99 each. What is the subtotal amount of the order?

(1) \$48.93
(2) \$197.93
(3) \$623.96
(4) \$596.00
(5) \$644.93

20. George drove 291.6 miles on 12 gallons of gas. How many miles per gallon did his car get?

(1) 1.2 miles per gallon
(2) 2.43 miles per gallon
(3) 12 miles per gallon
(4) 24.3 miles per gallon
(5) 29.16 miles per gallon

21. Mrs. Quan is ordering notebooks for her students. Each notebook costs \$1.29. Which is the correct expression to find the cost of 31 notebooks?

(1) \$1.29 ÷ 31
(2) \$1.29 + 31
(3) \$1.29 × 0.31
(4) \$1.29 × 31
(5) 31 ÷ \$1.29

22. Raphael is building shelves. He needs to cut pieces of wood 1.2 meters long. Which is the correct expression to find how many pieces with that length can he cut from a board 5 meters long?

(1) 5 ÷ 1.2
(2) 5 × 1.2
(3) 1.2 ÷ 5
(4) 5 ÷ 1.02
(5) Not enough information is given.

23. A kitchen is 398 centimeters wide. How many meters wide is the kitchen?

Mark your answer in the circles on the grid.

Check your answers on pages 297–298.

Math at Work

Service: Bank Teller

One important service industry is banking. Do you have an excellent eye for detail? Do you enjoy working with numbers and money? If so, you may be interested in becoming a bank teller.

Bank tellers handle money and checks presented to them by the bank's customers. Tellers must be able to count money quickly and accurately. Because they are handling money, they must be able to add and subtract decimals. They may be required to do these calculations with actual coins and bills or on special pieces of paper called deposit and withdrawal slips. Tellers use calculators and mental math to make and check these calculations.

Bank tellers must be able to record the money brought into the bank (deposits) and the money taken out of the bank (withdrawals). All transactions performed by the teller must balance at the end of each day. Because they work constantly with customers, tellers should have good communication skills and a pleasant appearance.

Look at the *Some Careers in Service* chart.

- Do any of the careers interest you? If so, which ones?
- What information would you need to find out more about those careers?

On a separate piece of paper, write some questions that you would like answered. You can find out more information about those careers in the *Occupational Outlook Handbook* at your local library or online.

Some Careers in Service

Cashier
operates cash register, handles money, stocks items and maintains store appearance

Fee Collector
welcomes visitors to museums and parks, collects fees, informs visitors of policies

Park Visitor Use Assistant
collects visitor fees, answers questions and provides assistance to visitors

Property Appraiser
calculates value of properties by examining their condition and comparing them to similar properties

Josepha is a bank teller. She often helps her customers correctly fill out deposit tickets like the one below. **Look at the deposit slip and the information. Then answer the questions.**

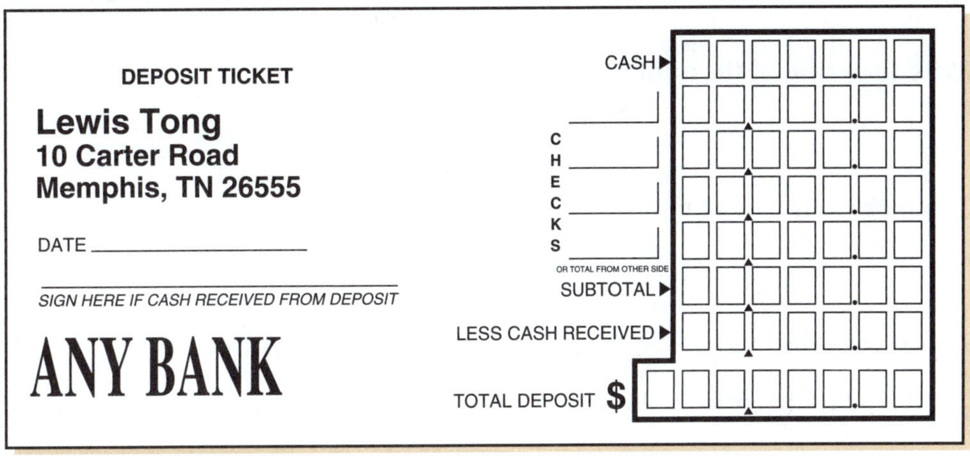

Mr. Tong presents Josepha with his deposit ticket and four checks for deposit. He has three checks in the amount of $345.23 each. His fourth check is for $476.55. He tells her he also wishes to receive $125.50 in cash from his deposit. Josepha tells Mr. Tong to write the amount of each check in the spaces beside the word *CHECKS* and the money he wants back beside *Less Cash Received.*

1. What is the total amount of Mr. Tong's checks?
   (1) $125.50
   (2) $345.23
   (3) $1,035.69
   (4) $1,386.74
   (5) $1,512.24

2. In which space should Mr. Tong enter the total amount of the four checks he wants to deposit?
   (1) Cash
   (2) Checks
   (3) Subtotal
   (4) Less Cash Received
   (5) Total Deposit

3. In the space marked *Total Deposit,* What amount should Mr. Tong enter?
   (1) $125.50
   (2) $345.23
   (3) $1,035.69
   (4) $1,386.74
   (5) $1,512.24

Check your answers on page 299.

Unit 3 Review: Decimals

Unit 3 Review: Decimals

Unit 3 Review Decimals

Write the value of the underlined digit in words.

1. $5.394\underline{6}$ ________________

2. $7.2\underline{8}13$ ________________

Compare each pair of numbers. Write >, <, or =.

3. 0.25 ________________ 0.025

4. 0.97 ________________ 0.970

Round each decimal to the given place value. Show your work.

5. 5.362 to the nearest hundredth

6. 7.351 to the nearest tenth

Solve each problem. Show your work.

7. Write 0.42 as a fraction. Reduce to lowest terms.

8. Write $\frac{19}{25}$ as a decimal.

Solve each problem. Show your work.

9. $\begin{array}{r} 3.925 \\ 4.6 \\ +3.26 \\ \hline \end{array}$

10. $\begin{array}{r} 25.36 \\ -17.5 \\ \hline \end{array}$

11. $\begin{array}{r} 8.4 \\ -3.257 \\ \hline \end{array}$

12. $4\overline{)37.6}$

13. $\begin{array}{r} 6.3 \\ \times\ 8 \\ \hline \end{array}$

14. $\begin{array}{r} 2.54 \\ \times\ 0.9 \\ \hline \end{array}$

15. $0.06\overline{)2.7}$

16. $0.016\overline{)7.488}$

17. $3.54 + 6.83 + 5.19 =$

18. $0.0357 \div 0.07 =$

19. $45.4 - 2.92 =$

20. $0.57 \times 0.05 =$

Solve each problem. Show your work.

21. Bryant bought a pair of athletic shoes for \$64.95 and a package of socks for \$5.99. Sales tax on his purchases was \$3.55. Bryant gave the clerk \$80.00 in cash. How much change should the clerk give Bryant?

22. What is the value of $9 \div (4 - 1) + 8$?

23. Freddie can order a dozen pens for \$3.12. How much would one pen cost at this rate?

24. Rounded to the nearest cent, what is the total cost for $\frac{2}{5}$ pound of cauliflower?

25. What is the total amount for the two items shown below?

Item Number	Description	Quantity	Unit Price	Amount
35D	Twin sheets	4	\$12.99	
26F	Down pillow	5	\$20.00	

Circle the best answer.

26. Auto Buys has tires on sale this week for \$42.50 each. Lin needs to buy four tires. Lin also gets the tires balanced for \$15.95. Tax amounted to \$14.87. Which is the correct expression to find the total cost to Lin?

(1) (4 × \$42.50) + \$15.95 + \$14.87
(2) 4 × (\$42.50 + \$15.95) + \$14.87
(3) 4 × (\$42.50 + \$15.95 + \$14.87)
(4) 4 × (\$42.50 + \$15.95 − \$14.87)
(5) 4 × (\$42.50 − \$15.95) + \$15.95 + \$14.87

Question 27 refers to the following graph.

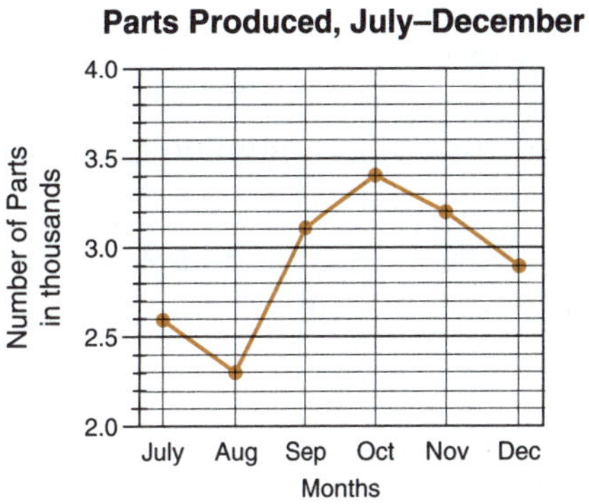

27. How many thousands of parts were produced during the last two months?

Mark your answer in the circles on the grid.

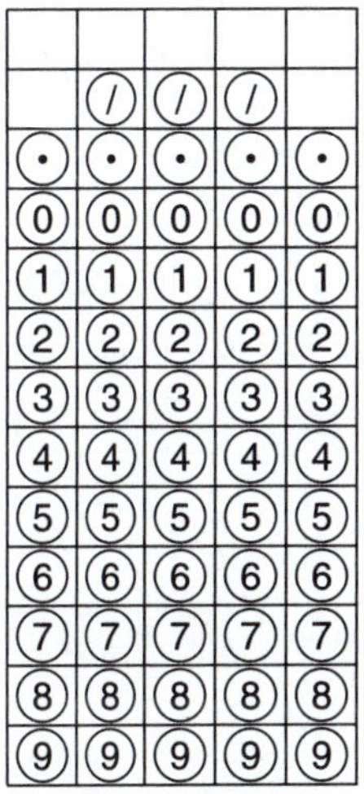

Math Extension

Measure the dimensions of two rooms in your home. Convert to metric amounts.

Check your answers on pages 299–300.

Mini-Test • Unit 3

Directions: This is a 15-minute practice test. After 15 minutes, mark the last number you finished. Then complete the test and check your answers. If most of your answers were correct but you did not finish, try to work faster next time.

Part I Directions: Choose the one best answer to each question. You MAY use your calculator.

1. A pest control service costs $16.50 per month or $168 per year. If you want the service for twelve months, how much would you save by paying the yearly rate?
   (1) $4.50
   (2) $14.00
   (3) $30.00
   (4) $33.00
   (5) $198.00

2. An electronics part must be no wider than 2.0145 centimeters. A manufacturer finds a part exactly 2.15 centimeters wide. How many centimeters too wide is the part?
   (1) 0.0005
   (2) 0.0355
   (3) 0.1355
   (4) 4.1645
   (5) Not enough information is given.

3. A bushel of wheat takes up 1.25 cubic feet of space. A truck can hold 20,000 cubic feet. How many bushels can the truck hold?
   (1) 1,600
   (2) 2,500
   (3) 7,500
   (4) 16,000
   (5) 25,000

4. Andy owns 10 shares of Cyber and 30 shares of Titan stocks. Cyber is worth $24.50 per share and Titan is worth $14.61 per share. How much are his shares worth in all?
   (1) $683.30
   (2) $782.20
   (3) $881.10
   (4) $1,564.40
   (5) Not enough information is given.

5. Keri weighed the four packages below using a digital scale.

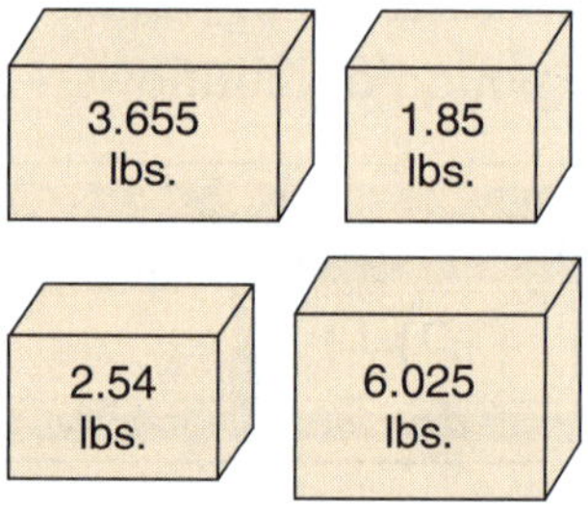

What is the total weight of the boxes in pounds?

Mark your answer in the circles in the grid.

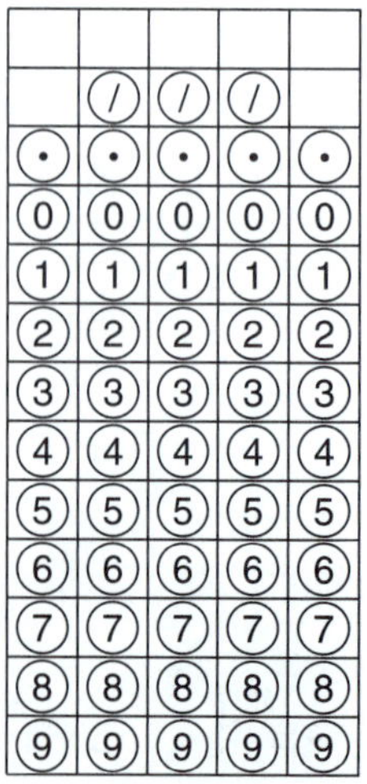

Mini-Test

Part II Directions: Choose the one best answer to each question. You MAY NOT use your calculator.

Questions 6 and 7 refer to the following graph.

Basketball Season Ticket Sales

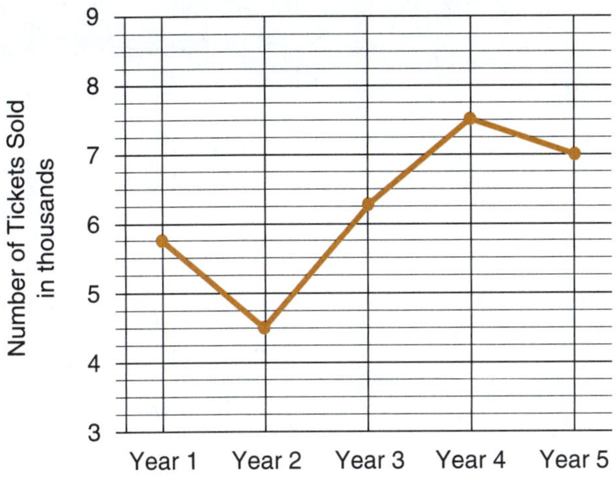

6. Which of the following expressions could be used to find out how many more season tickets were sold in Year 5 than in Year 1?
   (1) 7,250 − 5,750
   (2) 7,000 − 5,900
   (3) 7,500 − 4,500
   (4) 7,000 − 5,750
   (5) 7,250 − 6,250

7. In which year was the greatest increase in season ticket sales?
   (1) Year 2
   (2) Year 3
   (3) Year 4
   (4) Year 5
   (5) Not enough information is given.

8. What is the value of the expression $1.4 - 0.2 \times 6 + 2.5$?
   (1) 0.3
   (2) 2.7
   (3) 9.7
   (4) 10.2
   (5) 16.68

9. Jan buys 5 plants at the cost of \$3.59 each and a large plant for \$14.79. She gives the clerk \$40 in cash. Which is the correct expression to find out how much change Jan should receive?
   (1) \$40 − \$3.59 × 5 + \$14.79
   (2) (\$40 − \$3.59) × 5 + \$14.79
   (3) (\$40 − \$3.59 × 5) + \$14.79
   (4) \$40 − (\$3.59 × 5) + \$14.79
   (5) \$40 − (\$3.59 × 5 + \$14.79)

10. A large container holds 3 kiloliters of water. How many liters of water are in the container?
   (1) 30
   (2) 300
   (3) 3,000
   (4) 30,000
   (5) 300,000

11. Sandy expects a project to take 12.75 hours. She works 4.25 hours on Monday. How many more hours should the project take?

   Mark your answer in the circles in the grid.

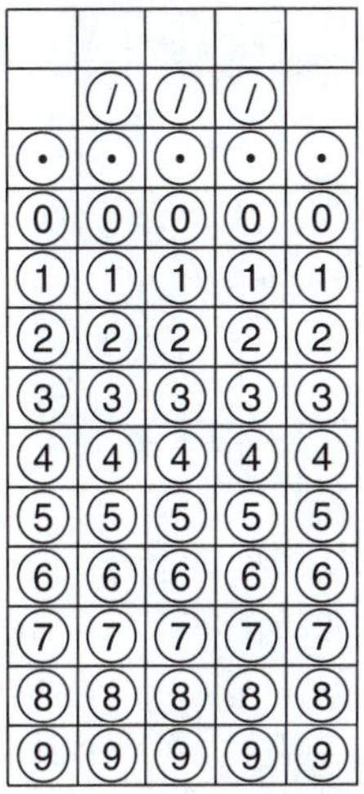

Check your answers on pages 300–301.

4 UNIT

Ratios, Proportions, and Percents

Ratios, proportions, and percents are used every day. For example, people often use common rates such as miles per gallon or dollars per pound. Percents are used to compute the amount you save at a sale, and ratios can be used to figure distances on a map.

What is the sales tax rate where you live? ____________________

What is today's chance of rain or snow in your area? ______________

What percent interest have you paid on a credit card or loan? ________

Thinking About Ratios, Proportions, and Percents

You may not realize how often you say, read, or use ratios, proportions, or percents as you go about your daily life. Think about your recent activities.

Check the box for each activity you have done recently.

- ☐ Did you look at the cost per ounce of a food item?
- ☐ Did you find the cost of two items when you knew the cost for one?
- ☐ Did you pay tax on an item?
- ☐ Did you increase or decrease the amounts in a recipe?
- ☐ Did you see a percent-off sale in a store?
- ☐ Did you pay interest on a loan or credit?
- ☐ Did you find the amount of a paycheck using your hourly pay rate?

Write other activities where you used ratios, proportions, or percents.

____

____

____

Previewing the Unit

In this unit you will learn:

- to write and use ratios and rates
- how to solve problems with ratios and rates
- how to write and use proportions
- to solve problems using a proportion
- how to use a map scale
- how to change percents to decimals and fractions
- how to change decimals and fractions to percents
- the three parts of a percent problem
- how to solve percent problems
- how to solve ratio, proportion, and percent problems on a calculator

Lesson 11	Ratios and Proportions	Lesson 15	Solving for the Base and Percent of Change
Lesson 12	Percent Basics	Lesson 16	Statistics and Probability
Lesson 13	Solving for the Part (p)		
Lesson 14	Solving for the Rate (r)		

LESSON 11

Ratios and Proportions

Finding Ratios and Rates

Ratios compare quantities or amounts. When you compare inches to inches or dollars to dollars, you are comparing like quantities. To compare 3 inches to 8 inches, you could write 3 to 8, 3:8, or $\frac{3}{8}$. The colon in 3:8 says this is a ratio.

Example Tanya works for a photo lab. During an 8-hour day, Tanya works 3 hours at the counter helping customers and 5 hours in the lab. What is the ratio of the time Tanya works at the counter to the time she works in the lab?

(1) $\frac{3 \text{ hours}}{8 \text{ hours}}$

(2) $\frac{3 \text{ hours}}{5 \text{ hours}}$

(3) $\frac{5 \text{ hours}}{8 \text{ hours}}$

(4) $\frac{8 \text{ hours}}{5 \text{ hours}}$

(5) $\frac{8 \text{ hours}}{3 \text{ hours}}$

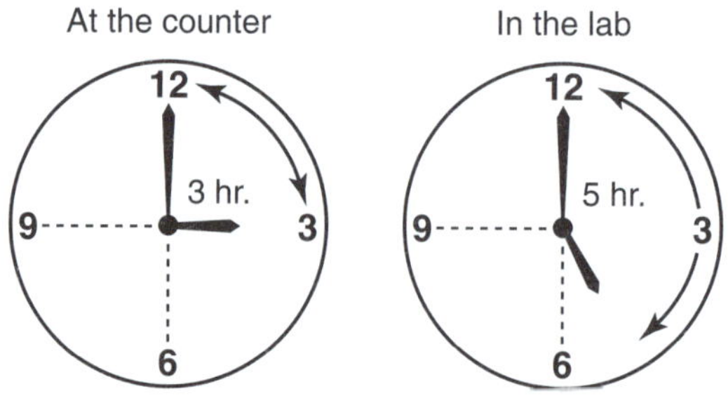

Answer **(2)** is correct. The ratio of hours working at the counter to hours working in the lab is 3:5 or $\frac{3}{5}$. The ratio shows that for every **3 hours** Tanya works at the counter, she works **5 hours** in the lab.

You can also compare unlike quantities or amounts. A ratio comparing unlike quantities is called a **rate.** A rate usually has a denominator of 1. It is very important to use the units of both the numerator and denominator in a rate.

Example Dean does a lot of city driving. He finds that he drives 15 miles on 1 gallon of gasoline. What is the rate of miles to gallons? Remember the order in which you are asked to compare. Here you are comparing miles to gallons.

(1) $\frac{1 \text{ mile}}{15 \text{ gallons}}$

(2) $\frac{1 \text{ gallon}}{15 \text{ miles}}$

(3) $\frac{15 \text{ gallons}}{15 \text{ miles}}$

(4) $\frac{1 \text{ mile}}{1 \text{ gallon}}$

(5) $\frac{15 \text{ miles}}{1 \text{ gallon}}$

Answer **(5)** is correct. The rate of miles to gallons is 15:1 or $\frac{15}{1}$. This means Dean can drive **15 miles** for every **one gallon** of gasoline. When writing rates, it is common to use the word *per* instead of *for every.* In this case you could write that Dean can drive 15 miles per gallon.

PRACTICE

Circle the best answer for each question.

1. Ned is making bread. The recipe calls for 1 cup of sugar and 6 cups of flour. Ned wants to know the ratio of the ingredients in case he decides to double the recipe. What is the ratio of cups of sugar to cups of flour?

(1) $\frac{\text{1 cup sugar}}{\text{3 cups flour}}$

(2) $\frac{\text{1 cup sugar}}{\text{6 cups flour}}$

(3) $\frac{\text{1 cup sugar}}{\text{12 cups flour}}$

(4) $\frac{\text{1 cup sugar}}{\text{36 cups flour}}$

(5) $\frac{\text{6 cups sugar}}{\text{1 cup flour}}$

2. The City Library has 1,055 fiction books and 1,267 nonfiction books checked out. What is the ratio of nonfiction books to fiction books checked out of the library?

(1) $\frac{\text{1,055 fiction}}{\text{1,055 nonfiction}}$

(2) $\frac{\text{1,055 fiction}}{\text{1,267 nonfiction}}$

(3) $\frac{\text{1,055 nonfiction}}{\text{1,267 fiction}}$

(4) $\frac{\text{1,267 nonfiction}}{\text{1,055 fiction}}$

(5) $\frac{\text{1,267 nonfiction}}{\text{1,267 fiction}}$

3. A 12-oz. package of wheat bran cereal costs $2.88, or $0.24 for each ounce. What is the rate of cost in dollars to weight in ounces?

(1) $\frac{\$0.24}{\text{1 oz.}}$

(2) $\frac{\$0.24}{\text{12 oz.}}$

(3) $\frac{\$2.88}{\text{1 oz.}}$

(4) $\frac{\text{1 oz.}}{\$0.24}$

(5) $\frac{\text{12 oz.}}{\$0.24}$

4. During a 40-hour week, Eiko earns $18 an hour as a copyeditor. What is the rate of money earned per hour?

(1) $\frac{\text{1 hr.}}{\$.18}$

(2) $\frac{\text{40 hr.}}{\$18}$

(3) $\frac{\text{40 hr.}}{\text{1 hr.}}$

(4) $\frac{\$18}{\text{1 hr.}}$

(5) $\frac{\$18}{\text{40 hr.}}$

5. Daksha works a total of 45 hours a week as a baker making cakes and cookies. She spends about 19 hours baking cakes. What is the ratio of hours baking cakes to hours baking cookies?

(1) $\frac{\text{19 hours}}{\text{45 hours}}$

(2) $\frac{\text{19 hours}}{\text{26 hours}}$

(3) $\frac{\text{26 hours}}{\text{45 hours}}$

(4) $\frac{\text{26 hours}}{\text{19 hours}}$

(5) Not enough information is given.

6. **Reason** Refer to the second example on page 162. Suppose Dean has two cars. The car discussed in the example is listed below as Car 1. His second car uses 10 gallons to drive 100 miles. Explain how you can find which car gives Dean the better rate of miles to gallons.

Car 1: 15 miles to 1 gallon

Car 2: ? miles to 1 gallon

Which car gets more miles per gallon?

Check your answers on page 301.

Working with Ratios and Rates

A ratio is a way to compare two quantities. A ratio that compares two unlike quantities is called a rate. You can write ratios in lowest terms the same way you can reduce fractions. Remember that the order in which you compare quantities is very important. The amount mentioned before the word *to* is always the numerator. The amount after the word *to* is always the denominator.

Example Clyde worked 8 hours on Thursday and 10 hours on Friday. What is the ratio of hours worked on Thursday to hours worked on Friday?

Step 1 Write the quantities in a ratio in the same order that you are asked to compare them. In this example, compare the hours worked on Thursday to the hours worked on Friday.

$\frac{\text{Thursday hours}}{\text{Friday hours}}$

$\frac{8}{10}$

Step 2 Reduce the ratio to lowest terms the same way you reduce a fraction. Divide the numerator and denominator by the same number. The ratio is in lowest terms when no number except 1 will divide both numbers evenly.

$\frac{8}{10} = \frac{8 \div 2}{10 \div 2} = \frac{4}{5}$

The ratio of hours worked on Thursday to hours worked on Friday is $\frac{4}{5}$ or 4:5.

TIP

When a ratio is an improper fraction, do *not* change it to a mixed number or a whole number. A ratio is always expressed as a fraction. The fraction $\frac{10}{1}$ equals the whole number 10. However, as a ratio, keep the number in the form $\frac{10}{1}$.

Example Nancy earned \$100 for 10 hours of work. What is the rate of her earnings to the hours she worked?

Step 1 Write the quantities in a rate in the same order that you are asked to compare them. Include units when comparing unlike quantities (amounts with different units of measurement).

$\frac{\text{earnings}}{\text{hours}} = \frac{\$100}{10 \text{ hr.}}$

Step 2 Reduce the rate to lowest terms.

$\frac{\$100}{10 \text{ hr.}} = \frac{\$100 \div 10}{10 \div 10} = \frac{\$10}{1 \text{ hour}}$

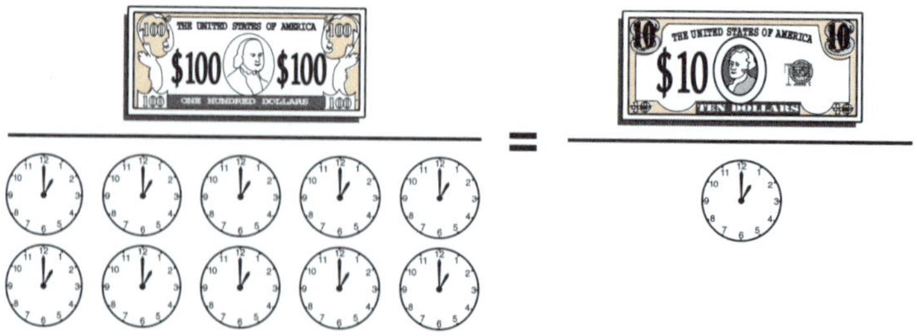

The rate of Nancy's earnings to the hours she worked is \$10/1 hour or \$10:1 hour.

PRACTICE

A. **Write each quantity as a ratio in lowest terms. Include units when comparing unlike quantities. The first one is done for you. The third one is started for you.**

1. 4 inches to 6 inches
   $\frac{4 \text{ inches}}{6 \text{ inches}} = \frac{4}{6} = \frac{2}{3}$
2. 16 employees to 21 employees
3. 2 gallons of paint to 400 square feet
   $\frac{2 \text{ gallons}}{400 \text{ square feet}}$
4. 2 pounds of apples to $1.50
5. 4 hours to 10 hours
6. 8 grams to 20 grams
7. 3 pairs of socks to $12
8. 12 women to 8 men
9. $280 to 20 hours
10. 40 books to 15 books

B. **Write the quantities as a ratio in lowest terms. The first one is done for you.**

11. It rained 9 days out of 31 days last month. What is the ratio of rainy days to total days in the month?
    $\frac{\textbf{9 rainy days}}{\textbf{31 days in the month}}$
    $\frac{9}{31}$
12. Mary's softball team won 12 games out of 18 games played. What is the ratio of games won to games played?
13. Midtown Motors sold 35 cars and 14 vans last week. What is the ratio of cars sold to vans sold?
14. Anwar drove 230 miles on 10 gallons of gasoline. What is the rate of miles to gallons?

C. **Solve.**

15. Today 15 of Ramona's customers paid with cash and 25 customers used a credit card. What is the ratio of cash customers to charge customers?
16. Kara earned $42 for 7 hours of work. What is the ratio of her earnings to the hours she worked?

Check your answers on pages 301–302.

Solving Proportions

A **proportion** is an equation comparing two equal ratios. The numbers in the proportion are called **terms.** A proportion has four terms. Sometimes one of the terms is not known. If you know three of the four terms, you can solve for the unknown term.

Example Find the value of the unknown term in this proportion: $\frac{10}{2} = \frac{?}{5}$

Step 1 Cross multiply to solve. Start by cross multiplying the numerator and denominator you know.

$\frac{10}{2} \searrow \frac{?}{5}$

$10 \times 5 = 50$

Step 2 Divide the result by the remaining number. In this case, divide 50 by the remaining term of 2. The answer, **25,** is the unknown term.

$50 \div 2 = 25$

Step 3 Check your answer by cross multiplying. If the results are equal, you know your answer is correct.

$\frac{10}{2} \times \frac{25}{5}$

$10 \times 5 = 50$

$2 \times 25 = 50$

TIP

If the numbers you are dividing do not divide evenly, write the remainder as a decimal. For example:

$\frac{3}{4} = \frac{?}{6}$

$3 \times 6 = 18$

$18 \div 4 = 4.5$

$\frac{3}{4} = \frac{4.5}{6}$

Another way to check a solved proportion is to recognize that it has the same value as two equivalent fractions. In the example above, $\frac{10}{2} = \frac{25}{5}$ because they both can be reduced to $\frac{5}{1}$.

$\frac{10}{2} = \frac{5}{1} \quad \frac{25}{5} = \frac{5}{1}$

Example Elena is a caterer. She uses 3 cups of ground coffee to serve 16 people. How many cups of ground coffee should she use to serve 80 people?

$$\frac{\text{(3 cups)}}{\text{16 people}} = \frac{\text{? cups}}{\text{80 people}}$$

Step 1 Write the ratio that has two known quantities. $\frac{\text{3 cups}}{\text{16 people}}$

Step 2 Write the second ratio with the unknown term. $\frac{\text{? cups}}{\text{80 people}}$

Step 3 Write a proportion using the two ratios. Be sure both ratios are set up in the same order: cups/people = cups/people

$\frac{\text{3 cups}}{\text{16 people}} = \frac{\text{? cups}}{\text{80 people}}$

$3 \times 80 = 240$

Step 4 Solve the proportion. $240 \div 16 = 15$

Step 5 Check your answer by cross multiplying. The cross products are equal: 240 = 240. Thus the ratios in the proportion are equal.

$\frac{3}{16} \times \frac{15}{80}$

$3 \times 80 = 240$

$16 \times 15 = 240$

PRACTICE

A. Solve each proportion. Show your work. The first one is done for you.

1. $\frac{3}{4} = \frac{?}{12}$

   $3 \times 12 = 36$

   $36 \div 4 = 9$ Thus, $\frac{3}{4} = \frac{9}{12}$

   Check: $\frac{3}{4} = \frac{9}{12}$

   $3 \times 12 = 36$

   $4 \times 9 = 36$

2. $\frac{4}{?} = \frac{12}{18}$

3. $\frac{30}{18} = \frac{10}{?}$

4. $\frac{7}{14} = \frac{?}{4}$

5. $\frac{20}{?} = \frac{4}{11}$

6. $\frac{?}{30} = \frac{10}{12}$

7. $\frac{?}{42} = \frac{10}{12}$

8. $\frac{5}{24} = \frac{4}{?}$

9. $\frac{5}{8} = \frac{?}{40}$

B. Use proportions to solve these problems. Show your work. The first one is started for you.

10. Toya gets paid an hourly wage. During a 40-hour week, Toya earns $240 in her job as a receptionist. This week she worked only 35 hours. How much did she earn?

    $\frac{40 \text{ hr.}}{\$240} = \frac{35 \text{ hr.}}{?}$

11. Kelsey swims 35 laps in 25 minutes. How many laps does he swim in 15 minutes?

12. Sarah is a photographer. For every 15 rolls of film she shoots, 5 are black-and-white. If Sarah uses 51 rolls of film, how many rolls are in black-and-white?

13. Dixie walks 45 minutes a day. How many hours will she walk in 8 days?

C. Solve. Show your work.

14. One meter is 39.37 inches. How many inches is 2.5 meters?

    ______

15. Noriko is in charge of the airport parking lots. Lot A has 4,000 spaces. Today there are 2,500 cars in Lot A. Lot B has 2,000 spaces. If both lots have the same ratio of cars to spaces, how many cars are in Parking Lot B? ______

Check your answers on page 302.

MEASUREMENT AND GEOMETRY

Using a Map Scale

Example Luis is a truck driver for Brown's Bakery. He needs to drive east from Lake City to Mt. Vernon, then south from Mt. Vernon to Hartford.

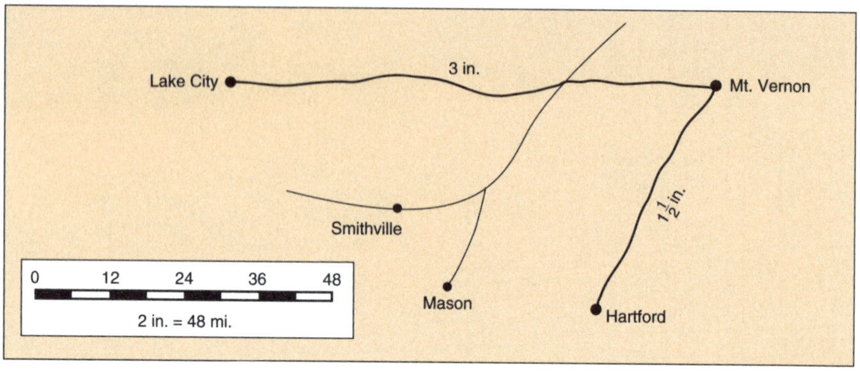

The map has a scale that shows you how to figure out distances. The scale shows the ratio of inches on the map to miles of actual distance.

On Luis' map, the scale is 2 inches equals 48 miles. The distance from Lake City to Mt. Vernon on the map is 3 inches. What is the actual distance in miles?

(1) 56
(2) 68
(3) 72
(4) 78
(5) 94

Answer **(3)** is correct. Set up a proportion comparing the map scale and a ratio with the distance shown on the map from Lake City to Mt. Vernon. Luis writes a proportion to solve for the unknown. The proportion is $\frac{2 \text{ in.}}{48 \text{ mi.}} = \frac{3 \text{ in.}}{? \text{ mi.}}$.

Set up the proportion.	Cross multiply.	Divide.
$\frac{2 \text{ in.}}{48 \text{ mi.}} = \frac{3 \text{ in.}}{? \text{ mi.}}$	$48 \times 3 = 144$	$144 \div 2 =$ **72 miles**

Example Mt. Vernon and Hartford are $1\frac{1}{2}$ inches apart on the map. What is the actual distance in miles from Mt. Vernon to Hartford?

(1) 12
(2) 24
(3) 25
(4) 36
(5) 48

Answer **(4)** is correct. Luis sets up a proportion and solves.

Set up the proportion.	Cross multiply.	Divide.
$\frac{2 \text{ in.}}{48 \text{ mi.}} = \frac{1\frac{1}{2} \text{ in.}}{? \text{ mi.}}$	$48 \times 1\frac{1}{2} = 72$	$72 \div 2 =$ **36 miles**

PRACTICE

Circle the best answer for each question.

Questions 1 through 3 refer to the following information.

Cheryl is a saleswoman for a computer company. She leaves her home in Riverton to drive to her first customer in Plainview. Then she continues on to Rock Falls.

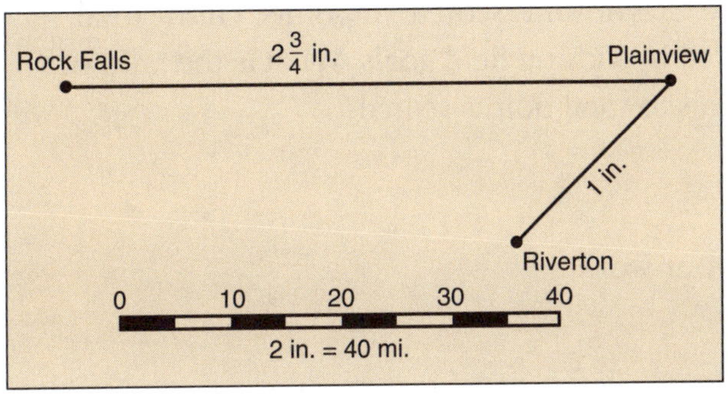

1. What is the actual distance from Riverton to Plainview in miles?
   (1) 1
   (2) 10
   (3) 20
   (4) 40
   (5) 80

2. What is the actual distance from Plainview to Rock Falls in miles?
   (1) 50
   (2) 55
   (3) 60
   (4) 110
   (5) 220

3. One of Cheryl's customers is in Hillsdale. Hillsdale is 60 miles north of Plainview. If Hillsdale were shown on this map, how many inches north of Plainview would it be?
   (1) $1\frac{1}{3}$
   (2) $1\frac{1}{2}$
   (3) 3
   (4) 6
   (5) 60

Questions 4 through 6 refer to the following information.

Jerome works for a recycling company. He drives from Mesa to Canyon City to Bluffton.

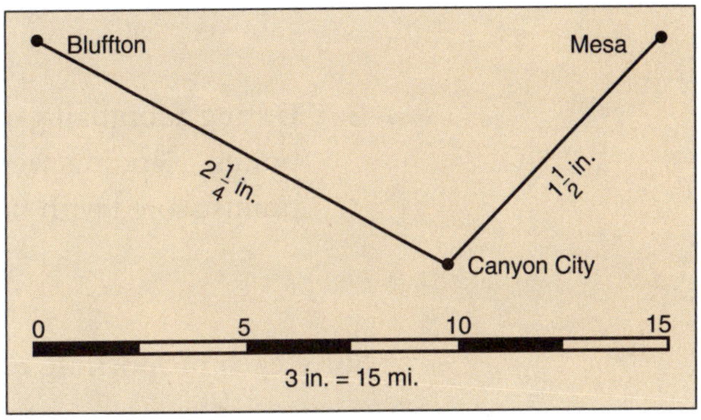

4. What is the actual distance from Mesa to Canyon City in miles?
   (1) $3\frac{1}{2}$
   (2) $5\frac{1}{2}$
   (3) $6\frac{1}{2}$
   (4) $7\frac{1}{2}$
   (5) $10\frac{1}{2}$

5. What is the actual distance from Canyon City to Bluffton in miles?
   (1) $4\frac{1}{2}$
   (2) $6\frac{2}{3}$
   (3) $6\frac{3}{4}$
   (4) $11\frac{1}{4}$
   (5) Not enough information is given.

6. **Reason** Assume Jerome is driving at a constant rate of 40 miles per hour. Explain about how long it would take Jerome to drive from Mesa to Bluffton through Canyon City.

Check your answers on page 302.

Lesson 11 Review

Write each ratio in lowest terms.

1. At Quality Products, 20 employees work in the office and 28 work in the warehouse. What is the ratio of employees who work in the office to employees who work in the warehouse?

2. During a football game, the Greyhounds scored 36 points. Of the total points, 15 points were scored by kicking field goals. What is the ratio of points scored with field goals to total points scored?

Solve each proportion. Show your work.

3. $\frac{6}{7} = \frac{?}{42}$

4. $\frac{12}{8} = \frac{30}{?}$

5. Andre drove 106 miles in 2 hours. If he can continue this speed, how far can he drive in 5 hours?

6. Of every 400 parts Tri-State Assembly makes, an average of 3 parts are defective. On Monday Tri-State Assembly made 1,600 parts. How many of these parts are likely to be defective?

Question 7 refers to the following ad.

CRAFTER'S WORLD		
Fabric paints	5 bottles for	\$2.25
Plastic canvas	3 sheets for	\$0.99
Yarn, 4 oz.	2 skeins for	\$2.86

7. Eva works at Crafter's World. A customer needs 5 sheets of plastic canvas to make holiday decorations. What price should Eva charge the customer?

PRACTICE

Use proportions to solve each problem. Show your work.

8. It takes Manuel 2 hours to mat and frame 5 pictures. How long will it take him to mat and frame 16 pictures?

9. Michela is planning to paint her house. The color she wants requires 3 parts blue to 4 parts gray paint. How many gallons of blue paint does she need to mix with 10 gallons of gray paint?

10. Darius drove 300 miles on 12 gallons of gasoline. His tank holds 16 gallons. How far could Darius drive on a full tank of gasoline?

11. At Pioneer Insurance 5 out of 9 employees are women. Pioneer Insurance employs 432 people. How many women work at Pioneer Insurance?

<u>Questions 12 through 14</u> refer to the following map.

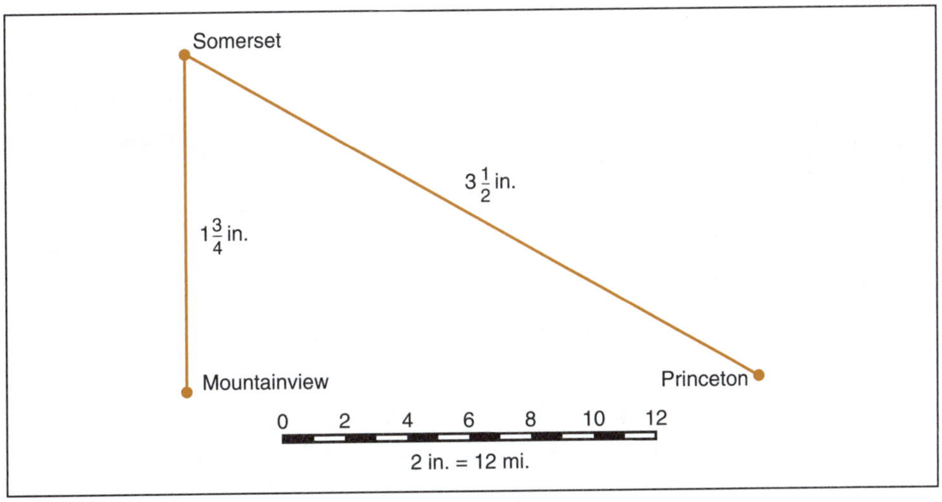

12. What is the actual distance from Mountainview to Somerset in miles?

13. What is the actual distance from Somerset to Princeton in miles?

14. Pine Ridge is 30 miles north of Somerset. If Pine Ridge were shown on this map, how many inches north of Somerset would it be?

Check your answers on pages 302–303.

LESSON 12

Percent Basics

The Meaning of Percent

Percent is another way to show part of a whole. The sign for percent is %.

Example Lana, a store clerk, is selling an AM/FM radio with a CD player for $100. The customer wants to make a down payment. Lana tells the customer that he must make a 25% down payment. She figures out that 25% of $100 is $25.

Percent means "for every 100" or "out of 100."

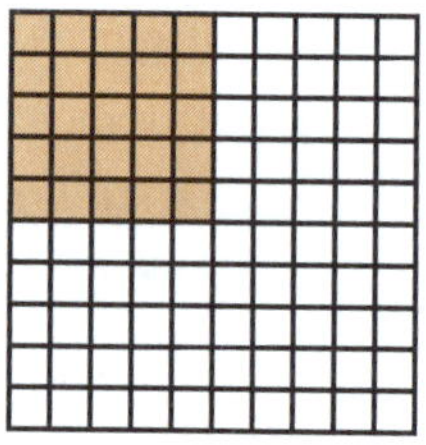

The figure is divided into 100 equal parts. Since 25 out of 100 parts are shaded, the shaded portion is 25% of the whole figure.

Percents can also be written as fractions or decimals.

The customer gives Lana $25 for the down payment. The customer was paying what fraction of the total cost for the radio?

(1) $\frac{1}{4}$ (4) $\frac{2}{3}$

(2) $\frac{1}{3}$ (5) $\frac{3}{4}$

(3) $\frac{1}{2}$

Answer **(1)** is correct. You can describe a percent as a fraction in lowest terms. $\frac{25}{100} = \frac{25 \div 25}{100 \div 25} = \frac{1}{4}$, so the customer was paying $\frac{1}{4}$ **of $100,** or **$25.**

Example You can also write a percent as a decimal. What is 25% written as a decimal? (Hint: 25% is also $\frac{25}{100}$, or 25 hundredths.)

(1) 0.0025
(2) 0.025
(3) 0.25
(4) 25.0
(5) 250.0

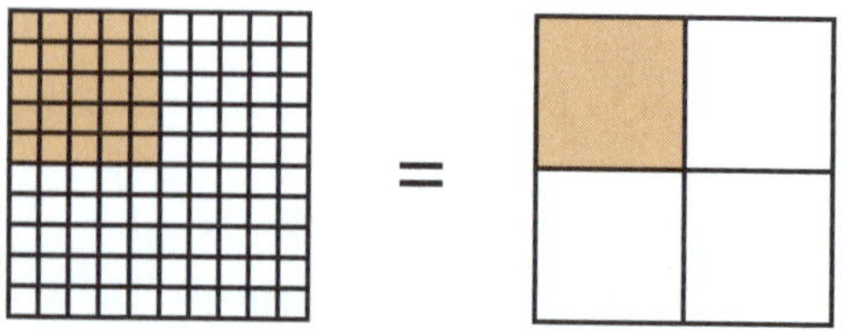

Answer **(3)** is correct. Since you know **25%** equals $\frac{25}{100}$, you can write the decimal **0.25** for 25%.

PRACTICE

Circle the best answer for each question.

Questions 1 through 3 refer to the following information.

Francisco is a bank teller. He counted the number of pennies in his bank drawer. The drawer held 30% of a dollar in pennies. The figure below shows this percent.

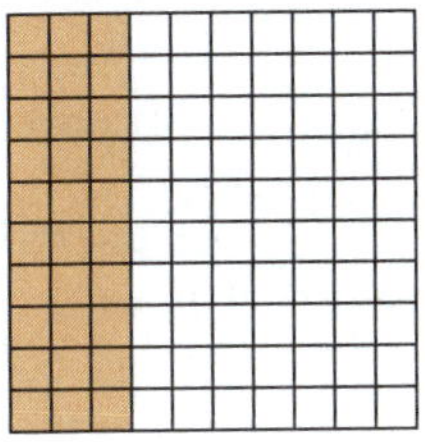

1. How many pennies does Francisco have in the drawer?

(1) 0.03
(2) 0.3
(3) 3
(4) 30
(5) 300

2. What fraction of a dollar does Francisco have in pennies?

(1) $\frac{1}{100}$
(2) $\frac{3}{100}$
(3) $\frac{3}{10}$
(4) $\frac{1}{3}$
(5) $\frac{3}{1}$

3. What dollar amount does Francisco have in pennies?

(1) $0.30
(2) $3.00
(3) $3.30
(4) $30.00
(5) $300.00

Questions 4 through 6 refer to the following information.

Gina is a mathematics teacher. She gave a test that had 100 questions. The highest score on the test was 90%. The figure at the right shows this percent.

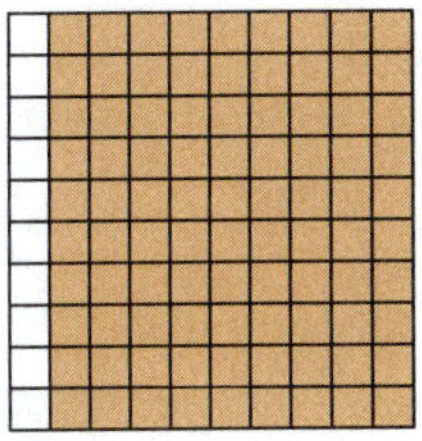

4. Marcus received the highest score. Gina wrote the scores on the tests as fractions. What fraction would you see on Marcus' test?

(1) $\frac{9}{100}$
(2) $\frac{90}{100}$
(3) $\frac{100}{90}$
(4) $\frac{10}{9}$
(5) $\frac{9}{1}$

5. Leah keeps track of all her tests in her notebook. She writes her scores as decimals so that she can average them at the end of the semester. Leah saw the fraction $\frac{85}{100}$ on her test. What decimal does Leah write in her notebook?

(1) 0.085
(2) 0.80
(3) 0.85
(4) 8.50
(5) Not enough information is given.

6. **Communicate**
Explain how you would find a fraction, a decimal, and a percent to describe the shaded portion of the figure to the right.

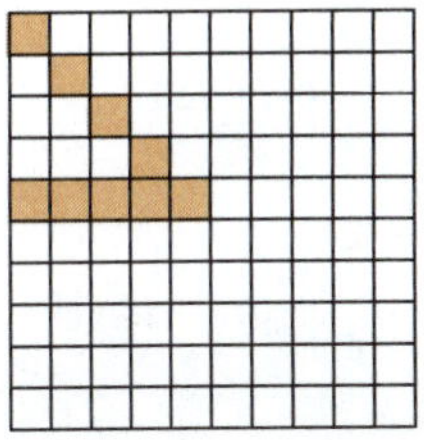

Check your answers on page 303.

Relating Percents, Fractions, and Decimals

You can write any percent as a decimal or a fraction. Since these numbers represent the same amount, only in different forms, we can say they are **equivalent.**

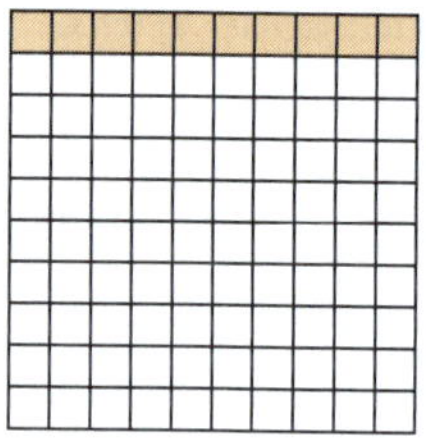

> **TIP**
> Percents, fractions, and decimals are different ways to show parts of a whole.

Example Write a fraction, a decimal, and a percent to describe the shaded portion of the figure at the right.

Step 1 Count the number of shaded parts. 10

Step 2 Count the total number of parts. 10 out of 100 parts are shaded. Write the fraction for "10 out of 100." Reduce. $\frac{10}{100} = \frac{1}{10}$

Step 3 Write the decimal. Hundredths is two places to the right of the decimal point. 0.10

Step 4 Write the percent. Percent means "out of 100," so 10 out of 100 would be 10%. 10%

PRACTICE

Write a fraction, a decimal, and a percent to describe the shaded portion of each figure. The first one is done for you.

1.

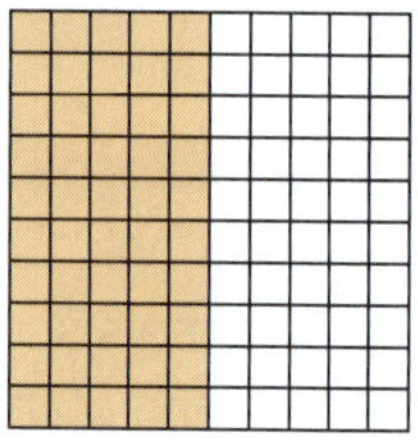

Fraction $\frac{50}{100} = \frac{1}{2}$

Decimal 0.50

Percent 50%

2.

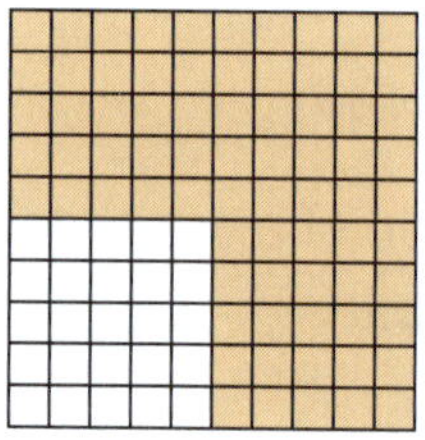

Fraction ____________

Decimal ____________

Percent ____________

3.

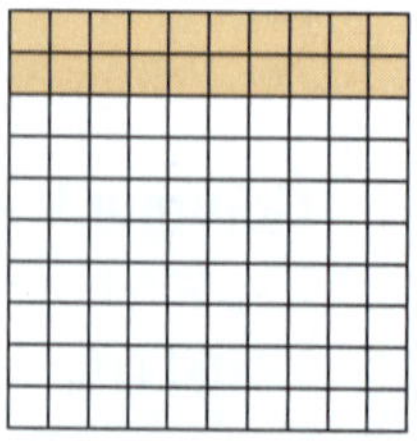

Fraction ____________

Decimal ____________

Percent ____________

4.

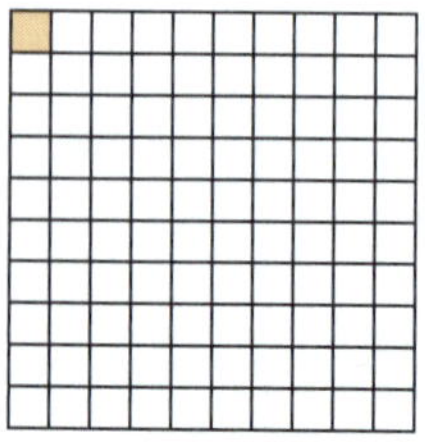

Fraction ____________

Decimal ____________

Percent ____________

 Check your answers on page 303.

Changing Percents and Decimals

You can change a percent to a decimal and a decimal to a percent.

TIP
To change a percent with a fraction to a decimal, change the fraction to its decimal form, drop the percent sign, then move the decimal point. For example, change $5\frac{3}{4}\%$ to a decimal.
First: $5\frac{3}{4}\% = 5.75\%$
Then: $5.75\% = 0.0575$

Example Change 150% to a decimal.

Step 1 Drop the percent sign. 150

Step 2 Write the number as a fraction with 100 as the denominator (since a percent is "out of 100"). Move the decimal point two places to the left (this is the same as dividing by 100). Add a zero as a placeholder if necessary. If a percent does not have a decimal point, it is understood to have one to the right of the ones place. $\frac{150}{100}$ 1.50. = 1.5

Example Change 0.875 to a percent.

Step 1 Multiply by 100. Move the decimal point two places to the right. 0.87.5

Step 2 Write the percent sign after the number. 87.5%

PRACTICE

Change each percent to a decimal. The first one is done for you.

1. 32% = 0.32
2. 4%
3. 250%
4. $6\frac{1}{2}\%$
5. 5.25%
6. 0.14%
7. $97\frac{1}{4}\%$
8. 100%

Change each decimal to a percent. The first one is done for you.

9. 2.60 = 260%
10. 0.03
11. 0.625
12. 0.0008
13. 4.635
14. 0.2
15. 5.86
16. 0.345

Check your answers on pages 303–304.

Changing Percents and Fractions

TIP

When you change $\frac{1}{3}$ and $\frac{2}{3}$ to decimals, the division will not work out evenly. Since they are used a lot, it is a good idea to memorize the equivalent decimal and percent for these fractions.

$\frac{1}{3} = 0.333...$

$\frac{1}{3} = 33\frac{1}{3}\%$

$\frac{2}{3} = 0.666...$

$\frac{2}{3} = 66\frac{2}{3}\%$

You can change a percent to a fraction and a fraction to a percent.

Example Change 75% to a fraction.

Step 1 Drop the percent sign. 75

Step 2 Write the number as a fraction with a denominator of 100 (since a percent is "out of 100"). $\frac{75}{100}$

Step 3 Reduce the fraction to lowest terms. $\frac{75}{100} = \frac{75 \div 25}{100 \div 25} = \frac{3}{4}$

Example Change $\frac{1}{8}$ to a percent.

Step 1 Change the fraction to a decimal. To do this, divide the denominator into the numerator.

$$\begin{array}{r} 0.125 \\ 8\overline{)1.000} \\ -\ 8 \\ \hline 20 \\ -16 \\ \hline 40 \\ -40 \\ \hline 0 \end{array}$$

Step 2 Change the decimal to a percent. $0.12.5 = 12.5\%$

PRACTICE

Change each percent to a fraction in lowest terms. The first one is started for you.

1. $60\% = \frac{60}{100}$ $\frac{60}{100} =$
2. 8%
3. 52%
4. 1%
5. 250%
6. 16%
7. 100%
8. 42%

Change each fraction to a percent. The first one is started for you.

9. $\frac{1}{2} = 2\overline{)1.0}$ → $0.5 =$ (-1.0, 0)
10. $\frac{3}{5}$
11. $\frac{7}{10}$
12. $\frac{3}{8}$
13. $\frac{1}{4}$
14. $\frac{5}{8}$
15. $6\frac{5}{10}$
16. $\frac{15}{5}$

For more practice changing percents and fractions, see Calculator Handbook page 251.

Lesson 12 Review

Change each percent to a decimal.

1. 37%
2. 4%
3. 225%
4. $6\frac{1}{2}\%$

Change each decimal to a percent.

5. 0.46
6. 0.08
7. 2.5
8. 0.375

Change each percent to a fraction.

9. 50%
10. 40%
11. 25%
12. 90%

Change each fraction to a percent.

13. $\frac{7}{10}$
14. $\frac{3}{4}$
15. $\frac{4}{5}$
16. $\frac{7}{8}$

Check your answers on pages 304–305.

LESSON 13

Solving for the Part (p)

Solving for the Part (*p*)

ALGEBRA TOPIC

The Percent Formula

Every percent problem has three elements: the base, the rate, and the part. For example: "30 is 50% of 60" or "50% of 60 is 30."

- **The base** ***(b)*** **is the whole amount. The base (here, 60) represents 100%. The base usually follows the word** ***of.***

- **The rate** ***(r)*** **is easy to find because it is always followed by the percent sign (%) or the word** ***percent*** **(here, 50%).**

- **The part** ***(p)*** **is part of the whole amount. The part (here, 30) usually precedes or follows the word** ***is.***

Example Chris earns $240 a week as a receptionist. Each week he puts $9.60 or 4% of his earnings into a savings account. Which number is the base?

(1) 4%
(2) 0.04
(3) $3.84
(4) $9.60
(5) $240

Answer **(5)** is correct. The base, or the whole amount, is **$240.**

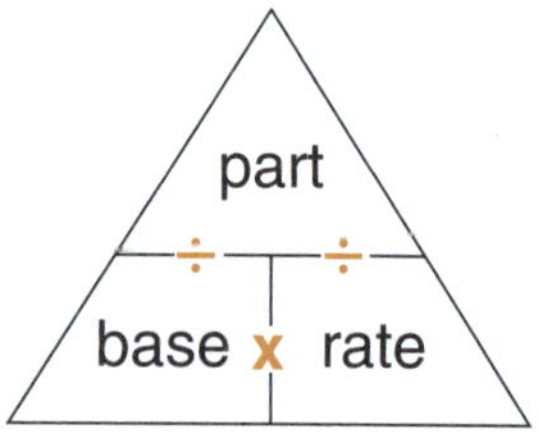

You can see the relationship of the three elements in the following formula. The triangle on the left will help you remember how these elements are related.

- ***base*** **×** ***rate*** **=** ***part*** **$240 × 4% = $9.60**

In percent problems, you will know two of the three elements. To solve, first identify the elements that you know. Then solve for the unknown element.

Example Karl makes repairs for a large apartment building. Velma, Karl's supervisor, estimates he will spend 20% of his 10-hour day fixing the garbage disposal in Apartment 3B. Karl wants to find how many hours the job should take. Which element does Karl need to find?

(1) the base
(2) the rate
(3) the part
(4) the proportion
(5) Not enough information is given.

Answer **(3)** is correct. Karl knows the base (his whole 10-hour day) and the rate (20%). He needs to find the **part** (hours to fix the garbage disposal). To solve, he needs to multiply the base (10) by the rate (20%).

PRACTICE

Circle the best answer for each question.

1. Hector works 8 hours a day in an office. He spends 20% of his time photocopying and filing. To find how many hours Hector spends on these tasks, which element do you need to find?
   (1) the base
   (2) the rate
   (3) the part
   (4) the proportion
   (5) Not enough information is given.

2. Margaret earns $550 a week. Her employer deducts 8% or $44 for social security. Which number is the rate?
   (1) $44
   (2) 8%
   (3) 8
   (4) $352
   (5) $550

3. Arturo's company employs 64 people. Of the employees, 16 are women. Arturo wants to know what percent of the employees are women. Which element does Arturo need to find?
   (1) the base
   (2) the rate
   (3) the part
   (4) the whole amount
   (5) Not enough information is given.

4. Joyce works 6 hours a day at a school cafeteria. She spends 50%, or 3 hours, of her workday preparing food. Which number is the base?
   (1) 6
   (2) 3
   (3) 50%
   (4) 18
   (5) 150

5. Mathew saw a sweatshirt originally priced at $24 on sale for 20% off the original price. Mathew wants to know how much he will save by buying the sweatshirt on sale. Which element does Mathew need to find?
   (1) the base
   (2) the rate
   (3) the part
   (4) the percent
   (5) Not enough information is given.

6. Fashions Plus has a 25% discount on sweaters. A customer saves $10 on one sweater. To find the regular price of the sweater, what element do you need to find?
   (1) the base
   (2) the rate
   (3) the part
   (4) the interest
   (5) Not enough information is given.

7. The Bluejays won 75%, or 12 games, of the 16 games they played this season. Which number is the part?
   (1) 0.75
   (2) 75%
   (3) 16
   (4) 12
   (5) Not enough information is given.

8. **Communicate** Given the following numbers, $500, 20%, and $100, write a percent word problem. Use the triangle on page 178 to help you label the three elements. Then show the relationship of the three numbers using the percent formula.

Check your answers on page 305.

Finding the Part (*p*)

Use the percent triangle to help you remember the formula. Cover the letter *p* when solving for the part. The base and the rate are connected by the multiplication sign. Solve the problem by multiplying the base times the rate.

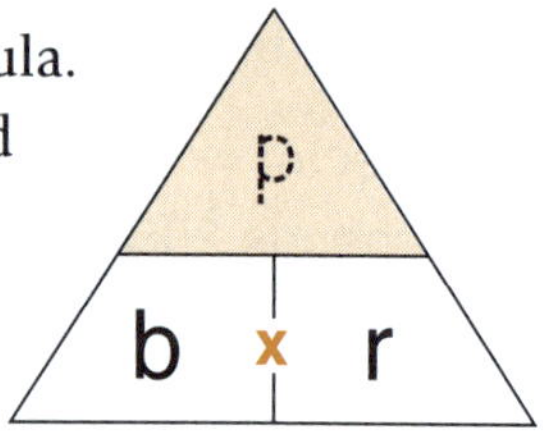

$p = b \times r$

> **TIP**
> When the rate is greater than 100%, the part will be greater than the base. Change the rate to a decimal as you would with rates less than 100%. For example, 150% becomes 1.5.

Example What is 25% of 800?

Step 1 Identify the rate and the base.
The rate is 25%.
The base is 800.

Step 2 Change the percent to a decimal.
25% = .25. = 0.25

Step 3 Use the percent formula, *part* = *base* × *rate*, to find the part. Multiply the base (800) by the rate (0.25) to solve for the part.

800
×0.25 ← 2 decimal places
4000
+1600
200.00 ← 2 decimal places

200 is 25% of 800.

You can use the % function on the CASIO *fx-260*SOLAR calculator to find the part. Enter the whole amount and then multiply by the percent. Since the % function is above the equals [=] key, press [SHIFT], then the [=] key. Do not change the percent to a decimal.

800 [×] 25 [SHIFT] [=] 200

Sometimes the rate contains a fraction.

Example What is $5\frac{1}{2}$% of $60?

Step 1 Identify the rate and the base.
The rate is $5\frac{1}{2}$%.
The base is $60.

Step 2 Change the fraction to a decimal.
$5\frac{1}{2}$% = 5.5%

Step 3 Change the percent to a decimal.
5.5% = 0.05.5 = 0.055

Step 4 Multiply the base ($60) by the rate (0.055) to find the part.

$60
×0.055 ← 3 decimal places
$3.300 ← 3 decimal places

$3.30 is $5\frac{1}{2}$% of $60.

On the CASIO *fx-260*SOLAR, you could use the % function to multiply by a percent without changing the percent to a decimal.

60 [×] 5.5 [SHIFT] [=] 3.3, or $3.30

PRACTICE

A. **Solve. Show your work. The first one is done for you.**

1. What is 35% of 220?

   35% = 0.35

   $$\begin{array}{r} 220 \\ \times 0.35 \\ \hline 1100 \\ +660 \\ \hline 77.00 \end{array}$$

2. What is 40% of $45?

3. What is 8% of 16?

4. What is 250% of 48?

5. What is $4\frac{1}{2}$% of $50?

6. What is $7\frac{1}{4}$% of 144?

B. **Solve. Show your work. The first one is done for you.**

7. Laura bought a pair of jeans that regularly cost $32 for 25% off the regular price. How much did Laura save?

   25% = 0.25

   $$\begin{array}{r} \$32 \\ \times 0.25 \\ \hline 160 \\ +64 \\ \hline \$8.00 \end{array}$$

8. The Jaguars won 80% of the 15 games they played this season. How many games did they win?

9. Of the 8,000 people at a baseball game, 5% were season ticket holders. How many people at the game were season ticket holders?

10. Electronic Buys sold 240 items last week. Of the items sold, 45% were kitchen appliances. How many kitchen appliances did Electronic Buys sell last week?

C. **Solve.**

11. Rick earns $32,500 per year. He spends 16% of his salary on rent. How much does Rick spend on rent a year? ______

12. A gallon of gasoline costs $1.32. Of that amount, 27% goes to state and local taxes. To the nearest cent, how much money per gallon goes to taxes? ______

Check your answers on pages 305–306.

ALGEBRA TOPIC

Simple Interest Formula

Interest is a charge that someone pays to borrow or to use someone else's money. The amount of interest charged is based on the **principal** or amount of money borrowed. An **interest rate** is used to figure out how much interest the borrower has to pay.

You can use a formula (similar to the percent formula) to find the amount of simple interest owed.

$i = p \times r \times t$, where *i* is the amount of interest (similar to the part), *p* is the principal (similar to the base), *r* is the interest rate (the rate), and *t* is the time in years over which the money is borrowed (time must be included because the rate includes time, usually % per year).

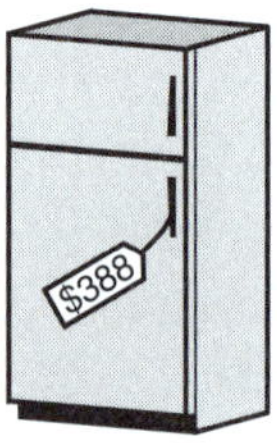

Example Gordon wants to buy a refrigerator. J & B Appliances has a used refrigerator advertised for $388. Gordon wants to pay $100 down on the refrigerator and the rest in monthly payments over 2 years. The store would charge an 18% interest rate on the remaining $288. Which is the correct expression to find how much simple interest Gordon will pay on the money he owes (the principal)?

(1) $388 + 0.18 + 2
(2) $388 × 0.18 × 2
(3) $388 × 100 × 2
(4) $288 × 0.18 × 2
(5) $288 + 0.18 + 2

Answer **(4)** is correct. Before finding the interest you must identify the principal. The principal is the amount owed. Gordon still owes $288 (he paid $100 when he bought the refrigerator). So $288 is the principal. The rate is 18%. The time is 2 years. Use the simple interest formula to find the interest Gordon owes. To multiply by the rate, first change the percent to a decimal: 18% = 0.18. = 0.18

$288 × 0.18 × 2 = $103.68, so Gordon will pay **$103.68** in interest in order to pay in installments over 2 years.

Example If Gordon buys the refrigerator, how much will he end up paying for the refrigerator including the original price and the interest?

(1) $184.32
(2) $339.84
(3) $391.68
(4) $443.52
(5) $491.68

Answer **(5)** is correct. Add the cost of the refrigerator ($388.00) and the interest ($103.68). **$388.00 + $103.68 = $491.68,** so the total cost will be **$491.68.**

PRACTICE

Circle the best answer for each question.

1. Dan borrowed \$650 for 1 year at a rate of 12% to make home improvements. Which is the correct expression to find the interest Dan paid?

(1) $\$650 + 0.12 + 1$
(2) $\$650 \times 0.12 \times 1$
(3) $\$650 \times 0.12 \times 0.01$
(4) $\$650 \times 12 \times 1$
(5) $\$650 \times 0.12 + 1$

2. The Davis family took out a loan of \$420 for 1 year at a rate of 18%. How much interest will the family owe?

(1) \$75.60
(2) \$76.60
(3) \$110.60
(4) \$326.00
(5) \$336.00

3. Donna took out a \$1,240 loan for college for 2 years. If the interest rate is $8\frac{1}{2}\%$, how much will Donna owe in interest?

(1) \$52.70
(2) \$105.40
(3) \$198.40
(4) \$210.80
(5) \$2,108.00

4. Mr. Martinez deposited \$1,600 in his savings account. If his savings account earns $5\frac{1}{4}\%$ interest, how much will Mr. Martinez earn in interest after 3 years?

(1) \$240
(2) \$252
(3) \$320
(4) \$400
(5) \$533

5. Morris bought a stove for \$560. He paid \$150 down. He will pay the rest at the end of 2 years at 16% interest. Which is the correct expression to find how much interest Morris will owe after 2 years?

(1) $\$150 \times 0.16 \times 2$
(2) $\$410 \times 0.16 \times 2$
(3) $\$410 \times 16 \times 2$
(4) $\$560 \times 0.16 \times 2$
(5) $\$710 \times 0.16 \times 2$

6. Recall the amount paid back is the total of the principal plus the interest. How much will Morris have to pay back?

(1) \$410.00
(2) \$524.10
(3) \$541.20
(4) \$608.00
(5) Not enough information is given.

7. Freddie got a personal loan of \$2,100 from a friend. He will pay the money back over 1 year at 12% interest. How much will Freddie pay back to his friend?

(1) \$252
(2) \$2,100
(3) \$2,112
(4) \$2,352
(5) Not enough information is given.

8. **Reason** Gina bought a used car for \$12,000. Gina paid \$2,000 as a down payment. She will pay the rest in monthly payments over 2 years at 15% interest. Explain how you would estimate the monthly payment. Then do the actual calculations. Show your work.

Check your answers on page 306.

DATA ANALYSIS

Circle Graphs

Circle graphs show a whole amount divided into parts. Circle graphs represent parts of a whole with amounts such as dollars, fractions, or percents.

The circle graph shown here shows Bill's gross pay of $1,100 per month. The sections of the graph tell what percent of his gross pay goes for deductions and net pay. The percents in a circle graph add to 100%.

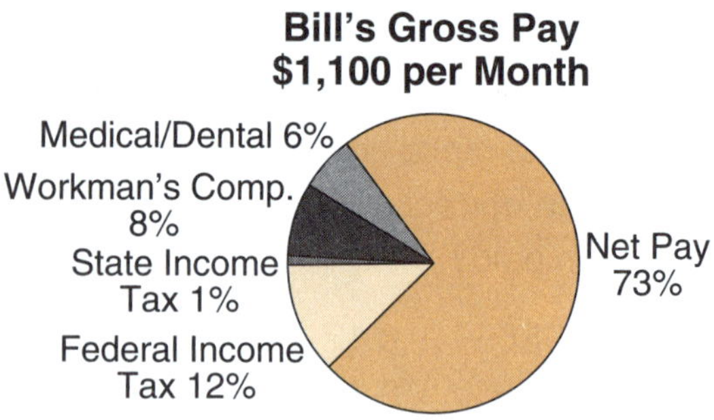

You can use this information to find the total percent for deductions, the dollar amounts for each deduction, and the amount of Bill's net pay.

Example What percent of Bill's gross pay goes towards deductions?

(1) 12%
(2) 13%
(3) 21%
(4) 27%
(5) 73%

Answer **(4)** is correct. Add the deduction percents:
12% + 1% + 8% + 6% = 27%.

Example How much of Bill's gross income goes for medical/dental insurance?

(1) $11.00
(2) $44.00
(3) $66.00
(4) $132.00
(5) $300.00

Answer **(3)** is correct. Multiply Bill's gross pay by the percent for medical/dental insurance: $1,100 × 6% = $1,100 × 0.06 = $66.00

Example What is the amount of Bill's net pay?

(1) $11.00
(2) $66.00
(3) $132.00
(4) $300.00
(5) $803.00

Answer **(5)** is correct. Multiply Bill's gross pay by the percent for his net pay:
$1,100 × 73% = $1,100 × 0.73 = $803.00

PRACTICE

Circle the best answer for each question. You may use a calculator.

Questions 1 through 3 refer to the graph below.

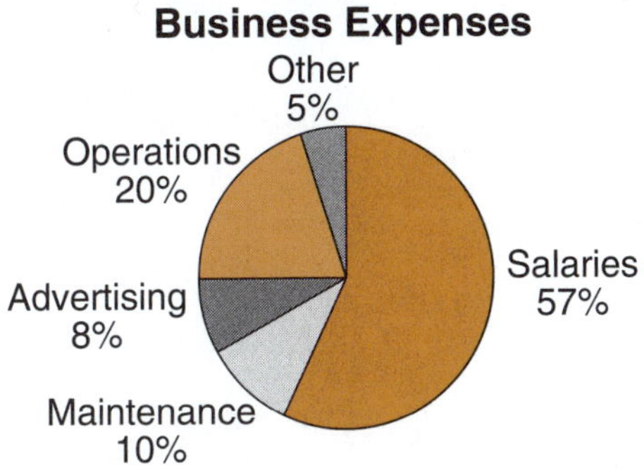

1. What percent of business expenses is used for maintenance, advertising, and operations combined?

   (1) 8%
   (2) 18%
   (3) 38%
   (4) 43%
   (5) 57%

2. If the company has $36,000 in business expenses each year, how much is spent on plant operations alone?

   (1) $1,800
   (2) $2,880
   (3) $3,600
   (4) $7,200
   (5) $10,800

3. What is the amount spent on salaries if the company has $36,000 in expenses?

   (1) $5,700
   (2) $6,400
   (3) $11,880
   (4) $15,480
   (5) $20,520

Questions 4 through 6 refer to the graph below.

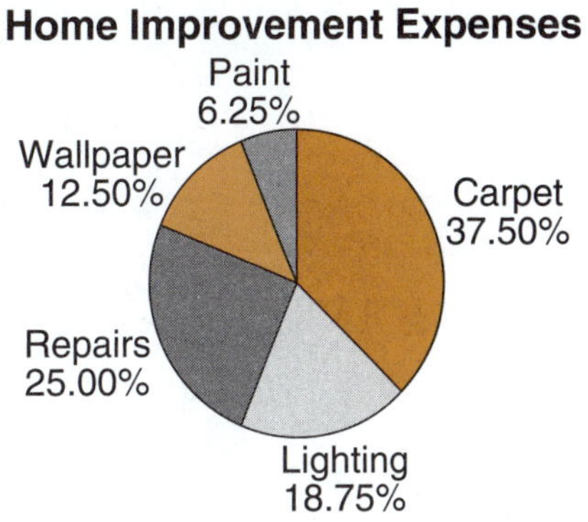

4. What percent of the Lansings' home improvement expenses was used for carpet and lighting combined?

   (1) 81.25%
   (2) 56.25%
   (3) 50.0%
   (4) 43.75%
   (5) 31.25%

5. If the Lansings spent a total of $1,600 on their home improvements, how much did they spend on carpet and lighting?

   (1) $900
   (2) $600
   (3) $300
   (4) $200
   (5) $100

6. How much was spent on wallpaper and paint if the improvements cost $1,600?

   (1) $600
   (2) $400
   (3) $300
   (4) $200
   (5) $100

Check your answers on page 306.

LESSON 13

Lesson 13 Review

Lesson 13 Review
Lesson 13 Review
Lesson 13 Re

Solve. Show your work.

1. What is 64% of 300?

2. What is 6% of 72?

3. What is 90% of $96?

4. What is 400% of $56?

5. What is 125% of 36?

6. What is $3\frac{1}{2}$% of 150?

Solve. Show your work.

7. The Mortons' health insurance covers 80% of the cost of medical care. John Morton had a series of lab tests that cost $145. How much of this cost was covered by his insurance?

8. Jill buys a coat that regularly sells for $125 for 33% off. How much does Jill save?

9. Aretha is recarpeting her house for $1,750. She pays $250 down. She will pay the rest in monthly payments over 2 years at 18% interest. How much will Aretha have to pay back?

10. In a survey, 3,000 adults were asked which newspaper they prefer to read. Fifty-five percent said they prefer a morning paper. How many adults prefer a morning paper?

11. Miguel puts 12% of his earnings in a savings account. This week Miguel earns $310. How much will he put in his savings account?

12. The Tallchiefs bought a refrigerator for $475 and a stove for $680. They paid $150 down. They will pay the rest in monthly payments over 1 year at 16% interest. What is the Tallchiefs' monthly payment?

Circle the best answer for each question.

Questions 13 through 15 refer to the following information.

Dayna is the manager of the clothing departments at Shop-Wise. She is scheduled to work 40 hours each week. The table shows what parts of her total week she spends working at different tasks.

Task	Percent of Time
Supervising	40%
Planning	25%
Selling	20%
Customer relations	15%

13. Which is the correct expression to find how many hours Dayna spends on customer relations each week?
    (1) 15 × 40%
    (2) 15 ÷ 40%
    (3) 40% × 0.15
    (4) 40 × 15%
    (5) 40 ÷ 15%

14. Which is the correct expression to find the total percentage of time Dayna spends on supervising and planning each week?
    (1) 40% × 25%
    (2) 40% + 25%
    (3) 40% − 25%
    (4) 40 × 40% × 25%
    (5) 40 × 40% + 25%

15. How many hours does Dayna spend selling each week?
    (1) 8
    (2) 10
    (3) 20
    (4) 40
    (5) 80

Questions 16 through 18 refer to the following graph.

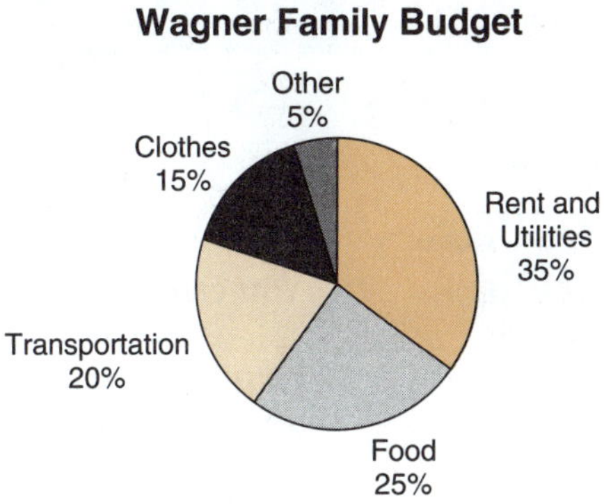

16. For which item do the Wagners budget the least amount?
    (1) Rent and utilities
    (2) Food
    (3) Transportation
    (4) Clothes
    (5) Other

17. The Wagners' total monthly income is $2,200. How much do they budget for rent and utilities?
    (1) $35
    (2) $350
    (3) $550
    (4) $570
    (5) $770

18. How much do the Wagners budget for food and clothing combined if their total monthly income is $2,200?
    (1) $330
    (2) $550
    (3) $880
    (4) $1,100
    (5) $1,200

Check your answers on page 307.

14 LESSON

Solving for the Rate (r)

ALGEBRA TOPIC

The Percent Formula

You can rewrite the percent formula to solve for the **rate** in percent problems.

Example Jan traveled a distance of 780 miles on her vacation. The first day of her trip she drove 234 miles. What percent of the total distance did she drive the first day?

Before you can solve for the rate, r, you must identify the base and the part. The base, b, is the total distance, 780 miles. The distance Jan drove the first day is part of the total distance. The part, p, is 234 miles.

Using the percent triangle, you can rewrite the percent formula to solve for the rate. Cover the letter r. The other elements, the part and the base, are separated by the division sign.

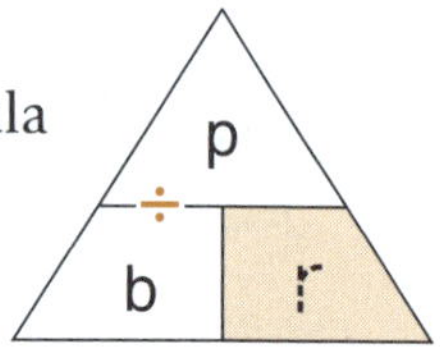

part ÷ base = rate

Which is the correct expression to find the percent of the total distance Jan drove the first day?

(1) 234 ÷ 780
(2) 234 × 780
(3) 780 ÷ 234
(4) 780 − 234
(5) 780 + 234

Answer **(1)** is correct. The total distance (780) is the base. The distance Jan drove on the first day (234) is the part. To find the rate, divide the part by the base. **234 ÷ 780 = 0.30 = 30%,** so Jan drove **30%** of the total distance the first day.

Example The Hillside Neighborhood Association has 450 members. In May 360 members attended the annual meeting. Which is the correct expression to find the percent of the association members who attended the meeting?

(1) 360 ÷ 450
(2) 450 ÷ 360
(3) 450 − 360
(4) 450 + 360
(5) 450 × 360

Answer **(1)** is correct. The total membership (450) is the base. The members at the meeting (360) is the part. Divide the part by the base to find the rate. **360 ÷ 450 = 0.8 = 80%,** so **80%** of the members attended the meeting.

PRACTICE

Circle the best answer for each question.

1. Mr. Cruz earns \$1,150 a month. He pays \$299 a month for rent. Which is the correct expression to find what percent of his income he pays for rent?

   (1) \$1,150 ÷ \$299
   (2) \$1,150 − \$299
   (3) \$299 ÷ \$1,150
   (4) \$299 + \$1,150
   (5) \$299 × \$1,150

2. Chiang bought a used car that cost \$4,500. She made a down payment of \$500. Which is the correct expression to find what percent of the total cost she put down on the car?

   (1) \$500 ÷ \$4,500
   (2) \$500 × \$4,500
   (3) \$500 + \$4,500
   (4) \$4,500 − \$500
   (5) \$4,500 ÷ \$500

3. A set of tires regularly sells for \$200. Lamont saved \$30 by buying the tires on sale. Which is the correct expression to find what percent of the regular price Lamont saved by buying the tires on sale?

   (1) \$200 − \$30
   (2) \$200 ÷ \$30
   (3) \$30 ÷ \$200
   (4) \$30 × \$200
   (5) \$30 + \$200

4. Computer Network employs 110 people. Of the employees, 65 are women. Which is the correct expression to find what percent of the employees are women?

   (1) 110 ÷ 65
   (2) 110 − 65
   (3) 65 ÷ 110
   (4) 65 + 110
   (5) 65 × 110

5. Ms. Chu earns \$380 a week. Her employer deducts \$45 out of her paycheck each week. Which is the correct expression to find what percent of Ms. Chu's earnings are taken out of her paycheck for deductions?

   (1) \$380 − \$45
   (2) \$380 ÷ \$45
   (3) \$45 + \$380
   (4) \$45 ÷ \$380
   (5) \$45 × \$380

6. Flavia earns \$420 a week. She shops for food once a week. Which is the correct expression to find what percent of Flavia's earnings is spent on food?

   (1) \$50 ÷ \$420
   (2) \$50 × \$420
   (3) \$420 − \$50
   (4) \$420 ÷ \$50
   (5) Not enough information is given.

7. There are 300 calories in a box of macaroni and cheese. Of the 300 calories, 117 calories come from fat. Which is the correct expression to find what percent 117 calories is of 300 calories?

   (1) 300 ÷ 117
   (2) 300 − 117
   (3) 117 + 300
   (4) 117 ÷ 300
   (5) Not enough information is given.

8. **Reason** To find the part in a percent problem you use the percent formula, *part = base × rate.* To find the rate in a percent problem, you rewrite the formula so that *rate = part ÷ base.* Explain in a sentence or two how the two formulas are related.

Check your answers on page 308.

Finding the Rate (*r*)

Use the percent triangle to help you find the rate, *r*, (percent). Cover the letter *r* because you need to solve for the rate. The other two elements, the part and the base, are separated by the division sign.

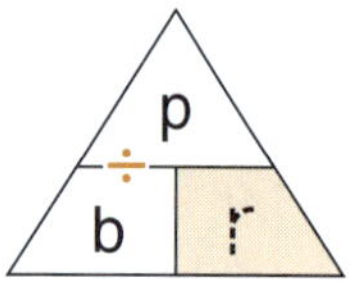

$p \div b = r$

Example What percent of 62 is 31?

Step 1 Identify the base and the part.

The base is 62.
The part is 31.

Step 2 Rewrite the percent formula to find the rate.

$p \div b = r$

Step 3 Divide the part (31) by the base (62) to find the rate.

$31 \div 62 = r$

```
    0.5
62)31.0
  -31 0
      0
```

Step 4 Change the decimal to a percent. Move the decimal point two places to the right.

$0.5 = 0.50. = 50\%$

31 is 50% of 62.

> **TIP**
>
> The base usually comes after the word *of*. The part often precedes or follows the word *is*.
>
> What percent *of* 20 *is* 10?
>
> The base is 20. The part is 10.

You can use the % function on the CASIO *fx-260SOLAR* calculator to find the rate. Enter the part first and then divide by the base. Press SHIFT, then the = key. Add the % symbol to the answer.

AC 31 ÷ 62 SHIFT = 50, or 50%

> **TIP**
>
> When the part is greater than the base, the rate will be greater than 100%. For example, what percent of 50 is 300?
>
> $300 \div 50 = 6$
>
> $6 = 6.00 = 600\%$

Example To the nearest whole percent, what percent of 16 is 9?

Step 1 Identify the base and the part.

The base is 16.
The part is 9.

Step 2 Rewrite the percent formula to find the rate.

$p \div b = r$

Step 3 Divide the part (9) by the base (16) to find the rate. Carry the division out to three decimal places.

$9 \div 16 = r$

```
   0.562
16)9.000
  -8 0
    1 00
  -   96
      40
     -32
       8
```

Step 4 Round the decimal to the nearest hundredth.

0.562 rounds to 0.56

Step 5 Change the decimal to a percent. Move the decimal point two places to the right.

$0.56 = 0.56. = 56\%$

9 is about 56% of 16.

A calculator will show some answers to several place values. You may need to round to the nearest whole percent.

AC 9 ÷ 16 SHIFT = 56.25, or 56.25%, which rounds to 56%

PRACTICE

A. **Solve. Show your work. Round your answers to the nearest whole percent. The first one is done for you.**

1. What percent of 76 is 19?

   $p \div b = r$

   $19 \div 76 = r$

   $$\begin{array}{r} 0.25 \\ 76\overline{)19.00} \\ -15\,2 \\ \hline 380 \\ -380 \\ \hline 0 \end{array} = 25\%$$

2. What percent of 360 is 36?

3. What percent of 45 is 180?

4. What percent of 95 is 57?

5. What percent of 200 is 4?

6. What percent of 6 is 15?

B. **Solve. Show your work. Round your answers to the nearest whole percent. The first one is started for you.**

7. Lyle answered 32 out of 40 questions on a test correctly. What percent of the questions did he answer correctly?

   $32 \div 40 = r$

8. Furniture Buys was having a sale on couches. One of the couches was marked down from $950 to $570. What percent of the original price was the sale price?

9. Bev saved $9 on a sweater that regularly sells for $36. What percent of the regular price did she save?

10. Dwayne's weekly gross pay is $460.00. Deductions of $73.60 are taken out of his paycheck each week. What percent of Dwayne's gross pay is taken out of his paycheck for deductions?

C. **Solve. Round your answers to the nearest whole percent.**

11. The Goto family earned a total annual income of $46,000. The family saved $1,100 for the year. What percent of their total income did they save?

    ______

12. Disk Company employs 74 people. Of the employees, 42 are women. What percent of the employees are women? ______

Check your answers on page 308.

PROBLEM SOLVING

Using Proportions to Solve Percent Problems

Up until now, you have solved percent problems using the percent formula working with the rate as a decimal. You can also solve a percent problem using the percent formula as a proportion. To solve a percent problem using a proportion, think of the rate (percent) as a ratio comparing an amount out of 100, or $\frac{?}{100}$.

The rate (percent) can be written as a fraction $= \frac{?}{100}$

The percent formula, $p \div b = r$, can be written as a proportion: $\frac{\text{part}}{\text{base}} = \frac{\text{rate}}{100}$.

TIP
Recall the steps for solving a proportion:
Step 1: Cross multiply
Step 2: Divide

Example Andrew's doctor told him to avoid foods in which more than 30% of the calories come from fat. Andrew decides to compare the fat content of a bag of microwave popcorn and a bag of potato chips. The labels from the products are shown below. Andrew knows that there are 9 calories in every gram of fat. He can find how many calories come from fat by multiplying the grams of fat by 9: $5 \times 9 = 45$.

MICROWAVE *LITE* POPCORN	
SERVING SIZE	4 CUPS
CALORIES	150
PROTEIN, grams	5
FAT, grams	5
CARBOHYDRATE, grams	21
SODIUM, milligrams	180

RANCH STYLE LIGHT CHIPS	
SERVING SIZE	14 CHIPS
CALORIES	130
PROTEIN, grams	1
FAT, grams	6
CARBOHYDRATE, grams	18
SODIUM, milligrams	110

Which is the correct proportion to find what percent 45 calories is of 150 calories?

(1) $\frac{100}{150} = \frac{45}{?}$
(2) $\frac{150}{45} = \frac{?}{100}$
(3) $\frac{45}{150} = \frac{?}{100}$
(4) $\frac{45}{150} = \frac{100}{?}$
(5) $\frac{45}{100} = \frac{?}{150}$

Answer **(3)** is correct. 45 calories is the part, and 150 calories is the base. The rate is unknown. Set up the proportion and solve for the unknown.
$\frac{45}{150} = \frac{?}{100}$, **? = 4,500 ÷ 150 = 30,** so 45 calories is **30%** of 150 calories.

Example Andrew wants to find the percent of calories that comes from fat in potato chips. First, he multiplies the 6 grams of fat in a serving by 9 to find 54 of the 130 calories come from fat. Using a proportion, what percent of 130 calories is 54 calories?

(1) 30%
(2) 36%
(3) 40%
(4) 42%
(5) 43%

Answer **(4)** is correct. Set up the proportion and solve for the unknown.
$\frac{54}{130} = \frac{?}{100}$, **? = 5,400 ÷ 130 = 41.5,** so about **42%** of the 130 calories is from fat.

PRACTICE

Circle the best answer for each question.

Questions 1 through 3 refer to the following label.

SPAGHETTI	
SERVING SIZE	2 OUNCES (DRY)
CALORIES	200
PROTEIN, grams	7
FAT, grams	1
CARBOHYDRATE, grams	41
SODIUM, milligrams	0

1. What two amounts do you need to know to find how many calories that come from fat are in a 2-ounce serving of spaghetti?

   (1) grams of fat and serving size
   (2) grams of fat and number of calories in a gram of fat
   (3) grams of fat and grams of protein
   (4) serving size and number of calories in a gram of fat
   (5) grams of protein and number of calories in a gram of fat

2. Recall there are 9 calories in 1 gram of fat. Which is the correct proportion to find what percent 9 calories is of 200 calories?

   (1) $\frac{200}{9} = \frac{?}{100}$
   (2) $\frac{63}{200} = \frac{?}{100}$
   (3) $\frac{9}{200} = \frac{?}{100}$
   (4) $\frac{9}{200} = \frac{100}{?}$
   (5) $\frac{200}{63} = \frac{?}{100}$

3. Using a proportion, what percent of 200 calories is 9 calories?

   (1) 0.045%
   (2) 4.5%
   (3) 9%
   (4) 22.2%
   (5) 222%

Questions 4 through 6 refer to the following label.

OAT CEREAL	
SERVING SIZE	1 OUNCE
CALORIES	120
PROTEIN, grams	4
FAT, grams	2
CARBOHYDRATE, grams	21
SODIUM, milligrams	290

4. Which is the correct proportion to find what percent 18 calories is of 120 calories?

   (1) $\frac{120}{18} = \frac{?}{100}$
   (2) $\frac{18}{120} = \frac{?}{100}$
   (3) $\frac{18}{100} = \frac{?}{120}$
   (4) $\frac{18}{120} = \frac{100}{?}$
   (5) $\frac{18}{100} = \frac{120}{?}$

5. Using a proportion, what percent of 120 calories is 18 calories?

   (1) 15%
   (2) 21%
   (3) 64%
   (4) 66%
   (5) Not enough information is given.

6. **Communicate** Andrew was told to avoid foods in which over 30% of the calories come from fat. Should Andrew avoid spaghetti or oat cereal? Look at the labels above. Explain your answer in a sentence or two. Use percents in your explanation.

Check your answers on pages 308–309.

DATA ANALYSIS

Budgets and Circle Graphs

Example Ella works for the Conrad Chemical Company. She is helping her boss, Paula, collect information on the company's yearly budget. Paula gives Ella the following circle graph.

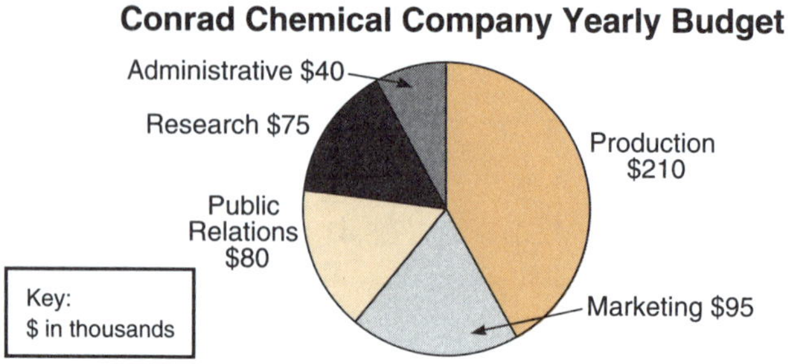

Which is the correct expression to find what percent of the total budget is spent on marketing?

(1) $95 × $500
(2) $95 ÷ $500
(3) $500 ÷ $95
(4) $500 − $95
(5) $500 + $95

Answer **(2)** is correct. First, find the total dollar amount for all the areas on the graph: $210 + $95 + $80 + $75 + $40 = $500. To find the rate (percent), divide the part ($95) by the base ($500).

$95 ÷ $500 = 0.19 = 19%, so the amount spent on marketing is **19%** of the company budget.

Note: Because the key states that the dollar amounts are in thousands, you would multiply each amount by $1,000 to find the actual dollar amounts. The company's yearly budget is $500,000, and $95,000 is spent on marketing. This is still 19% of the total budget.

Example The amount in the section labeled *Production* is $210. The amount in the *Research* section is $75. What is the total percent spent on research and production?

(1) 47%
(2) 54%
(3) 57%
(4) 63%
(5) 75%

Answer **(3)** is correct. First, add the amounts from the two sections on the graph: $210 + $75 = $285. To solve for the rate, divide the part ($285) by the base ($500). **$285 ÷ $500 = 0.57 = 57%**, so the amount spent on research and production is **57%** of the total budget.

PRACTICE

Circle the best answer for each question.

Questions 1 through 3 refer to the following graph.

The Morenos' Monthly Budget

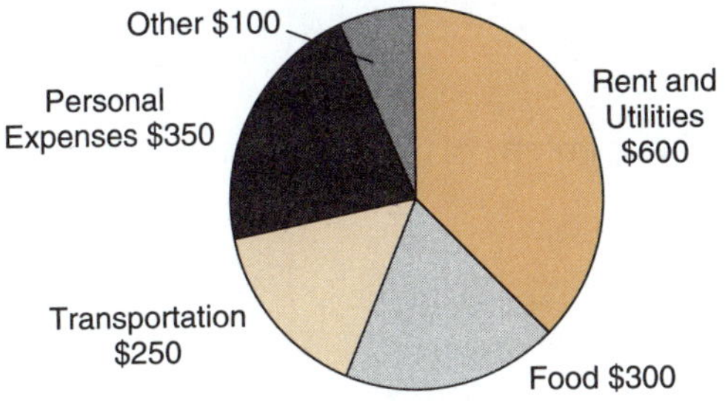

1. What is the total amount of the Morenos' monthly budget?
   (1) $100
   (2) $360
   (3) $1,500
   (4) $1,600
   (5) $5,760

2. To the nearest whole percent, what percent of the Morenos' budget is for rent and utilities?
   (1) 4%
   (2) 17%
   (3) 38%
   (4) 60%
   (5) 600%

3. To the nearest whole percent, what percent of the budget is for personal expenses and other expenses?
   (1) 4%
   (2) 22%
   (3) 28%
   (4) 45%
   (5) 450%

Questions 4 through 6 refer to the following graph.

Projected Sales of Team-Wear, Inc.

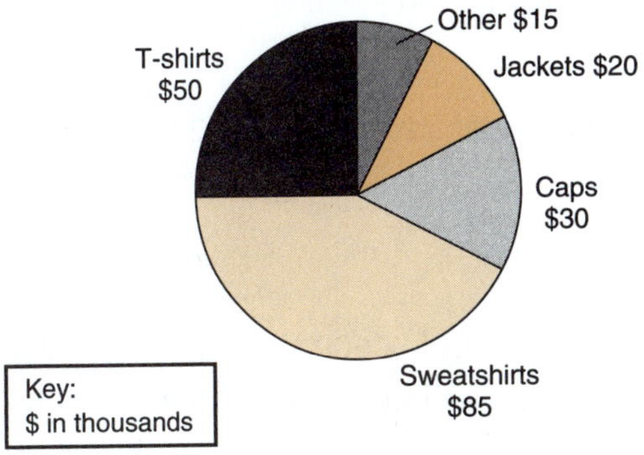

4. What are Team-Wear, Inc.'s total projected sales?
   (1) $65
   (2) $135
   (3) $200
   (4) $2,000
   (5) $200,000

5. To the nearest whole percent, what percent of the sales came from T-shirts and sweatshirts?
   (1) 25%
   (2) 43%
   (3) 50%
   (4) 68%
   (5) Not enough information is given.

6. **Reason** How much larger is the percent of sales from caps than from jackets? Explain how you would find this difference. Then answer the question.

Check your answers on page 309.

Lesson 14 Review

Solve. Show your work.

1. What percent of 10 is 16?

2. What percent of 72 is 54?

3. What percent of 300 is 12?

4. What percent of 15 is 75?

5. Sedika took a 600-mile trip. She drove 240 miles before stopping for lunch. What percent of the total trip did she drive before lunch?

6. Curtis paid $80 for a jacket that originally cost $96. What percent of the original price did Curtis pay?

7. Of the 220 employees at Adams Department Store, 125 work part-time. What percent of the employees work part-time?

8. Sarah works 40 hours per week as assistant manager of a hotel. She spends an average of 12 hours each week responding to guest comments and complaints. What percent of her time does Sarah spend responding to guest comments and complaints?

Questions 9 through 11 Refer to the following graph.

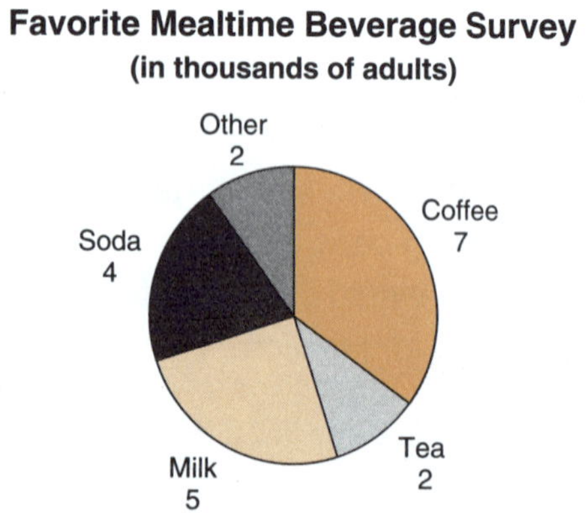

9. What is the total number of adults surveyed?

10. What percent of those surveyed prefer milk?

11. What percent of those surveyed prefer coffee or tea?

Circle the best answer for each question.

Questions 12 through 14 refer to the following label.

YOGURT	
SERVING SIZE	8 OUNCES
CALORIES	180
PROTEIN, grams	11
FAT, grams	5
CARBOHYDRATE, grams	23
SODIUM, milligrams	125

12. There are two amounts you need to know to find how many calories that come from fat are in an 8-ounce serving of yogurt. One amount you need is grams of fat. What is the other amount?
    (1) milligrams of sodium
    (2) grams of protein
    (3) number of calories in a gram of fat
    (4) number of servings
    (5) number of calories in a serving

13. Recall that there are 9 calories in 1 gram of fat. Which is the correct proportion to find what percent 45 calories is of 180 calories?
    (1) $\frac{180}{45} = \frac{?}{100}$
    (2) $\frac{45}{100} = \frac{?}{180}$
    (3) $\frac{45}{180} = \frac{100}{?}$
    (4) $\frac{45}{180} = \frac{?}{100}$
    (5) $\frac{45}{100} = \frac{180}{?}$

14. Using a proportion, what percent of 180 calories is 45 calories?
    (1) 15%
    (2) 25%
    (3) 45%
    (4) 65%
    (5) Not enough information is given.

Questions 15 through 17 refer to the following payroll check stub.

Tony Marquez	
Gross pay	\$500.00
Deductions:	
Federal income tax	63.40
State income tax	6.60
Credit union	25.00
Total deductions	\$ 95.00
Net pay	\$405.00

15. Which is the correct expression to find the rate of Tony's gross pay deducted for his credit union account?
    (1) \$25.00 × \$500.00
    (2) \$25.00 ÷ \$500.00
    (3) \$500.00 − \$25.00
    (4) \$500.00 × \$25.00
    (5) \$500.00 ÷ \$25.00

16. What percent of Tony's gross pay is deducted for taxes?
    (1) 7%
    (2) 12%
    (3) 14%
    (4) 35%
    (5) 70%

17. Tony wants to contribute 4% of his gross pay to his company's savings plan. How much would Tony contribute each pay period?
    (1) \$4.00
    (2) \$20.00
    (3) \$16.20
    (4) \$125.00
    (5) \$200.00

Check your answers on pages 309–310.

LESSON 15

Solving for the Base and Percent of Change

Solving for the Base and Percent of Change

ALGEBRA TOPIC

The Percent Formula

You can rewrite the percent formula to solve for the **base** in percent problems.

Example Ellen works at a furniture store. She needs to price a new sofa 45% higher than the original cost. But Ellen has a problem. She does not know what the store paid for the sofa. Her boss tells her that she should add $292.50 to the cost. If $292.50 represents 45% of the cost, how can Ellen find the cost?

In this example, Ellen knows that the rate is 45%. The cost is the whole amount, or base. She needs to solve for the base. Ellen covers the letter *b* on the percent triangle, which shows to divide the part by the rate to solve for the base.

part ÷ rate = base

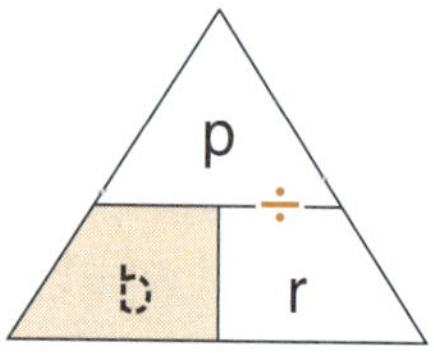

Which is the correct expression to find the cost of the sofa?

(1) $292.50 ÷ 45
(2) $292.50 + 45
(3) $292.50 ÷ 45%
(4) $292.50 × 45%
(5) $292.50 − 45%

Answer **(3)** is correct. To find the base, divide the part ($292.50) by the rate (45%). First change the rate from a percent to a decimal: **45% = 0.45**
$292.50 ÷ 0.45 = $650, so the base or the original price (cost) of the sofa is $650. Ellen will price the sofa at $942.50 ($650 + $292.50).

Example A bookstore manager adds 20% to the price she pays for each hardcover book sold in the store. If the added cost to a book is $5, which is the correct expression to find the cost of the book?

(1) $5 × 0.20
(2) $5 ÷ 0.20
(3) $5 − 20
(4) $5 ÷ 20
(5) $5 + 20

Answer **(2)** is correct. Divide the part ($5) by the rate (20%) to find the base. Remember to first change the percent to a decimal (20% = 0.20).
$5 ÷ 0.20 = $25, so $25 is the base, the original cost of the book.

PRACTICE

Circle the best answer for each question.

1. Clara puts 6% of her gross pay in the credit union each pay period. This pay period she put $17.52 in the credit union. Which is the correct expression to find Clara's gross pay?

   (1) 0.06 ÷ $17.52
   (2) $17.52 ÷ 0.06
   (3) $17.52 − 6
   (4) $17.52 + 6
   (5) $17.52 × 6

2. Michael found that 75% of the people who come into his hardware store make a purchase. Today Michael made 270 sales. Which is the correct expression to find how many people came into the hardware store?

   (1) 270 × 0.75
   (2) 270 ÷ 0.75
   (3) 270 + 75
   (4) 270 − 75
   (5) 270 ÷ 75

3. Dustin answered 22 questions correctly on a test. His score was 88%. Which is the correct expression to find how many questions were on the test?

   (1) 22 + 88
   (2) 22 × 0.88
   (3) 22 ÷ 0.88
   (4) 0.88 ÷ 22
   (5) 88 − 22

4. This week Curtis spent 45% or 18 hours of his time taking inventory. Which is the correct expression to find how many hours Curtis worked this week?

   (1) 18 ÷ 45
   (2) 18 ÷ 0.45
   (3) 18 × 0.45
   (4) 45 − 18
   (5) 45 + 18

5. Lenora made 20% of the sales at Super Sporting Goods this week. She made 85 sales. Which is the correct expression to find how many sales Super Sporting Goods had this week?

   (1) 85 − 20
   (2) 85 ÷ 20
   (3) 85 + 20
   (4) 85 ÷ 0.20
   (5) 85 × 0.20

6. Transport Movers need to load boxes onto moving vans. They have 3 vans. By noon, the movers loaded 129 boxes. Which is the correct expression to find the number of boxes the movers need to load?

   (1) 129 ÷ 0.3
   (2) 129 − 3
   (3) 129 × 0.33
   (4) 129 ÷ 3
   (5) Not enough information is given.

7. Carmen attached the following sale tag to a pair of pants. Which is the correct expression to find the regular price for the pair of pants?

   (1) $13.60 × 0.40
   (2) $13.60 ÷ 40
   (3) $13.60 − 0.40
   (4) $13.60 ÷ 0.40
   (5) Not enough information is given.

8. **Reason** To find the part in a percent problem, you used the percent formula, *part* = *base* × *rate.* To find the rate, you rewrote the formula so that *rate* = *part* ÷ *base.* Explain in a sentence or two how to rewrite the formula to find the base.

Check your answers on page 310.

WORKPLACE MATH

Selling at a Discount

Example Paul works at Discount Fashions, a clothing store. The store buys clothes in large quantities at low prices and sells the clothes at a discount. Sometimes customers return clothes to the store. If the tags have been taken off, Paul must print out new tags and attach them to the clothing.

Paul needs to print a tag for a jacket. From the top of the tag, he knows that the discount rate is 25%, and the amount of customer savings is $15.50. Paul needs to find the regular price.

Which is the correct expression to find the regular price of the jacket?

(1) 75% ÷ $15.50
(2) 25% ÷ $15.50
(3) $15.50 × 75%
(4) $15.50 × 25%
(5) $15.50 ÷ 25%

Answer **(5)** is correct. The regular price of the jacket is the base. The customer savings, $15.50, is the part of the base. The rate of discount is 25%. Using the percent formula, Paul divides the part ($15.50) by the rate (25%).

First change the percent to a decimal. 25% = 0.25

$15.50 ÷ 0.25 = $62.00, so the regular price of the jacket is **$62.00.**

Example Next, Paul needs to find the sale price of the jacket. Which is the correct expression to find the sale price of this jacket?

(1) $62.00 + $15.50
(2) $62.00 − $15.50
(3) $62.00 × 25%
(4) $62.00 ÷ 25%
(5) $15.50 − 25%

Answer **(2)** is correct. To find the sale price of the jacket, subtract the amount of customer savings ($15.50) from the regular price ($62.00).

$62.00 − $15.50 = $46.50, so the sale price of the jacket is **$46.50.**

PRACTICE

Circle the best answer for each question.

1. Which is the correct expression to find the regular price of this sweater?

(1) $7.20 × 30%
(2) $7.20 ÷ 30%
(3) $7.20 × 70%
(4) $7.20 + 30%
(5) 70% ÷ $7.20

2. Which is the correct expression to find the regular price of this coat?

(1) 45% × $54.00
(2) 45% ÷ $54.00
(3) $54.00 ÷ 45%
(4) $54.00 − 45%
(5) $54.00 ÷ 55%

3. Which is the correct expression to find the sale price of these jeans?

(1) $12.60 × 35%
(2) $12.60 ÷ 35%
(3) $36.00 ÷ 35%
(4) $36.00 + $12.60
(5) $36.00 − $12.60

4. Which is the correct expression to find the regular price of this sweatshirt?

(1) $5.60 × 40%
(2) $5.60 ÷ 40%
(3) $5.60 × 60%
(4) $5.60 ÷ 60%
(5) 40% ÷ $5.60

5. Which is the correct expression to find the amount you save on this dress?

(1) 25% ÷ $64.00
(2) $64.00 × 25%
(3) $64.00 ÷ 25%
(4) $64.00 × 75%
(5) Not enough information is given.

6. **Communicate** Explain how to find the discount rate for this skirt. Next, explain how you would find the sale price of the skirt. No calculations are necessary.

Check your answers on page 311.

Finding the Base (*b*)

Use the percent triangle to find the base. Cover the letter *b*, because you need to solve for the base. The other two elements, the part and the rate, are separated by a division sign. Divide the part by the rate to find the base.

TIP

To divide by a decimal, move the decimal point to the right to make the decimal into a whole number. Then move the decimal point the same number of places to the right in the number you are dividing.

$.4\overline{).32} = 4\overline{)3.2}$ (quotient .8)

Example The number 28 is 20% of what number?

The rate is 20%.
The part is 28.

Step 1 Identify the rate and the part. You know that 20% is the rate because of the percent sign. The number 28 is the part.

Step 2 Rewrite the percent formula to find the base.
$p \div r = b$

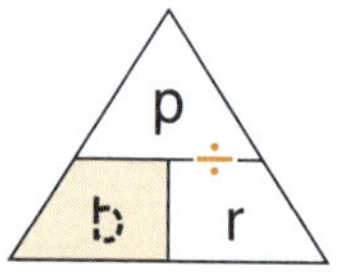

$b = p \div r$

Step 3 First change the percent to a decimal. Divide the part (28) by the rate (20%) to find the base.

$20\% = .20. = 0.20$

The number 28 is 20% of 140.

You can also use the % function on the CASIO *fx-260*SOLAR to find the base. Enter the part first and press the [÷] key. Enter the rate, then press [SHIFT] and the [=] key.

[AC] 28 [÷] 20 [SHIFT] [=] 140

You can also solve these types of percent problems by using a proportion.

Example The number 50 is 125% of what number?

Step 1 Identify the elements.

The part is 50.
The rate is 125%.
The base is unknown.

Step 2 Write the percent as a ratio.

$125\% = \frac{125}{100}$

Step 3 Set up the proportion.

$\frac{\text{part}}{\text{base}} = \frac{\text{rate}}{100}$

$\frac{50}{?} = \frac{125}{100}$

Step 4 Cross-multiply.

$50 \times 100 = 5{,}000$

Step 5 Divide by the remaining number.

$5{,}000 \div 125 = 40$

The number 50 is 125% of 40.

Remember, when the rate is greater than 100%, the part will be greater than the base.

To use the CASIO *fx-260*SOLAR to solve, enter the part first and then press the [÷] key. Enter the rate, then press [SHIFT] and the [=] key.

[AC] 50 [÷] 125 [SHIFT] [=] 40

PRACTICE

A. Solve. Show your work. The first one is done for you.

1. 24 is 25% of what number?

   25% = 0.25

   $$0.25\overline{)24.00}$$

   96.

   −22 5

   150

   −150

   0

2. 36 is 40% of what number?

3. $9 is 75% of what number?

4. 57 is 300% of what number?

5. 40 is 8% of what number?

6. 35 is 175% of what number?

B. Solve. Show your work. The first one is started for you.

7. For a sale, Earl's Furniture has discounted all items by 30%. The tag on a chair shows that a customer can save $22.50. What is the regular price of this chair?

   30% = 0.30

   $0.30\overline{)\$22.50}$

8. Ladonna's softball team won 14 games this season. The team won 70% of the games played. How many games did Ladonna's team play this season?

9. Vern spent 35% of his income last month for rent and utilities. He spent $490 for rent and utilities. What was Vern's income last month?

10. Elvia deposited money in a money market fund that pays 5% annual interest. She earned $34.25 interest in one year. How much did she deposit?

C. Solve these problems using a calculator.

11. Attendance at this year's opening game was 16,313. This number is 110% of last year's attendance. What was last year's attendance?

    ______

12. At CR Fashions, 31% of the weekly sales are of children's clothes. This week's sales of children's clothes is $2,356. What are the total sales for the week? ______

Check your answers on page 311.

Finding the Percent of Change

> **TIP**
> You can also use a proportion to solve a percent of change problem. Set up the proportion and solve for the rate of change.
>
> $\frac{\text{amt. of change}}{\text{original amt.}} = \frac{?}{100}$

A **percent of change** problem is a type of percent problem. To solve the problem, identify the base as the original amount and find the part as the amount of change (or the difference between the new amount and the original amount). Then solve for the rate—the percent of increase or decrease.

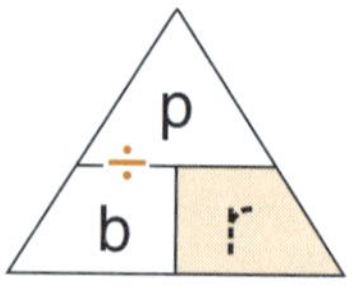

$r = p \div b$

rate	=	part	÷	base
percent of change	=	amount of change	÷	original amount

Finding the Percent of Increase

Example In January, Warren's landlord raised the rent from \$450 to \$486 per month. What is the percent of increase in Warren's rent?

Step 1 Identify the original amount (the base). The base is \$450.

Step 2 Identify the new amount. The new amount is \$486.

Step 3 Find the amount of change (the part) by subtracting the original amount (\$450) from the new amount (\$486). \$486 − \$450 = \$36 = part

Step 4 Use the percent formula. Divide the part (\$36) by the base (\$450) to find the rate of change. \$36 ÷ \$450 = 0.08

Step 5 Change the decimal to a percent. 0.08. = 8%

There was an 8% increase in Warren's monthly rent.

Finding the Percent of Decrease

Example Chris's Deli lowered the price of its fruit cup from \$4.00 to \$3.20. What is the percent of decrease?

Step 1 Identify the original price (the base). The base is \$4.00.

Step 2 Identify the new amount. The new amount is \$3.20.

Step 3 Find the amount of change (the part) by subtracting the new amount (\$3.20) from the original amount (\$4.00). \$4.00 − \$3.20 = \$0.80 = part

Step 4 Use the percent formula. Divide the part (\$0.80) by the base (\$4.00) to find the rate. \$0.80 ÷ \$4.00 = 0.20

Step 5 Change the decimal to a percent. 0.20. = 20%

There was a 20% decrease in the price of the fruit cup.

PRACTICE

A. Use the percent formula to find the percent of increase or decrease. Show your work. The first one is done for you.

	Original Amount	New Amount	Percent of Change
1.	$575	$690.00	20%

$$\begin{array}{r} \$690.00 \\ -575.00 \\ \hline \$115.00 \end{array} \qquad \begin{array}{r} 0.2 \\ \$575\overline{)\$115.00} \\ -115.00 \\ \hline 0 \end{array}$$

	Original Amount	New Amount	Percent of Change
2.	$8.80	$6.60	______
3.	25 pounds	15 pounds	______

B. Use proportions to find the percent of increase or decrease. Show your work. The first one is started for you.

	Original Amount	New Amount	Percent of Change
4.	4,800 people	5,376 people	______

$$\frac{\text{amount of change}}{\text{original amount}} = \text{rate of change}$$

$$\frac{576}{4{,}800} = \frac{?}{100}$$

	Original Amount	New Amount	Percent of Change
5.	70 gallons	63 gallons	______
6.	$24.00	$22.56	______

C. Solve these problems using a calculator.

7. Last year Michael's car insurance premium was $750. His car insurance premium this year is $1,125. What is the percent of increase in the premium? ______

8. A bus ticket from Mesa to Beaverton was $28.00. To attract more customers, the bus company reduced the fare to $23.80. What is the percent of decrease in the fare? ______

Check your answers on page 311.

DATA ANALYSIS

Using a Double Bar Graph

Example Acorn Business Systems has offices in four states. The company sells office machines to other businesses. The following graph shows company sales by state for the first and second quarters. A quarter is three months.

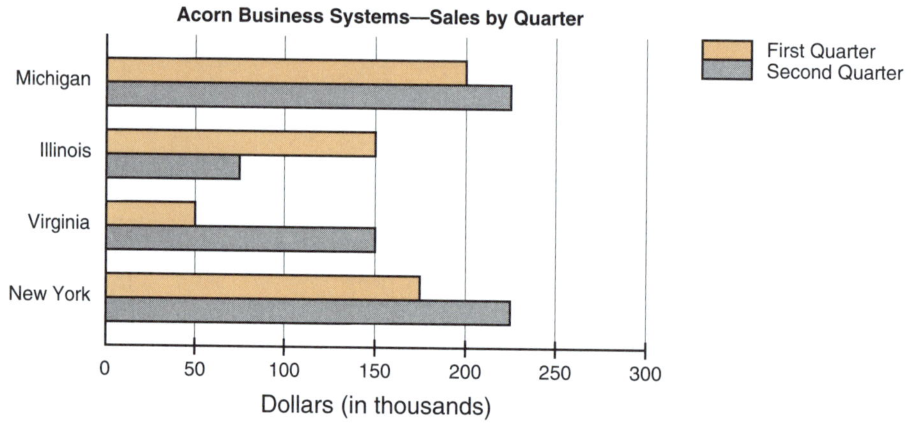

This graph is called a double bar graph because there are two bars for each state.

From the key, you know that the orange bar shows the first-quarter sales, and the gray bar shows the second-quarter sales.

Jorge Ruiz needs to find the percent of change in sales from the first quarter to the second quarter for Virginia. He reads the graph for Virginia. The orange bar reaches the mark labeled 50. The gray bar reaches the mark labeled 150.

What is the percent of increase from 50 to 150?

(1) $66\frac{2}{3}\%$
(2) 75%
(3) 100%
(4) 200%
(5) 250%

Answer **(4)** is correct. Jorge subtracts to find the part: 150 − 50 = 100
To find the rate, he divides the part (100) by the base (50).
100 ÷ 50 = 2 = 200%, so there was a **200%** increase in sales in Virginia from the first quarter to the second quarter.

Example Next, Jorge finds the percent of change in sales for New York. He reads the graph for New York. The orange bar represents 175. The gray bar represents 225. To the nearest whole percent, what is the percent of increase from 175 to 225?

(1) 22%
(2) 29%
(3) 33%
(4) 50%
(5) 72%

Answer **(2)** is correct. Jorge subtracts to find the part: 225 − 175 = 50
Then he divides the part (50) by the base (175) to find the rate.
$\frac{50}{175}$ **= 0.285, which rounds to 0.29 = 29%,** so there was about a **29%** increase in sales in New York from the first quarter to the second quarter.

PRACTICE

Circle the best answer for each question.

Questions 1 through 7 refer to the following graph.

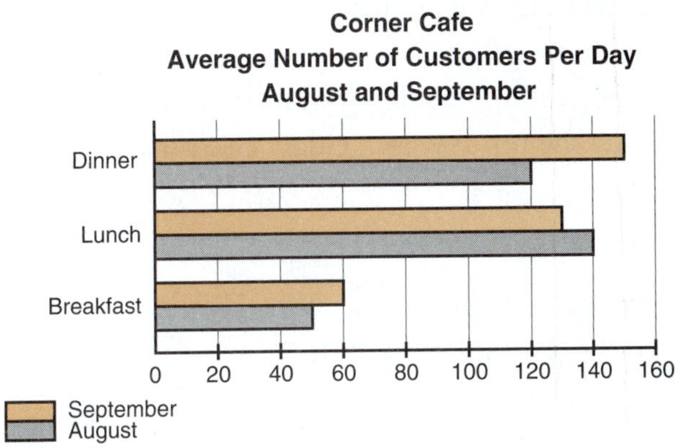

1. How many more customers did the Corner Cafe have each day for dinner in September than in August?
   (1) 15
   (2) 20
   (3) 30
   (4) 40
   (5) 270

2. To the nearest whole percent, what is the percent of increase in the number of customers each day for dinner from August to September?
   (1) 13%
   (2) 20%
   (3) 25%
   (4) 44%
   (5) 56%

3. To the nearest whole percent, what percent of the total customers each day during September were breakfast customers?
   (1) 11%
   (2) 18%
   (3) 25%
   (4) 40%
   (5) 60%

4. To the nearest whole percent, what is the percent of increase in the number of customers each day for breakfast from August to September?
   (1) 12%
   (2) 17%
   (3) 20%
   (4) 25%
   (5) 83%

5. To the nearest whole percent, what is the percent of decrease in the number of customers each day for lunch from August to September?
   (1) 7%
   (2) 8%
   (3) 14%
   (4) 93%
   (5) Not enough information is given.

6. What was the total number of average customers at the Corner Cafe during the months of August and September?
   (1) 270
   (2) 310
   (3) 340
   (4) 650
   (5) Not enough information is given.

7. **Communicate** Explain how to find the percent of increase in the total number of customers each day from August to September. Include the numbers you would use for the part, base, and rate. Do not do any calculations.

Check your answers on page 312.

Filling in Standard Grids for Percent Problems

Some questions require you to fill in the answer to a percent problem on a grid. To fill in an answer on a standard grid:

- You can begin in any column as long as your answer will fit.
- Write the answer in the boxes on the top row.
- A decimal point or fraction bar should be entered in its own column.
- Fill in the correct bubbles using the top row as a guide. Use only one circle in each column.
- Leave blank any unused column.

Example Randy earns a 5% commission, or bonus, based on a percent of the amount he sells. If he sells \$85 worth of clothing, how much money did he earn in commission on the sale?

Step 1 This problem is looking for the part (the commission). Identify the base and the rate.

$b = \$85$, $r = 5\%$, or 0.05

Step 2 Write the percent formula for finding the part. Multiply:

$b \times r = p$

$\$85 \times 0.05 = \$4.25$

Step 3 Write the answer in the top row of the grid. Do not write the \$ sign.

Step 4 Fill in the matching circle below the digits and decimal point. Leave unused columns blank.

Example A pair of jeans is on sale for 50% off the original price. If the sale price was \$21.25, what was the original price of the jeans?

Step 1 This problem is looking for the base (the original price). Identify the part and the rate.

$p = \$21.25$, $r = 50\%$, or 0.5

Step 2 Write the percent formula for finding the base. Divide:

$p \div r = b$

$\$21.25 \div 0.5 = \$42.50$

Step 3 Write the answer in the top row of the grid.

Step 4 Fill in the matching circle below the digits and decimal point.

PRACTICE

Solve each problem. Then fill in your answer on a grid.

1. Roberto Montoya earns 25% sales commission for his job. If he sells $1,320 in merchandise, how many dollars in commission will he make?

	/	/	/	
.	.	.	.	.
0	0	0	0	0
1	1	1	1	1
2	2	2	2	2
3	3	3	3	3
4	4	4	4	4
5	5	5	5	5
6	6	6	6	6
7	7	7	7	7
8	8	8	8	8
9	9	9	9	9

2. May Quan earned a commission of $425 last week. If her rate of commission is 25%, what was the dollar amount of the merchandise she sold?

	/	/	/	
.	.	.	.	.
0	0	0	0	0
1	1	1	1	1
2	2	2	2	2
3	3	3	3	3
4	4	4	4	4
5	5	5	5	5
6	6	6	6	6
7	7	7	7	7
8	8	8	8	8
9	9	9	9	9

3. Ben lent his brother $3,500 for one year. If his brother pays 5% interest, what dollar amount will Ben receive for interest at the end of one year?

	/	/	/	
.	.	.	.	.
0	0	0	0	0
1	1	1	1	1
2	2	2	2	2
3	3	3	3	3
4	4	4	4	4
5	5	5	5	5
6	6	6	6	6
7	7	7	7	7
8	8	8	8	8
9	9	9	9	9

4. Janet's volleyball team won 27 games this year. Her team won 90% of the games played. How many games did Janet's team play?

	/	/	/	
.	.	.	.	.
0	0	0	0	0
1	1	1	1	1
2	2	2	2	2
3	3	3	3	3
4	4	4	4	4
5	5	5	5	5
6	6	6	6	6
7	7	7	7	7
8	8	8	8	8
9	9	9	9	9

 Check your answers on pages 312–313.

15 LESSON

Lesson 15 Review

Solve. Show your work.

1. 27 is 30% of what number?

2. 60 is 3% of what number?

3. $40.80 is 68% of what number?

4. 4 is 20% of what number?

5. $35 is 125% of what number?

6. 36 is 75% of what number?

Circle the best answer for each question.

7. The Rileys' health insurance pays 80% of all medical expenses. The insurance paid $120 for Mona Riley's X rays. Which is the correct expression to find the charge for the X rays?
   (1) 20% × $120
   (2) 80% ÷ 120
   (3) $120 ÷ 20%
   (4) $120 × 80%
   (5) $120 ÷ 80%

8. Consumer's Summary reported that 9,610 people surveyed prefer to drive a sedan. This number represents 62% of those surveyed. Which is the correct expression to find the number of people surveyed?
   (1) 9,610 ÷ 38%
   (2) 9,610 × 62%
   (3) 9,610 ÷ 62%
   (4) 38% × 9,610
   (5) 62% ÷ 9,610

9. Last year the Hassans paid $216.00 each month for health insurance. This year, their monthly premium is $233.28. What is the percent of increase in their monthly premium?
   (1) 7%
   (2) 8%
   (3) 17%
   (4) 21%
   (5) 93%

10. Stan works as a data entry operator. His current wage is $6.50 per hour. Because of Stan's excellent performance, his boss gave him a raise. His new hourly wage will be $7.15. What is the percent of increase in Stan's wages?
   (1) 8%
   (2) 9%
   (3) 10%
   (4) 65%
   (5) 91%

Solve. Show your work.

Questions 11 and 12 refer to the following part of a price tag.

11. What is the regular price of the shirt?

12. What is the sale price of the shirt?

Questions 13 and 14 refer to the following graph.

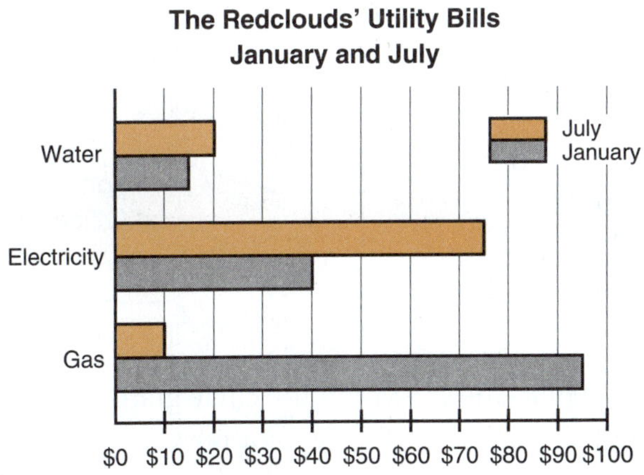

13. To the nearest whole percent, what is the percent of decrease in the gas bill from January to July?

14. To the nearest whole percent, what is the percent of increase in the electric bill from January to July?

Questions 15 and 16 refer to the following graph.

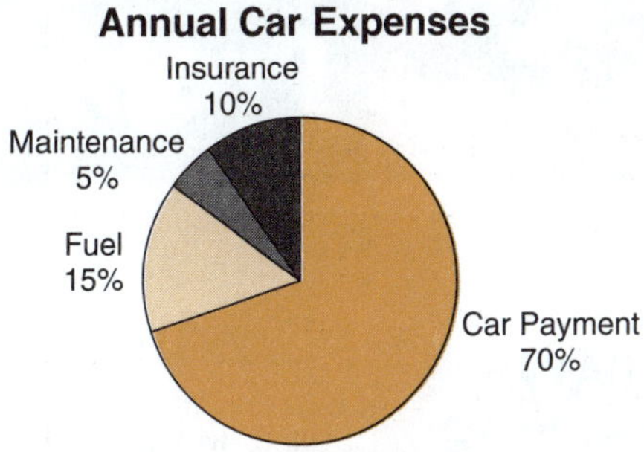

15. Jim's annual car expenses are about $3,500 each year. How much does he pay for fuel each year?
(1) $175
(2) $350
(3) $525
(4) $700
(5) $1,500

16. If Jim's annual expenses are $3,500, how much are his monthly car payments? Round to the nearest cent.
(1) $2,450.00
(2) $525.35
(3) $291.67
(4) $204.17
(5) Not enough information is given.

17. If Trudy earns 18% commission on her sales, what will she earn if she sells $12,000 in merchandise?

Mark your answer in the circles in the grid.

 Check your answers on pages 313–314.

LESSON 16

Statistics and Probability
Statistics and Probability

Statistics and Probability

DATA ANALYSIS

Basing Probability on Statistics

Many businesses gather information, or **data,** in order to make decisions. **Statistics** involves organizing and analyzing that data. Data can be gathered by taking a **survey.** When you gather information from an entire group, this group is called the **population.** Sometimes it is impractical to survey an entire population. In this case, gather information from a smaller group, or **sample,** that represents the population.

One uses statistics to make predictions based on the **probability** of something happening. Probability is the study of **chance** or **outcome.** Probability can be expressed as a fraction, a decimal, or a percent.

Example 1 Glenn works at Employment Op. He took a sample survey to find out how many people earn a living working from home. The circle graph at the right shows the results of his survey.

Preferred Place of Work

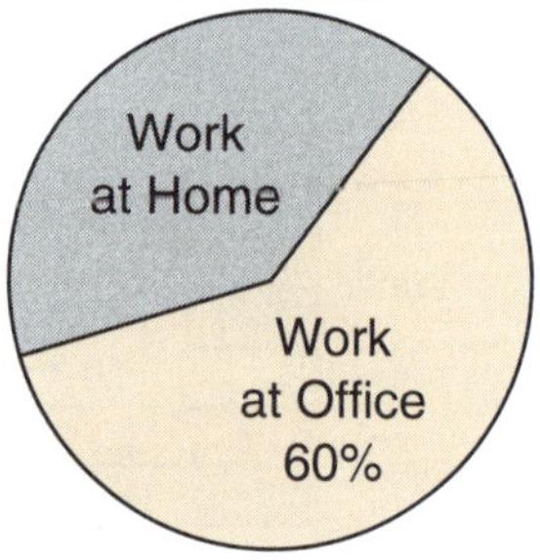

What is the probability that someone surveyed works at an office?

(1) 0 or 0%
(2) $\frac{2}{5}$ or 40%
(3) $\frac{3}{5}$ or 60%
(4) $\frac{4}{5}$ or 80%
(5) 1 or 100%

Answer **(3)** is correct. Based on the percents shown on the circle graph, the probability that someone surveyed works at an office is $\frac{3}{5}$ **or 60%.**

Example 2 Of 800 people surveyed, 52% named baseball as their favorite sport. What is the probability that the next person surveyed would choose a favorite sport other than baseball?

(1) 0%
(2) 48%
(3) 50%
(4) 52%
(5) 100%

Answer **(2)** is correct. To find the probability, subtract the percent of people who named baseball as their favorite sport (52%) from the total percent of people surveyed (100%). **100% − 52% = 48%.**

PRACTICE

Circle the best answer for each question.

Questions 1 and 2 refer to the following information and circle graph.

Neal is in charge of ordering inventory at Sporting Goods. He surveyed the customers in order to make decisions on what inventory to order. The results are shown below.

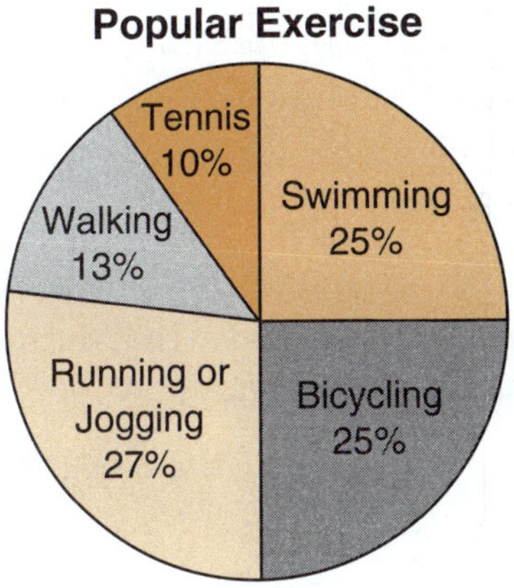

1. Of the 500 adults surveyed, 27% enjoyed running or jogging. How many people enjoy running or jogging as their favorite sport?

   (1) 5
   (2) 45
   (3) 90
   (4) 135
   (5) 450

2. According to the graph, what is the probability that the next customer Neal surveys would be interested in running or walking shoes?

   (1) $\frac{1}{4}$ or 25%
   (2) $\frac{2}{5}$ or 40%
   (3) $\frac{1}{2}$ or 50%
   (4) $\frac{3}{4}$ or 75%
   (5) 1 or 100%

Questions 3 and 4 refer to the following information and circle graph.

Frank works at Auto Dealer. He took a survey of 60 people and asked them to name the kind of car they drive. The results are shown below.

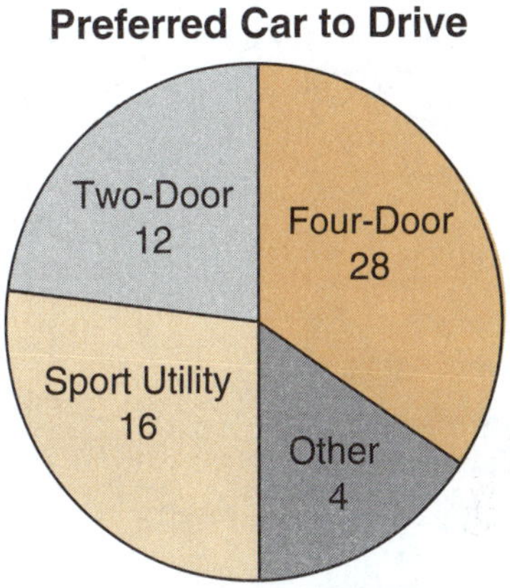

3. What is the probability that the next person surveyed drives a two-door?

   (1) 10%
   (2) 20%
   (3) 30%
   (4) 40%
   (5) 50%

4. What is the probability that the next person surveyed drives a four-door or a two-door?

   (1) 0%
   (2) 20%
   (3) 46%
   (4) $66\frac{2}{3}\%$
   (5) Not enough information is given.

5. **Reason** Frank's survey is based on a sample population of about 30,000 potential customers. What percent of the population does the sample of 60 people surveyed represent? What would make the survey a stronger predictor of probability?

Check your answers on page 314.

Facts about Probability

Probability uses numbers to show how likely it is that an event will happen. When you cannot control how an event will turn out, the **outcome** or result is left to **chance.**

You can express probability by using the numbers from 0 to 1 or by using percents from 0% to 100%. A probability of 1 or 100% means that you are sure that a certain outcome will occur. A probability of 0 or 0% means that a certain outcome cannot possibly happen.

TIP
The name that Wanda picks is left to **chance.** In other words, Scott and Wanda cannot control how the drawing will turn out, and they cannot predict with certainty what will happen.

Example Scott is organizing the holiday office party. He writes the names of all ten employees on separate cards. Then he puts the ten cards in a box. Each person picks a name and buys that co-worker a gift. Wanda chooses first. What is the probability that Wanda will pick the name of one of the ten office employees?

Step 1 Decide if the possible outcomes are equally likely to occur. *Yes, each of the ten outcomes is equally likely to occur.* This means that each card has an equal chance of being chosen.

Step 2 Decide if there are any other possibilities. *There is no other possibility.*

The probability is **1 or 100%** that Wanda will pick a card with the name of an employee on it. There is no other possibility.

PRACTICE

Write each probability using the numbers 0 or 1 and using percents 0% or 100%. Explain your answer. The first one is done for you.

1. Read the example problem again. What is the probability that the card Wanda picks will be blank?

   0 or 0%
   None of the cards is blank, so there is no possibility that the card she chooses will be blank.

2. A box contains 15 blue marbles. What is the probability of picking a blue marble from this box?.

3. Toshiro's sock drawer contains 11 pairs of socks: 3 pairs of blue socks, 4 pairs of black socks, and 4 pairs of white socks. What is the probability that he will pick a pair of brown socks?

4. Mr. Hernandez is a meteorologist. He told his television viewing audience that there is a 100% chance of rain tomorrow. According to Mr. Hernandez, what is the probability that it will not rain tomorrow?

 Check your answers on page 314.

Finding the Probability of an Outcome

You can write probabilities that are greater than 0 but less than 1 either as a fraction, decimal, or a percent.

Probability of an outcome $= \dfrac{\text{number of ways the outcome can occur}}{\text{number of all possible outcomes}}$

Example Ten co-workers decide to exchange gifts for the holidays. They each write their name on a card and place the cards in a box. Then each person will choose a name. Jesse picks first. What is the probability that Jesse will pick his own name?

Step 1 Find the number of all possible outcomes. Because there are 10 cards in the box, there are 10 equally likely outcomes.

There are 10 possible outcomes

Step 2 Find the number of ways the outcome (Jesse picking his own name) can occur. Since Jesse's name is written on only one of the cards, there is only one way this outcome can occur.

There is one way the outcome can occur.

Step 3 Write the probability as a fraction in lowest terms.

$\frac{1}{10}$

Step 4 To write the probability as a percent, change the fraction $(\frac{1}{10})$ to a decimal (0.1). Then change the decimal (0.1) to a percent (10%).

$1 \div 10 = 0.1 = 10\%$

The probability Jesse will pick his own name is 10%.

TIP
Remember to reduce the fraction to lowest terms.
$\frac{2}{10} = \frac{1}{5}$

PRACTICE

Write each probability as a fraction and as a percent. The first one is done for you.

Questions 1 through 4 refer to the following information.
A number cube is a cube with six sides. Each side is shown below.

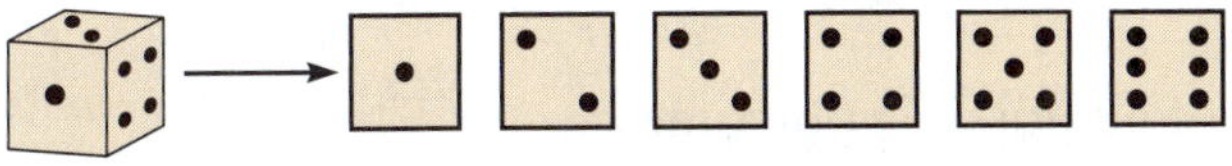

1. If you roll a number cube, what is the probability that you will roll a 4?

   Fraction: $\frac{1}{6}$ Percent: $16\frac{2}{3}\%$

2. If you roll a number cube, what is the probability that you will roll a 9?

   Fraction: ________ Percent: ________

3. If you roll a number cube, what is the probability that you will roll a number less than 7?

   Fraction: ________ Percent: ________

4. If you roll a number cube, what is the probability that you will roll an even number?

   Fraction: ________ Percent: ________

Check your answers on page 314.

More than One Chance for the Same Outcome

Sometimes there is more than one way for an outcome to occur.

Example On a TV game show, a player spins a wheel to win a prize. The wheel has eight equal sections. Since the sections are the same size, there is an equal chance that the wheel will stop on any one section.

Only one section is marked *New Car.* A player has a 1 in 8 chance of winning a new car. The probability that the wheel will stop on the section marked *New Car* is $\frac{1}{8}$ or 12.5%. What is the probability that the wheel will stop on $100?

TIP
Four sections are marked *$100.*

Step 1 Find the number of all possible outcomes. There are eight sections on the wheel. — 8

Step 2 Find the number of ways this particular outcome can occur. There are four ways to win $100. — 4

Step 3. Write the probability of the outcome. — $\frac{\text{number of ways an outcome can occur}}{\text{number of possible outcomes}} =$ $\frac{4}{8} = \frac{1}{2}$

Step 4: Change the fraction to a percent. — $\frac{1}{2} = 50\%$

The probability of a player winning $100 is $\frac{1}{2}$ or 50%.

Example A box contains 12 marbles. Nine marbles are black. Three marbles are white. If you take one marble from the box without looking, what is the probability that you will choose a white marble?

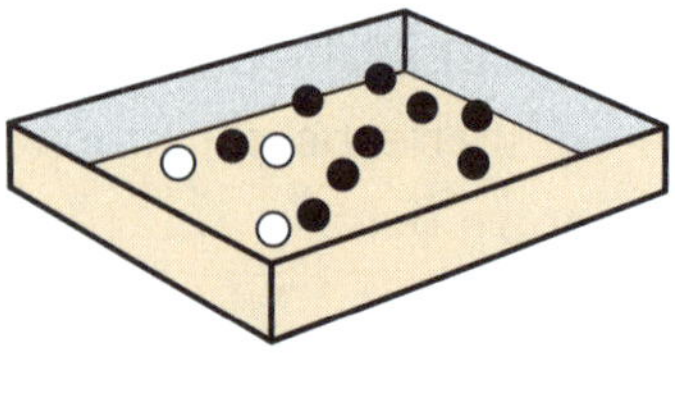

Step 1 Find the number of all possible outcomes. There are 12 marbles in the box. — 12

Step 2 Find the number of ways this particular outcome can occur. There are 3 chances of choosing a white marble. — 3

Step 3 Write the probability of the outcome. — $\frac{3}{12} = \frac{1}{4}$

Step 4 Change the fraction to a percent. — $\frac{1}{4} = 25\%$

The probability of choosing a white marble is $\frac{1}{4}$ or 25%.

PRACTICE

A. Write the probability for each problem. Show your work. The first one is done for you.

Questions 1 through 4 refer to the following information. The carnival game spinner shown contains 10 equal sections.

1. What is the probability that the wheel will stop on a shaded section?

$$\frac{\text{number of shaded sections}}{\text{total number of sections}} = \frac{3}{10} = 30\%$$

2. What is the probability that the wheel will stop on an even number (2, 4, 6, 8, 10)?

3. What is the probability that the wheel will stop on a number greater than 4?

4. What is the probability that the wheel will stop on a number less than 12?

B. Write the probability. Show your work. The first one is started for you.

Questions 5 through 7 refer to the following information. A box contains 8 marbles: 3 black, 3 gray, and 2 white. You take one marble out of the box without looking.

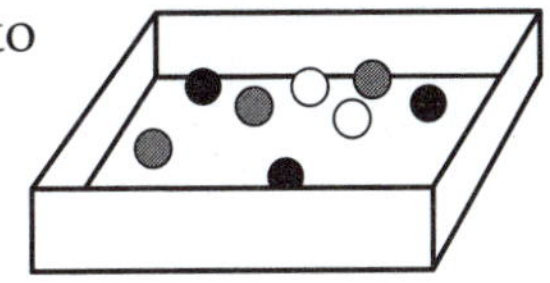

5. What is the probability you will choose a gray marble?

$$\frac{\text{number of gray marbles}}{\text{number of marbles}} =$$

6. What is the probability that you will choose a black or a gray marble?

7. What is the probability that you will not choose a black marble?

C. Find the probability. Show your work.

8. A box has 20 marbles: 6 black, 4 white, 5 blue, and 5 gray. What is the probability you will choose a white marble?

9. Out of every 500 pairs of jeans a company manufactures, 35 are defective. Out of the next batch of 500 pairs of jeans, what is the probability that the next pair of jeans is defective?

Check your answers on pages 314–315.

Finding a Dependent Probability

Sometimes the chance of something happening depends on another outcome. To find a **dependent probability**, consider the other outcomes.

Example M & M Furniture is having a drawing for its employees. The names of the six employees with the best attendance records will be written on cards. Two names will be drawn, and the employees whose names are drawn will win a trip. Of the six employees in the drawing, two work in shipping and four work in sales. What is the probability that the first name drawn will be an employee from the shipping department?

Step 1 Find the number of all possible outcomes. There are six employees in the drawing. — 6

Step 2 Find the number of ways this outcome can occur. Two of the employees work in shipping. — 2

Step 3 Write the probability of the outcome. — $\frac{\text{ways outcome can occur}}{\text{possible outcomes}} = \frac{2}{6} = \frac{1}{3}$

Step 4 Change the fraction to a percent. — $\frac{1}{3} = 33\frac{1}{3}\%$

The probability that the first name drawn will be an employee from the shipping department is $33\frac{1}{3}\%$.

Example The first card drawn was from shipping. This card was not placed back in the drawing. Then a second name was drawn. What is the probability that the second name drawn will be an employee from sales?

Step 1 Find the number of possible outcomes. There are five names (cards) left in the drawing. — 5 cards left

Step 2 Find the number of ways an outcome can occur. If the first name drawn was from shipping, then there are 4 names from sales and 1 name from shipping left. — 4 people in sales

First card drawn	These cards are left:				
Shipping	Shipping	Sales	Sales	Sales	Sales

Step 3 Write the probability of the outcome: — $\frac{\text{4 people in sales}}{\text{5 cards left}}$

Step 4 Change the fraction to a percent. — $\frac{4}{5} = 80\%$

The probability that the second name drawn will be an employee from sales is **80%**.

If the first name drawn has been from sales, then there would be 3 names from sales and 2 names from shipping left. That means the probability the second name drawn would be an employee from sales would have been $\frac{3}{5}$ or **60%**.

PRACTICE

A. Write the probability for each problem. Show your work. The first one is done for you.

Questions 1 through 3 refer to the following information.

Five friends are planning a fishing trip. To decide who will drive, they place five cards the same size in a bag. *Drive* is written on one card. The other cards are blank. Each person will draw a card until someone draws the card with *Drive* on it.

Drive				

1. What is the probability that the first card drawn will have *Drive* written on it?

   $\frac{\text{number of cards with Drive}}{\text{number of cards}} = \frac{1}{5} = 20\%$

2. The first card drawn is blank. What is the probability that the second card drawn will have *Drive* written on it?

3. The first card drawn is blank. The second card drawn is blank. What is the probability that the third card drawn will also be blank?

B. Write the probability. Show your work. The first one is started for you.

Questions 4 through 6 refer to the following information.

As part of Food Basket's grand opening, nine currency bills are placed in a box; six \$20 bills and three \$100 bills. Each hour, a customer will be selected to draw one bill from the box without looking.

\$20	\$100	\$20	\$20	\$100
\$20	\$20	\$100	\$20	

4. What is the probability that the first bill drawn will be a \$100 bill?

   $\frac{\text{number of \$100 bills}}{\text{total number of bills}} =$

5. The first bill drawn is a \$100 bill. What is the probability that the second bill drawn will also be a \$100 bill?

6. The first bill drawn is a \$100 bill. The second bill drawn is a \$20 bill. What is the probability that the third bill drawn will be a \$100 bill?

C. Solve the following problems using a calculator.

Questions 7 and 8 refer to the following information.

Amy's department is having a picnic. At the picnic there is going to be a raffle. There is one grand prize and five first-place prizes. Of the 25 people who bought a ticket, 8 are men and 17 are women. Tickets are drawn and not replaced.

7. What is the probability that the first ticket drawn will be Amy's ticket?

   ______

8. If the first three tickets drawn belonged to women, what is the probability that the fourth ticket drawn will belong to a man? ______

Check your answers on page 315.

Lesson 16 Review

Solve each problem. Show your work.

Questions 1 through 3 refer to the graph.

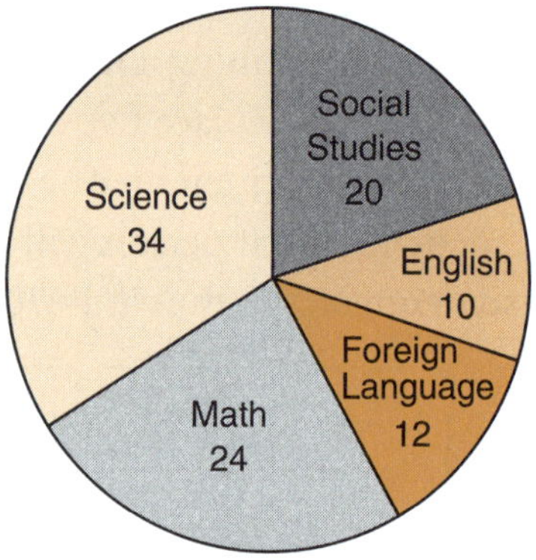

1. What is the total number surveyed?

2. What percent of those surveyed preferred social studies?

3. What is the probability that someone will name English as a preferred subject?

Write each probability using the numbers 0 or 1 and using percents 0% or 100%.

4. A drawer contains 12 multicolored scarves. No other items are in the drawer. What is the probability of choosing a multicolored scarf from this drawer?

5. The forecast is for 100% chance of snow today. According to the forecast, what is the probability that it will not snow today?

Write each probability as a fraction and as a percent.

Questions 6 through 9 refer to the following information.
A box contains 5 marbles: a red, a green, a blue, an orange, and a yellow marble. You take one marble out of the box without looking. After a marble is taken, it is put back in the box.

6. What is the probability you will choose a green marble?

   Fraction: __________ Percent: __________

7. What is the probability you will choose a marble that is not blue?

   Fraction: __________ Percent: __________

8. What is the probability you will choose an orange or a yellow marble?

   Fraction: __________ Percent: __________

9. What is the probability you will choose a black marble?

   Fraction: __________ Percent: __________

Questions 10 through 17 refer to the following information.
A box contains 10 marbles: 6 black, 1 gray, and 3 white.
You take one marble out of the box without looking.
After a marble is taken, it is put back in the box.

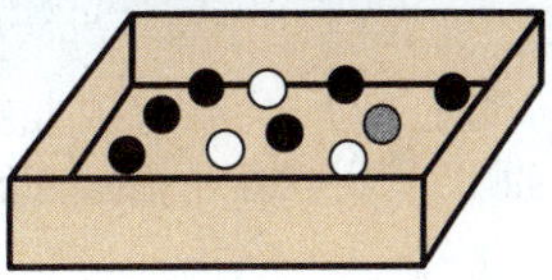

Circle the best answer for each question.

10. What is the probability that you will choose a black marble?

(1) $\frac{2}{5}$ or 40%
(2) $\frac{3}{5}$ or 60%
(3) $\frac{2}{3}$ or $66\frac{2}{3}$%
(4) 1 or 100%
(5) $\frac{3}{2}$ or 150%

11. What is the probability that you will choose a green marble?

(1) 0 or 0%
(2) $\frac{1}{10}$ or 10%
(3) $\frac{1}{9}$ or $11\frac{1}{9}$%
(4) $\frac{9}{10}$ or 90%
(5) 1 or 100%

12. What is the probability that you will choose a gray marble?

(1) 0 or 0%
(2) $\frac{1}{10}$ or 10%
(3) $\frac{1}{9}$ or $11\frac{1}{9}$%
(4) $\frac{9}{10}$ or 90%
(5) 1 or 100%

13. What is the probability that you will not choose a white marble?

(1) 0 or 0%
(2) $\frac{3}{10}$ or 30%
(3) $\frac{2}{5}$ or 40%
(4) $\frac{3}{5}$ or 60%
(5) $\frac{7}{10}$ or 70%

Solve. Show your work. Assume after each marble is drawn, it is not put back in the box.

14. The first marble drawn is gray. A second marble is drawn. What is the probability that it is white?

15. The second marble drawn is white. A third marble is drawn. What is the probability that it is also white?

16. The third marble drawn is black. A fourth marble is drawn. What is the probability that it is also black?

17. The fourth marble drawn is black. A fifth marble is drawn. What is the probability that it is white?

Check your answers on pages 315–316.

Math at Work

Hospitality: Cook

Almost everybody enjoys a tasty meal. Many people like having that meal prepared by a professional cook or chef and served in a restaurant.

Cooks have the ability to combine a variety of ingredients into dishes to be enjoyed by diners in a wide range of settings. Cooks work in casual and elegant restaurants, commercial kitchens, bakeries, grocery stores, and school cafeterias.

Many cooks are responsible for planning menus; creating new recipes; buying meats, spices, produce and other foodstuffs; and preparing a variety of foods to be served. Often restaurant cooks help set the prices of the dishes on the menu. Other cooks follow the directions of the head chef. They follow recipes, cook the foods, and present them in an appetizing display.

Measuring ingredients, following recipes, and adjusting recipes are some of the ways cooks apply math skills. Cooks must be able to use their knowledge of percents, ratio, and proportion.

Look at the *Some Careers in Hospitality* chart.

- Do any of the careers interest you? If so, which ones?
- What information would you need to find out more about those careers? On a separate piece of paper, write some questions that you would like answered. You can find more information about those careers in the *Occupational Outlook Handbook* at your local library or online.

Some Careers in Hospitality

Assistant Cook
prepares recipe ingredients for head cook by measuring, chopping, peeling, and cleaning

Hotel Desk Clerk
checks in and checks out hotel guests, calculates bills, answers questions, and assists guests

Restaurant Host
welcomes and seats diners, makes and checks reservations, and monitors number of diners in each section

Waitperson
takes food orders, serves customers, calculates bills, collects money, and makes change

Use the recipe below to answer the questions that follow.

Roasted Red Pepper Pesto
Serves 6 people

Ingredients

4 red peppers, halved, seeded, and roasted
3 garlic cloves
1 cup Parmesan cheese, shredded
$\frac{1}{3}$ cup pine nuts, toasted
$\frac{1}{2}$ cup olive oil
$\frac{1}{4}$ teaspoon salt
$\frac{1}{4}$ teaspoon freshly ground black pepper

Grind garlic and pine nuts in food processor until finely minced. Add all remaining ingredients except olive oil. Grind until mixture becomes a coarse puree. Keep processor running and drizzle in olive oil. Process until all ingredients are combined in a fine puree.

1. In the recipe, what is the ratio of the number of garlic cloves to the number of cups of cheese?
   (1) 4:3
   (2) 3:4
   (3) 4:1
   (4) 3:1
   (5) 1:3

2. Danielle needs to prepare the recipe above for 8 people. Which of the following expressions should she use to find the number of red peppers she should use?
   (1) $\frac{6}{8} = \frac{4}{?}$
   (2) $\frac{8}{6} = \frac{4}{?}$
   (3) $\frac{4 + 6}{4 + 8} = \frac{6 + 8}{? + 8}$
   (4) $\frac{6 \times 8}{4} = ?$
   (5) Not enough information is given.

3. Based on the recipe, which percentage represents the approximate amount of a garlic clove that each person would get in an individual serving?
   (1) 10%
   (2) 25%
   (3) 33%
   (4) 50%
   (5) 300%

Check your answers on page 316.

Unit 4 Review Ratios, Proportions, and Percents

Solve each problem. Show your work.

1. $\frac{8}{20} = \frac{?}{35}$

2. What is 450% of 72?

3. What percent of 320 is 8?

4. $4.05 is 9% of what amount?

Find the percent of increase or decrease. Round your answer to the nearest percent. Show your work.

5. Original amount $12.50 | New Amount $17.50 | Percent of Change ________

6. Original amount 45 pounds | New Amount 15 pounds | Percent of Change ________

7. Original amount 60 gallons | New Amount 50 gallons | Percent of Change ________

Solve. Show your work.

Donita bought a washing machine that cost $520. She paid $100 down. She will pay the rest in monthly payments over 2 years at 18% interest.

8. How much interest will Donita pay?

9. What will Donita's monthly payment be?

10. Auto Needs has oil on sale at 3 quarts for $2.85. Minh wants to buy 8 quarts. How much will he pay for the oil?

11. Carlos sold $4,527 in merchandise. He earns a 22% commission rate. What is the amount of commission Carlos earned?

Circle the best answer for each question.

Questions 12 and 13 refer to the following information.

A bag contains 3 nickels and 2 dimes. You choose one coin without looking.

12. What is the probability that you will choose a nickel?
   (1) $\frac{1}{5}$ or 20%
   (2) $\frac{2}{5}$ or 40%
   (3) $\frac{3}{5}$ or 60%
   (4) $\frac{4}{5}$ or 80%
   (5) 1 or 100%

13. The first coin you choose is a nickel. You do not put it back in the bag. You choose another coin without looking. What is the probability that the second coin you choose will also be a nickel?
   (1) 0 or 0%
   (2) $\frac{2}{5}$ or 40%
   (3) $\frac{1}{2}$ or 50%
   (4) $\frac{3}{4}$ or 75%
   (5) 1 or 100%

14. The Wilsons purchased a car for $11,500. If the sales tax rate is $7\frac{1}{2}$%, what was the tax on the car?
   (1) $86.25
   (2) $770.00
   (3) $805.00
   (4) $862.50
   (5) $1,533.33

15. Which is the correct expression to find the regular price of this set of dishes?

Discount House Dishes	
Discount	25%
You Save	$40.00

   (1) 25% × $40.00
   (2) 25% ÷ $40.00
   (3) $40.00 ÷ 25%
   (4) $40.00 − 25%
   (5) $40.00 + 25%

Questions 16 and 17 refer to the following graph.

Preferred Reading Survey
(in hundreds of adults)

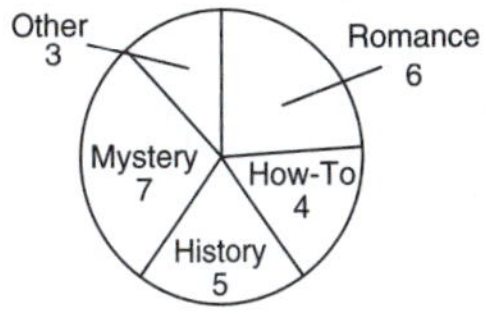

16. What percent of the adults in the survey preferred reading mysteries? Write the answer. ______

17. What is the probability that the next person surveyed prefers romance or history? Write your answer as a fraction.

Mark your answer in the circles in the grid.

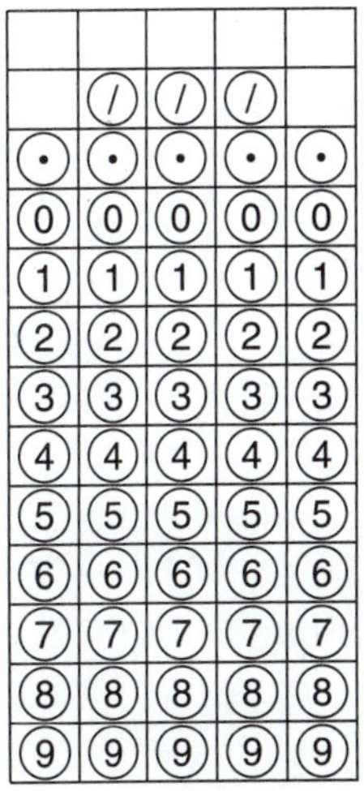

Math Extension

For the next week, keep track of how much money you spend on a hobby or favorite pastime. What percent of your weekly pay or allowance does this amount represent?

Check your answers on pages 316–317.

Mini-Test • Unit 4

Directions: This is a 15-minute practice test. After 15 minutes, mark the last number you finished. Then complete the test and check your answers. If most of your answers were correct, but you did not finish, try to work faster next time.

Part I Directions: Choose the one best answer to each question. You MAY use your calculator.

1. Lila's car used 15 gallons of gasoline to travel 600 miles. How many gallons are needed to drive 1,000 miles?

   (1) 17.5
   (2) 20
   (3) 22.25
   (4) 25
   (5) 28.5

2. At Capitan Pass, this year's snowfall is predicted to be 220 inches. So far this year, 154 inches has fallen. This is what percent of the predicted snowfall?

   (1) 30%
   (2) 70%
   (3) 95%
   (4) 115%
   (5) 143%

3. Renata borrowed $2,800 for a car. The interest on the loan is $12\frac{1}{2}$%. What is the total amount of interest she will pay back in 3 years?

   (1) $1,050
   (2) $3,150
   (3) $3,850
   (4) $4,850
   (5) $5,950

Question 4 refers to the following graph.

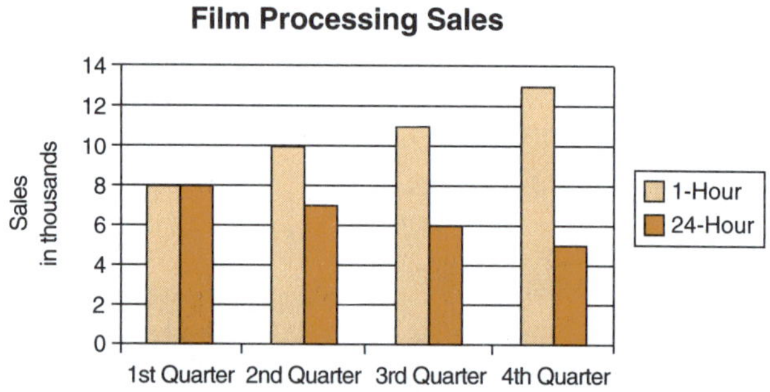

4. What was the percent of increase in 1-hour sales from the 1st quarter to the 4th quarter?

   (1) 25%
   (2) 37.5%
   (3) 50%
   (4) 62.5%
   (5) 75%

5. Samuel paid $47.92 sales tax on his new computer. If the sales tax rate was 8%, what did the computer cost in dollars?

   Mark your answer in the circles in the grid.

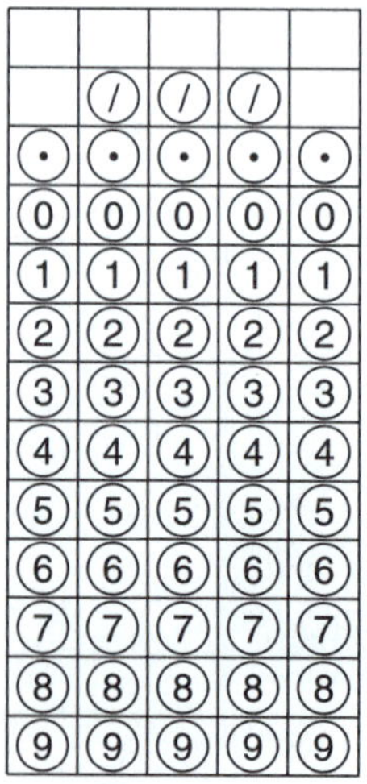

Part II Directions: Choose the one best answer to each question. You MAY NOT use your calculator.

6. Jake's Furnishings is having a year-end clearance sale of 60% off. How much will Tanya save on a dining room table that originally sold for $350?
   (1) $58
   (2) $60
   (3) $91
   (4) $140
   (5) $210

Questions 7 and 8 refer to the following graph.

Number of Moviegoers by Age

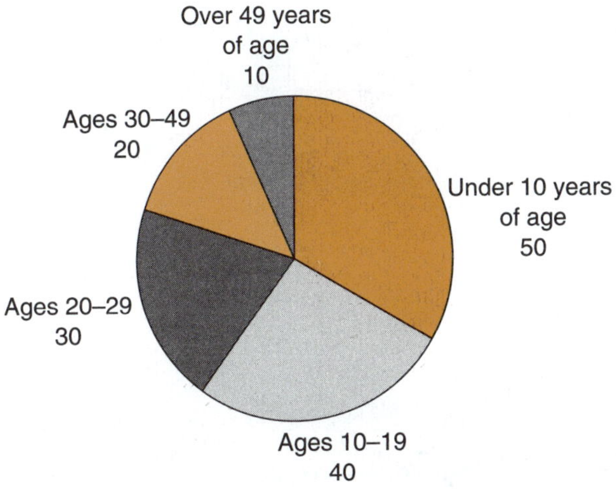

7. What percent of the moviegoers were ages 20 to 29?
   (1) 60%
   (2) 40%
   (3) 30%
   (4) 20%
   (5) 10%

8. What percent of the moviegoers were under age 20?
   (1) 40%
   (2) 50%
   (3) 60%
   (4) 80%
   (5) 90%

Question 9 refers to the following information.

The spinner below is made up of 6 equal sections.

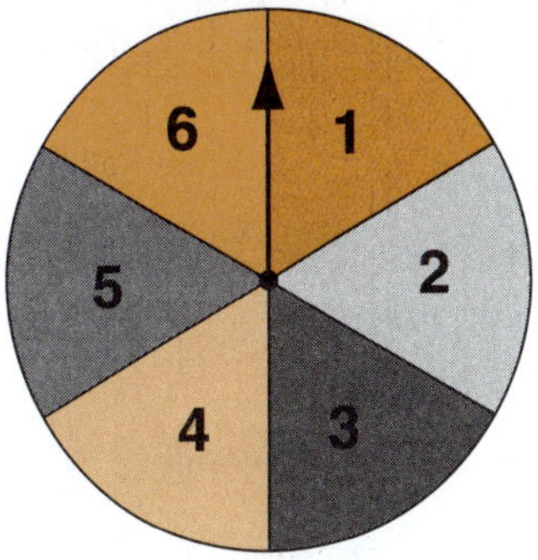

9. What is the probability that the spinner will stop on a number less than 5?
   (1) $\frac{1}{6}$
   (2) $\frac{1}{3}$
   (3) $\frac{1}{2}$
   (4) $\frac{2}{3}$
   (5) $\frac{5}{6}$

10. If 3 nights at a hotel cost $225, how many nights at that rate could a guest spend at the hotel for $375?

    Mark your answers in the circles in the grid.

Check your answers on pages 317–318.

ANSWER SHEET

Posttest
Mathematics

Name: ______________________ Class: ________________ Date: __________

Part 1

1 ① ② ③ ④ ⑤

2 ① ② ③ ④ ⑤

3 ① ② ③ ④ ⑤

4 ① ② ③ ④ ⑤

5

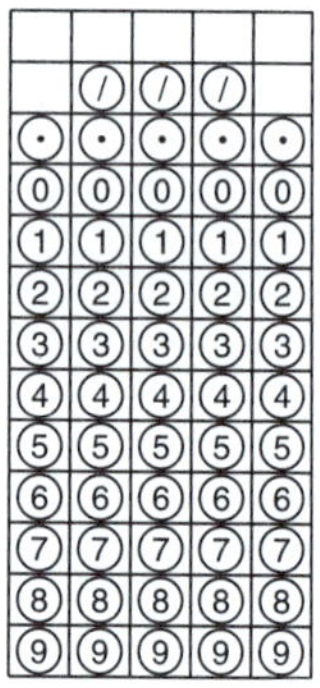

6 ① ② ③ ④ ⑤

7 ① ② ③ ④ ⑤

8 ① ② ③ ④ ⑤

9 ① ② ③ ④ ⑤

10

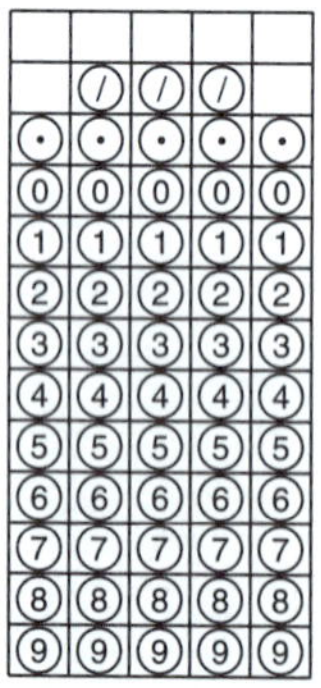

11 ① ② ③ ④ ⑤

12 ① ② ③ ④ ⑤

13 ① ② ③ ④ ⑤

Part 2

14 ① ② ③ ④ ⑤

15 ① ② ③ ④ ⑤

16 ① ② ③ ④ ⑤

17 ① ② ③ ④ ⑤

18

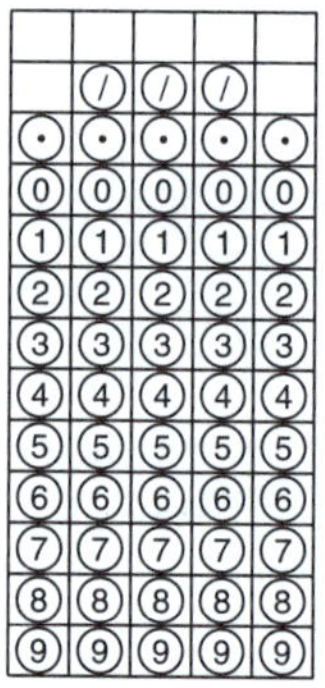

19 ① ② ③ ④ ⑤

20 ① ② ③ ④ ⑤

21 ① ② ③ ④ ⑤

22 ① ② ③ ④ ⑤

23 ① ② ③ ④ ⑤

24

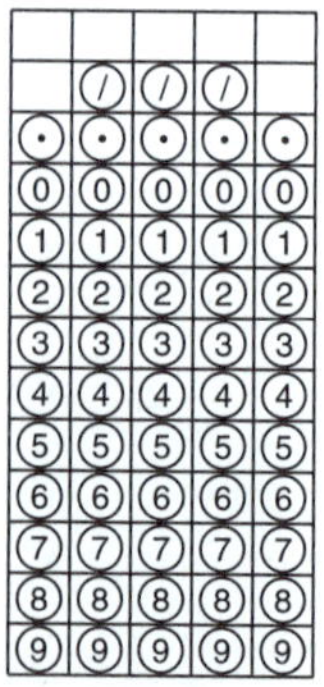

25 ① ② ③ ④ ⑤

POSTTEST

Directions

This is a 40-minute practice test. After 40 minutes, mark the last number you finished. Then complete the test and check your answers. If most of your answers were correct but you did not finish, try to work faster next time.

The Posttest consists of 25 questions, and has two parts. You will be allowed to use a calculator on Part I of the posttest, but you are not required to use it. In fact, you may be able to solve some questions more quickly without it. Do not use a calculator on the questions in Part II.

Record your answers on the answer sheet on page 228. You may photocopy this page. Most questions have five answer choices. To record your answer, fill in the numbered circle on the answer sheet that corresponds to the answer you select for each question in the Posttest. A few do not have choices. For these problems, write your answers on the special grids on your answer sheet.

After you complete the Posttest, check your answers on pages 318–320. Then use the chart on page 240 to identify the math skills that you need to practice more.

EXAMPLE

Brad paid his grocery bill of $34.57 with two $20.00 bills. How much change will he receive?

(1) $7.33
(2) $6.43
(3) $6.33
(4) $5.43
(5) $5.33

(On Answer Sheet)

①②③●⑤

The correct answer is **$5.43.** Therefore, answer space 4 would be marked on the answer sheet.

EXAMPLE

Lorie buys $5\frac{3}{4}$ pounds of ground beef. She puts $2\frac{1}{2}$ pounds in a separate package for freezing. How many pounds are left?

There are **$3\frac{1}{4}$ pounds** left. When entering answers in alternate grids, remember to fill in only one circle in each column, begin in any column that allows the entire answer to be entered, and enter a mixed number, such as $3\frac{1}{4}$, as an improper fraction $\frac{13}{4}$ or as a decimal (3.25).

If you do not use the answer sheet provided, mark your answers on each test page by circling or writing the correct answer for each question.

Go on to the next page.

FORMULAS

You may use the following formulas as you take the PreGED Mathematics Posttest.

AREA of a:	
square	Area = side^2
rectangle	Area = length × width
PERIMETER of a:	
square	Perimeter = 4 × side
rectangle	Perimeter = 2 × length + 2 × width
triangle	Perimeter = side 1 + side 2 + side 3
VOLUME of a:	
cube	Volume = side^3
rectangle	Volume = length × width × height
MEASURES OF CENTRAL TENDENCY	mean = [stack fraction – make sure subscripts are legible] $x_1 + x_2 + \ldots x_z/z$; where the x's are the values for which a mean is desired, and x is the total number of values of x. median = the middle value of an odd number of ordered scores and halfway between the two middle values of an even number of ordered scores.
SIMPLE INTEREST	interest = principal × rate × time
DISTANCE	distance = rate × time
TOTAL COST	total cost = (number of units) × (price per unit)

Adapted with permission from GEDTS.

Part I

Directions: Choose the one best answer for each question. You MAY use your calculator.

Question 1 refers to the following table.

Northbrook Hotel
Banquet Reservations for June 15

Room	Number of Guests
Ballroom	424
Terrace	208

1. Banquet guests are seated 8 to a table. How many tables must be set up for the banquets on June 15?

(1) 26
(2) 53
(3) 79
(4) 632
(5) Not enough information is given.

2. Wayne sees the following advertisement.

The store's regular price for a dress shirt is $18.95. If Wayne buys three dress shirts on sale, how much will he save off the regular price for three shirts?

(1) $8.15
(2) $10.35
(3) $18.50
(4) $27.55
(5) $37.90

3. Ray sees the following price tag on a sweater at Discount Fashions. Part of the tag has been torn off.

Which is the correct expression to find the original price of the sweater?

(1) $\$15.20 \times 0.4$
(2) $\$15.20 \div 0.4$
(3) $\$15.20 \times 0.04$
(4) $\$15.20 \div 0.04$
(5) $\$15.20 + 0.4$

4. Gloria delivers newspapers. She drives $2\frac{1}{5}$ miles on Broad Street, $3\frac{9}{10}$ miles through a housing division, and $1\frac{1}{2}$ miles on Hilltop Road. How many miles does Gloria drive on her route?

(1) $6\frac{1}{2}$

(2) $6\frac{3}{5}$

(3) $6\frac{11}{17}$

(4) $7\frac{1}{2}$

(5) $7\frac{3}{5}$

5. Martin is cooking a rice dish for a community shelter. A large box of rice holds 3.7 pounds. If Martin uses one-half the contents of the box, how many pounds of rice will he use?

Mark your answer in the circles in the grid on your answer sheet.

Questions 6 and 7 refer to the following graph.

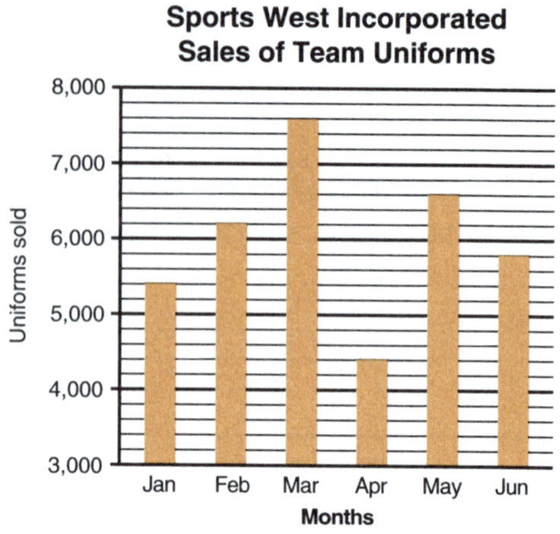

6. The company goal is to sell at least 5,800 uniforms per month. In which month or months did the company meet its goal?

(1) June
(2) January and April
(3) February and June
(4) February, March, May, and June
(5) Not enough information is given.

7. What was the percent of increase in uniform sales from April to May?

(1) 5%

(2) $33\frac{1}{3}$%

(3) 50%

(4) 100%

(5) 150%

8. What is the value of the following expression?

$144 + 18 \div 3 - (13 + 9 \times 2)$

(1) 10
(2) 23
(3) 106
(4) 119
(5) 155

9. The spinner below is divided into ten equal sections.

Suppose you spin the spinner only once. What is the probability that you will spin the number 5?

(1) $\frac{1}{5}$
(2) $\frac{3}{10}$
(3) $\frac{1}{2}$
(4) $\frac{3}{5}$
(5) Not enough information is given.

10. Angela has $4\frac{3}{4}$ yards of fabric. She plans to use $\frac{2}{3}$ of the fabric to make a skirt. How many yards of fabric will she use to make the skirt?

Mark your answer in the circles in the grid on your answer sheet.

11. A spool holds 25.8 meters of electrical wire. How many pieces measuring 1.7 meters in length can be cut from the spool?

(1) 14
(2) 15
(3) 16
(4) 24
(5) 43

Questions 12 and 13 refer to the following chart.

Net Profit from School Fundraisers

	Last year	This Year
Candy Sales	$2,400	$1,800
Jazz Concert	$6,400	$5,400
Spring Carnival	$7,200	$10,800
TOTAL FUNDS	$16,000	$18,000

12. Last year, what was the ratio (in lowest terms) of dollars raised from Candy Sales to dollars raised from the Spring Carnival?

    (1) 1:3
    (2) 1:6
    (3) 3:1
    (4) 3:20
    (5) Not enough information is given.

13. Matt plans to make a circle graph of this year's data. The section that represents the net profit from the Jazz Concert will be what fraction of the whole circle?

    (1) $\frac{2}{3}$
    (2) $\frac{3}{5}$
    (3) $\frac{3}{5}$
    (4) $\frac{3}{10}$
    (5) $\frac{1}{5}$

Part II

Directions: Choose the one best answer for each question. You MAY NOT use your calculator.

Question 14 refers to the following information.

A cereal company mixes the ingredients listed below to make a batch of Cranberry Crunch cereal. Recall that there are 16 ounces in a pound.

$2\frac{2}{3}$ pounds cranberries

$2\frac{1}{8}$ pounds almonds

2 pounds 2 ounces oat flakes

15 ounces wheat squares

14. Which is the best estimate for the total pounds of ingredients in the batch of cereal?

(1) 2
(2) 5
(3) 8
(4) 10
(5) 12

15. Mike has two boards. One is 2 yards long and the other is 24 inches long. Which is the correct expression to find the total length, in feet, of the boards?

(1) $(2 + 24) \times 12$
(2) $(2 + 24) \times 36$
(3) $(2 \times 36) + (24 \div 12)$
(4) $(2 \times 12) + (24 \times 12)$
(5) $(2 \times 3) + (24 \div 12)$

Questions 16 and 17 refer to the following graph.

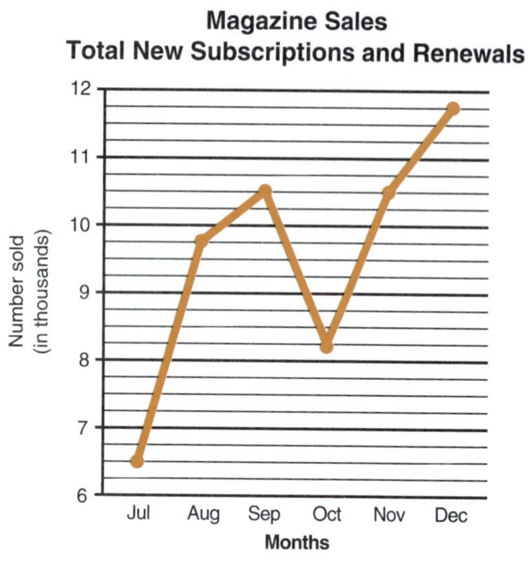

16. Altogether, how many new subscriptions and renewals were sold in November and December?

(1) 21,250
(2) 22,000
(3) 22,250
(4) 23,000
(5) 32,250

17. In August, what percent of the number sold were new subscriptions?

(1) 9.75%
(2) 15%
(3) 50%
(4) 75%
(5) Not enough information is given.

Question 18 refers to the following information.

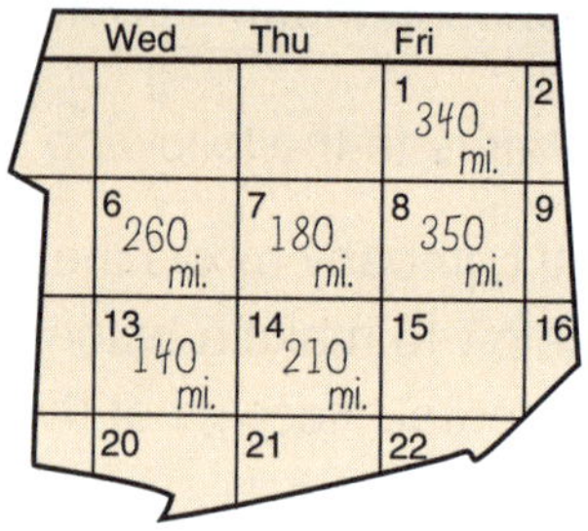

18. Nancy uses her own car to make deliveries on Wednesdays, Thursdays, and Fridays. She keeps a calendar in the car so that she can write down the number of miles she drives each day. What is the median number of miles she drove during the 6-day period shown on the calendar?

Mark your answer in the circles in the grid on your answer sheet.

Question 19 refers to the following figure.

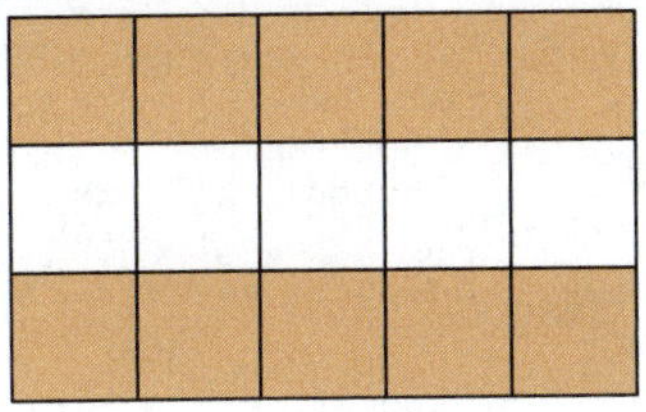

19. Which of the following is an equivalent fraction for the shaded portion of the figure?

(1) $\frac{1}{4}$
(2) $\frac{1}{3}$
(3) $\frac{1}{2}$
(4) $\frac{2}{3}$
(5) $\frac{3}{4}$

20. Samuel has a wooden dowel that is 3.5 meters in length. What is the correct expression to find the length of the dowel in centimeters?

(1) $3.5 \times 1{,}000$
(2) 3.5×100
(3) 3.5×10
(4) $3.5 \div 100$
(5) $3.5 \div 1{,}000$

Question 21 refers to the following diagram.

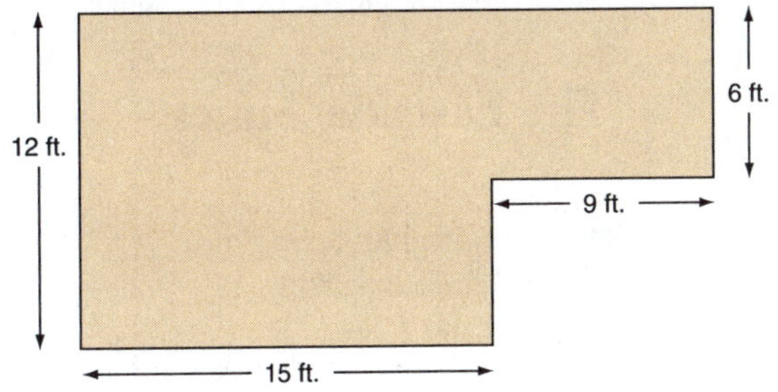

21. Denise plans to tile the room shown in the diagram. What is the area of the room in square feet?

(1) 42
(2) 72
(3) 234
(4) 288
(5) Not enough information is given.

Question 22 refers to the following information.

> **The Eastside Times**
>
> Classified Ad Special
> run your advertisement
> for 7 days!
>
> Only $10 for
> 3 lines of print.

22. Ron wants to place an ad in the Eastside Times to sell some furniture. His ad is 5 lines long. Which is the correct proportion to find the cost of placing Ron's ad in the newspaper for 7 days?

(1) $\frac{10}{5} = \frac{?}{3}$
(2) $\frac{10}{3} = \frac{5}{?}$
(3) $\frac{3}{7} = \frac{5}{?}$
(4) $\frac{3}{10} = \frac{7}{?}$
(5) $\frac{3}{10} = \frac{5}{?}$

23. Which number is the least common denominator for the following set of fractions?

$\frac{3}{4}$ $\frac{2}{5}$ $\frac{1}{6}$

(1) 12
(2) 24
(3) 30
(4) 60
(5) 120

Question 24 refers to the following graph.

School Board Election Results

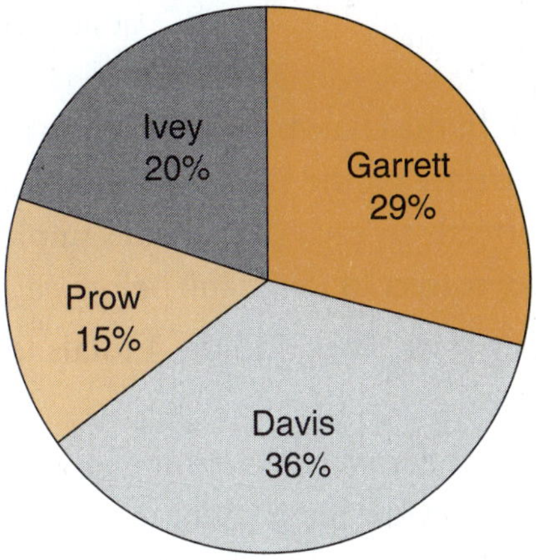

24. In an election, four candidates ran for the local school board. If 2,400 people voted in the election, how many votes went to Ivey, Garrett, and Davis?

    Mark your answer in the circles in the grid on your answer sheet.

25. A manufacturer pays 8.5 cents for a special fastener used in the company's circuit boards. How much will the manufacturer pay for 3,000 fasteners?

    (1) $2,550.00
    (2) $255.00
    (3) $25.50
    (4) $2.55
    (5) $0.26

Posttest Evaluation Chart

This chart will help you determine your strengths and weaknesses in basic mathematics skills.

Directions

Check your answers on pages 318–320. In the chart below, circle the number of each question that you answered correctly on the Posttest. Count the number of questions you answered correctly in each row. (For example, in the Whole Numbers row, write the number correct in the blank before *out of 6*.)

Complete this process for the remaining rows. Then add the 4 totals to get your Total Correct for the whole Posttest.

- If you answered fewer than 20 questions correctly, determine in which areas you need further practice. Go back and review the content in those areas. Page numbers for specific instruction appear in the left-hand column.
- If you answered 20 or more questions correctly, your instructor may decide that you are ready to go on to Steck-Vaughn's *GED Mathematics* book.

Skill Area	Computation	Data Analysis	Measurement	Total Correct
Whole Numbers (Pages 10–69)	1, 8	**6, 18**	15, 21	______ out of 6
Fractions (Pages 70–113)	4, 10, 19, 23	**13**	14	______ out of 6
Decimals (Pages 114–155)	2, 5, 11, 25	**16**	20	______ out of 6
Ratios, Proportions and Percents (Pages 156–227)	3, 12, 22	**7, 9, 17, 24**		______ out of 7

Total Correct for Posttest ______ out of 25

Boldfaced numbers indicate questions based on charts, diagrams, graphs, and maps.

Adding Whole Numbers

Practice this example of adding using your calculator.

Example	279 + 36	
Step 1	Estimate to know the approximate answer.	280 + 40 = 320
Step 2	First clear the display. Then enter each number. Press an operation key between the numbers. Here, press the plus sign between numbers to be added.	[AC] 279 [+] 36
Step 3	Press the equals sign at the end of the problem to get the total.	[=] 315
Step 4	Compare your calculator answer with your estimated answer. Based on your estimate, does your calculator answer make sense?	Yes, because 315 is close to 320.
Step 5	You can also check your exact answer by entering the numbers you added in a different order to see that you get the same total.	[AC] 36 [+] 279 [=] 315

It's Important to Know: When you add numbers on a calculator, you can enter the numbers in any order.
Example The numbers in the problem 20 + 8 + 43 could also be entered as 8 + 20 + 43 **or** 8 + 43 + 20 **or** 43 + 20 + 8, and so on.

PRACTICE

Follow the process shown above to solve these problems using a calculator.

1.

SOSA'S APPLIANCES INVOICE	
Oven	$499
Tax	$ 29
Delivery	$ 25
Installation	$ 75
TOTAL	

Estimated answer: ________

Exact answer: ________

Checked answer: ________

2.

McHENRY PLUMBING INVOICE	
Fixtures	$780
Pipes	$345
Faucets	$290
Connectors	$ 85
Labor	$320
TOTAL	

Estimated answer: ________

Exact answer: ________

Checked answer: ________

Check your answers on page 320.

Subtracting Whole Numbers

Practice this example of subtracting using your calculator.

Example 346 − 88

Step 1	Estimate to know the approximate answer.	350 − 90 = 260
Step 2	First clear the display. Then enter the number you are subtracting from. Use the minus sign between numbers. Then enter the number you are subtracting.	[AC] 346 [−] 88
Step 3	Press the equals sign at the end of the problem to find the answer.	[=] 258
Step 4	Compare your calculator answer with your estimated answer. Based on your estimate, does your calculator answer make sense?	Yes, because 258 is close to 260.
Step 5	To check your exact answer add your calculator answer to the number you subtracted. You should get the amount you started with.	[AC] 258 [+] 88 [=] 346

It's Important to Know: When you subtract numbers on a calculator, you must enter the numbers in order of the whole amount minus the number being subtracted.
Example The numbers in the problem 547 − 95 must be entered in that order: first 547, then the minus sign, then 95.

PRACTICE

Follow the process shown above to solve these problems using a calculator.

	1st Half	Final
Bluebirds	39	89
Redwings	47	116

1. By how many points were the Redwings leading at the end of the first half?
Estimated answer: ____________
Exact answer: ____________
Checked answer: ____________

2. By how many points did the Redwings beat the Bluebirds?
Estimated answer: ____________
Exact answer: ____________
Checked answer: ____________

3. 14,899 people attended this game. Paid attendance was 13,968. How many people didn't have to pay for their tickets?
Estimated answer: ____________
Exact answer: ____________
Checked answer: ____________

 Check your answers on pages 320–321.

Multiplying Whole Numbers

Practice this example of multiplying using your calculator.

Example 83×17

Step 1	Estimate to know the approximate answer.	$80 \times 20 = 1{,}600$
Step 2	First clear the display. Then enter each number. Use the multiplication sign between numbers to be multiplied.	AC 83 × 17
Step 3	Press the equals sign at the end of the problem to get the product.	= 1411
Step 4	Compare your calculator answer with your estimated answer. Based on your estimate, does your calculator answer make sense?	Yes, because 1,411 is close to 1,600.
Step 5	You can also check your exact answer by entering the numbers you multiplied in a different order to see that you get the same product.	AC 17 × 83 = 1411

It's Important to Know: When you multiply numbers on a calculator, you can enter the numbers in any order.
Example The numbers in the problem $2 \times 8 \times 4$ could also be entered as $8 \times 2 \times 4$ **or** $8 \times 4 \times 2$ **or** $4 \times 2 \times 8$, and so on.

PRACTICE

Follow the process shown above to solve these problems using a calculator.

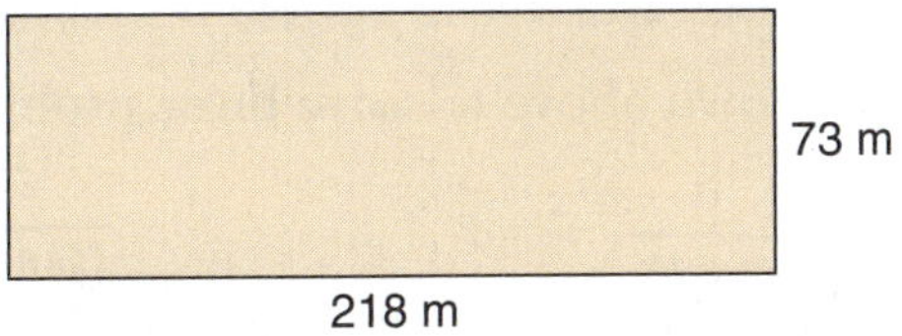

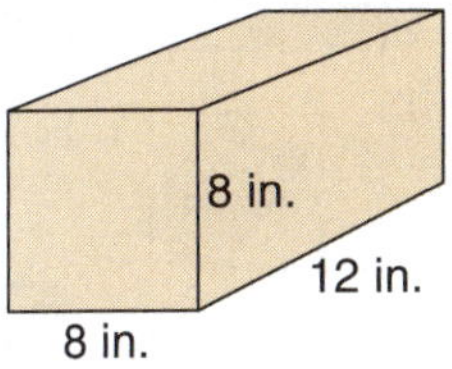

1. What is the area of the rectangle?
   Estimated answer: ____________
   Exact answer: ____________
   Checked answer: ____________

2. What is the volume of the box?
   Estimated answer: ____________
   Exact answer: ____________
   Checked answer: ____________

Check your answers on page 321.

Dividing Whole Numbers

Practice this example of dividing using your calculator.

Example 256 ÷ 4

Step 1 Estimate to know the approximate answer. — 250 ÷ 5 = 50

Step 2 First clear the display. Then enter the number to be divided. Use the division sign between numbers. Then enter the number you are dividing by. — [AC] 256 [÷] 4

Step 3 Press the equals sign at the end of the problem to get the answer. — [=] 64

Step 4 Compare your calculator answer with your estimated answer. Based on your estimate, does your calculator answer make sense? — Yes, because 64 is close to 50.

Step 5 You can also check your exact answer by multiplying your calculator answer by the number you divided by. You should get the same amount you started with. — [AC] 64 [×] 4 [=] 256

> **It's Important to Know:** When you divide numbers on a calculator, you must enter the numbers in order of the whole amount then the number you are dividing by.
>
> **Example** The numbers in the problem 96 ÷ 6 must be entered in that order: first 96, then the division sign, then 6.

PRACTICE

Follow the process shown above to solve these problems using a calculator.

1. How much will each lottery each winner receive?

   Estimated answer: ____________

   Exact answer: ____________

   Checked answer: ____________

2. How much grant money will each of the 6 counties receive?

   Estimated answer: ____________

   Exact answer: ____________

   Checked answer: ____________

 Check your answers on page 321.

Exponents and Square Roots

Calculators are very useful when you are solving problems with exponents and square roots. For example, to find the value of 32^2, you need to multiply 32×32. To find the value of 8^3, you need to multiply $8 \times 8 \times 8$. You can solve both of these problems quickly using the [x^2] and [x^y] keys. The square root function is located above the [x^2] key on most scientific calculators.

Example	Find the value of 32^2.	
Step 1	Estimate an approximate answer.	$30 \times 30 = 900$
Step 2	Use the calculator. First clear the display. Then enter the number and press x^2.	[AC] 32 [x^2] 1024
Step 3	Compare your calculator answer with the estimated answer. Does the answer make sense?	Yes, 900 is reasonably close to 1,024.

Example	Find the value of 8^3.	
Step 1	Estimate an approximate answer. Find the cubes of one number greater than 8 and one that is lower than 8. For example, find 10^3 and 5^3.	$10 \times 10 \times 10 = 1{,}000$ $5 \times 5 \times 5 = 125$
Step 2	Use the calculator. First clear the display. Enter the number, press x^y, and enter the exponent.	[AC] 8 [x^y] 3
Step 3	Press the equals sign.	[=] 512
Step 4	Compare your calculator answer with the estimated answer. Does the answer make sense?	Yes, the answer should be less than 1,000 and more than 125.

Example	Find the value of $\sqrt{75}$.	
Step 1	Estimate an approximate answer. Think: $\sqrt{64} = 8$ and $\sqrt{81} = 9$.	The answer must be between 8 and 9.
Step 2	Use the calculator. Enter the number. Then press [SHIFT] and the square root key.	[AC] 75 [SHIFT] [$\sqrt{\ }$] 8.660254038
Step 3	Compare your calculator answer with the estimate. Does the answer make sense?	Yes, because 8.66 is between 8 and 9.

PRACTICE

Follow the steps shown above to solve these problems using a calculator.

1. $5^5 =$ ______
2. $2.5^2 =$ ______
3. $\sqrt{625} =$ ______
4. $\sqrt{289} =$ ______

Check your answers on page 321.

Adding Decimals

Practice this example of adding using your calculator.

Example 45.67 + 24.069

Step 1 Estimate to know the approximate answer. — 45 + 25 = 70

Step 2 First clear the display. Then enter each number. Use the decimal point key to enter the decimal point at the correct place in each decimal number. Use the plus sign between numbers to be added. — [AC] 45 [.] 67 [+] 24 [.] 069

Step 3 Press the equals sign at the end of the problem to get the total. — [=] 69.739

Step 4 Compare your calculator answer with your estimated answer. Based on your estimate, does your calculator answer make sense? — Yes, because 69.739 is close to 70.

Step 5 You can also check your exact answer by entering the numbers you added in a different order to see that you get the same total. — [AC] 24 [.] 069 [+] 45 [.] 67 [=] 69.739

It's Important to Know: Whether you're adding whole numbers or decimals on a calculator, you can enter the numbers in any order.

Example The problem 20.4 + 8.02 + 43.9 could also be entered as 8.02 + 20.4 + 43.9 **or** 8.02 + 43.9 + 20.4 **or** 43.9 + 20.4 + 8.02 and so on.

PRACTICE

Follow the process shown above to solve these problems using a calculator.

1.

Oil Change — Every 3,500 Miles

Oil Changed at: 16,632.8 Miles

Next Oil Change: ______ Miles

Estimated answer: ____________

Exact answer: ____________

Checked answer: ____________

2.

INVOICE

Filter	$5.88
4 qt. oil	4.25
O-ring	0.60
Parts Total	

Estimated answer: ____________

Exact answer: ____________

Checked answer: ____________

 Check your answers on page 321.

Subtracting Decimals

Practice this example of subtracting using your calculator.

Example	36.78 − 18.9	
Step 1	Estimate to know the approximate answer.	40 − 20 = 20
Step 2	First clear the display. Then enter the number you are subtracting from. Use the decimal point key to enter the decimal point at the correct place in each decimal number. Use the minus sign between numbers to be subtracted. Now enter the number you are subtracting.	[AC] 3 6 [.] 7 8 [−] 1 8 [.] 9
Step 3	Press the equals sign at the end of the problem to find the answer.	[=] 17.88
Step 4	Compare your calculator answer with your estimated answer. Based on your estimate, does your calculator answer make sense?	Yes, because 17.88 is close to 20.
Step 5	You can also check your exact answer by adding your calculator answer to the number you subtracted. You should get the total amount you started with.	[AC] 1 7 [.] 8 8 [+] 1 8 [.] 9 [=] 36.78

It's Important to Know: Whether you're subtracting whole numbers or decimals on a calculator, you must enter the numbers in order of the whole amount minus the number being subtracted.

Example The numbers in the problem 54.7−9.5 must be entered in that order: first 54.7, then the minus sign, then 9.5.

PRACTICE

Follow the process shown above to solve these problems using a calculator.

May 10 — Odometer Reading	21,383.7
May 15 — Odometer Reading	21,734.6
May 20 — Odometer Reading	21,862.2

Start With:	$20.00
Spend:	$11.75 (Gasoline)
	$1.39 (Large Fries)
	$1.01 (Soda)

1. How many miles were driven between May 10 and May 15?

   Estimated answer: ____________

   Exact answer: ____________

   Checked answer: ____________

2. How much change is left?

   Estimated answer: ____________

   Exact answer: ____________

   Checked answer: ____________

Check your answers on page 321.

Multiplying Decimals

Practice this example of multiplying using your calculator.

Example 7.3 × 8.17

Step 1 Estimate to know the approximate answer. — 7 × 8 = 56

Step 2 First clear the display. Then enter each number. Use the decimal point key to enter the decimal point at the correct place in each decimal number. Use the multiplication sign between numbers to be multiplied. — [AC] 7 [·] 3 [×] 8 [·] 17

Step 3 Press the equals sign at the end of the problem to get the product. — [=] 59.641

Step 4 Compare your calculator answer with your estimated answer. Based on your estimate, does your calculator answer make sense? — Yes, because 59.641 is close to 56.

Step 5 You can also check your exact answer by entering the numbers you multiplied in a different order to see that you get the same product. — [AC] 8 [·] 17 [×] 7 [·] 3 [=] 59.641

TIP

When entering a decimal such as 0.8 or 0.56 on a calculator, you do not need to type in the zero that is before the decimal point.

It's Important to Know: Whether you're multiplying whole numbers or decimals on a calculator, you can enter the numbers in any order.

Example The numbers in the problem 2.67 × 8.1 × 0.04 could also be entered as 8.1 × 2.67 × 0.04 **or** 8.1 × 0.04 × 2.67 **or** 0.04 × 2.67 × 8.1 and so on.

PRACTICE

Follow the process shown above to solve these problems using a calculator.

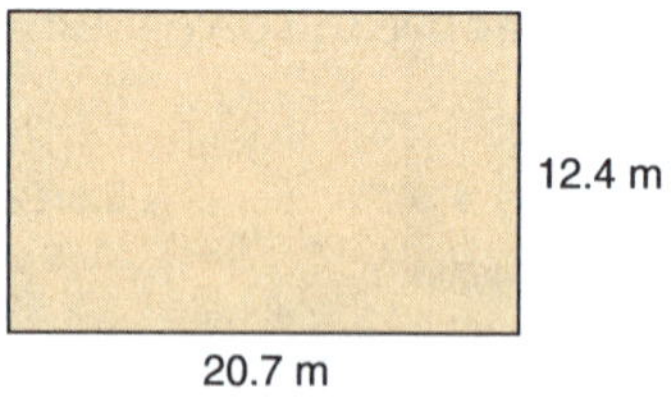

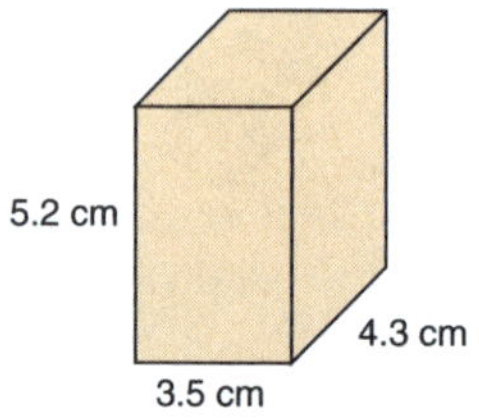

1. What is the area of the rectangle shown here?
   Estimated answer: ____________
   Exact answer: ____________
   Checked answer: ____________

2. What is the volume of the box?
   Estimated answer: ____________
   Exact answer: ____________
   Checked answer: ____________

 Check your answers on page 321.

Dividing Decimals

Practice this example of dividing using your calculator.

Example	25.6 ÷ 0.8	
Step 1	Estimate to know the approximate answer.	26 ÷ 1 = 26
Step 2	First, clear the display. Then enter each number. Use the division sign between numbers to be divided.	[AC] 25 [.] 6 [÷] [.] 8
Step 3	Press the equals sign at the end of the problem to get the answer.	[=] 32
Step 4	Compare your calculator answer with your estimated answer. Based on your estimate, does your calculator answer make sense?	Yes, because 32 is close to 26.
Step 5	You can also check your exact answer by multiplying your calculator answer by the number you divided by. You should get the same amount you started with.	[AC] 32 [×] [.] 8 [=] 25.6

It's Important to Know: Whether you're dividing whole numbers or decimals on a calculator, you must enter the numbers in order of the whole amount then the number you are dividing by.

Example The numbers in the problem 43.28 ÷ 0.4 must be entered in that order: first 43.28, then the division sign, then 0.4.

PRACTICE

Follow the process shown above to solve these problems using a calculator.

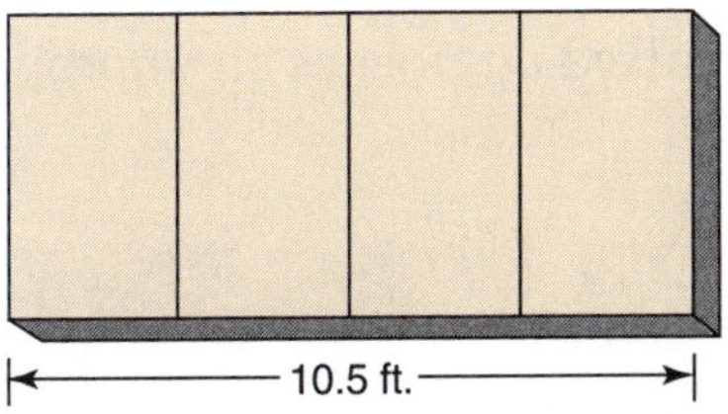

RECEIPT:	
Bakery	$15.83
Plates	$ 3.25
Napkins	$ 2.15
Subtotal	$21.23
Tax	$ 1.27
Total	$22.50

1. If the board were cut into 4 equal pieces, how long would each piece be?

   Estimated answer: ____________

   Exact answer: ____________

   Checked answer: ____________

2. If three people split the cost of these birthday items, how much would each person pay?

   Estimated answer: ____________

   Exact answer: ____________

   Checked answer: ____________

Check your answers on page 321.

Solving Fraction Problems

Scientific calculators have the ability to perform operations with fractions without first changing them to decimals. You can enter fractions and mixed numbers on a scientific calculator using the key marked (a b/c). Note the *d/c* above the (a b/c) key. Press (a b/c) after the whole number and again after the numerator. The *d/c* function is used to change a mixed number to an improper fraction. Access this function by pressing the (SHIFT) key.

Example $3\frac{5}{8} + \frac{3}{4}$

Step 1 Estimate an approximate answer.

$3\frac{5}{8}$ is almost 4 and $\frac{3}{4}$ is almost 1. Add: 4 + 1 = 5.

Step 2 Use the calculator. First clear the display. Then enter the problem.

(AC) 3 (a b/c) 5 (a b/c) 8 (+) 3 (a b/c) 4

Step 3 Press the equals key. The display means $4\frac{3}{8}$. The ⌟ symbol separates the parts of the mixed number.

4⌟3⌟8.

Step 4 If a decimal answer is needed, press (a b/c) again. The fraction will be displayed as a decimal.

(a b/c) 4.375

Step 5 Check your work. Compare your calculator answer with the estimated answer. Does the answer make sense?

Yes, $4\frac{3}{8}$, or 4.375, is close to 5.

If you are answering a grid item, use the *d/c* function to change a mixed number to an improper fraction.

Example Change $5\frac{2}{3}$ to an improper fraction.

Step 1 Use the fraction key to enter the mixed number.

5 (a b/c) 2 (a b/c) 3

Step 2 Use the *d/c* function. Do not press the equals key. The display will immediately show the mixed number as an improper fraction.

(SHIFT) (d/c a b/c)

The display reads 17⌟3., which means $\frac{17}{3}$.

PRACTICE

Use your calculator to solve these problems.

1. $\frac{5}{6} + 1\frac{1}{2} + \frac{1}{3} =$ ______
2. $7\frac{3}{5} \times 3\frac{1}{6} =$ ______
3. $10\frac{2}{9} - 3\frac{2}{3} =$ ______
4. $3\frac{4}{5} \div \frac{1}{4} =$ ______

Change each mixed number to an improper fraction.

5. $8\frac{2}{5} =$ ______
6. $12\frac{3}{4} =$ ______
7. $3\frac{4}{7} =$ ______
8. $6\frac{11}{12} =$ ______

 Check your answers on page 321.

Converting Fractions, Decimals, and Percents

Calculators are tools that help simplify converting between fractions, decimals, and percents. You have already seen how to convert fractions to decimals using a scientific calculator. You can also change a fraction to a decimal using division.

Example Change $\frac{3}{4}$ to a decimal.

Step 1 First, clear the display. Then enter the numerator (3). Divide the numerator by the denominator (4). Use the division sign between numbers to be divided. AC 3 ÷ 4

Step 2 Press the equals sign at the end of the problem to get the answer. = .75

You can then change 0.75 to a percent. Multiply by 100 and add a percent sign: 0.75 = 75%.

PRACTICE

Complete the chart below to show equivalent fractions, decimals, and percents. Use your calculator and the process shown above when needed.

Fraction	Decimal	Percent
$\frac{1}{10}$		10%
$\frac{1}{8}$		
	.2	
		25%
$\frac{1}{3}$	.33	$33\frac{1}{3}\%$
$\frac{2}{5}$		
		50%
	.6	
$\frac{2}{3}$	.67	$66\frac{2}{3}\%$
		75%
$\frac{4}{5}$		
	.9	
1	1	100%

Check your answers on page 321.

Solving Percent Problems

Use your calculator to practice finding the part in this percent example.

Example What is 40% of 60?

Step 1	First, clear the display. Then enter the whole (60). Press the multiplication sign.	AC 60 ×
Step 2	Enter the number that represents the percent (40) and press SHIFT % (=). Your answer will appear in the display.	40 SHIFT % (=) 24. 40% of 60 is **24**

Use your calculator to practice finding the percent in this percent example.

Example What percent is 50 of 200?

Step 1	First, clear the display. Then enter the part (50). Press the division sign.	AC 50 ÷
Step 2	Enter the whole, or base (200). Then press SHIFT and % (=).	200 SHIFT % (=) 25, which means 25%
Step 3	The answer is automatically displayed as a percent. Write the answer with a percent sign.	50 is **25%** of 200

Use your calculator to practice finding the whole in this percent example.

Example $30 is 25% of what amount?

Step 1	First, clear the display. Then enter the part (30). Press the division sign.	AC 30 ÷
Step 2	Enter the number part of the percent (25). Then press SHIFT and % (=).	25 SHIFT % (=) 120
Step 3	Write the answer with a dollar sign.	$30 is 25% of **$120**

PRACTICE

Follow the examples shown above to solve these problems using a calculator.

1. Purchase $35.00
   Tax at 6.25%

   What is the dollar amount of the tax on this purchase? (Hint: Round the amount of the tax to the nearest cent.)

   ______

2. 5 defective parts
   500 parts tested for defects

   Based on the number of defective parts and the number of parts tested, what percent of the parts are defective?

   ______

 Check your answers on page 322.

Solving Multi-step Problems

Apply what you have learned in this handbook to problems that require more than one step.

Practice this example using your calculator.

Example A square plot of land measuring 24 feet on each side is to be fenced in. Two gates are going to be installed: one 3-foot gate and one 6-foot gate. How much fencing is needed?

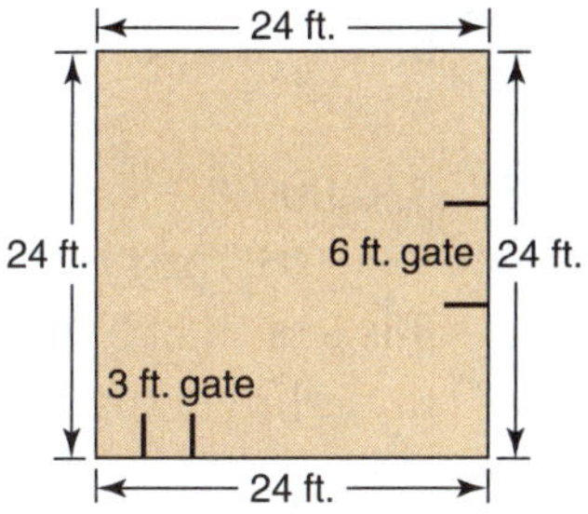

Step 1 Estimate to know the approximate answer.

25 × 4 = 100
100 − 10 = 90

Step 2 First, clear the display. Then enter the length of one side. Multiply that number by the number of sides to be fenced in (4). This will find the length of fencing needed <u>without</u> the two gates. Press the equals sign at the end of the problem to get the product.

AC 24 × 4 = 96

Step 3 Subtract the measurements of the gates to find the amount of fencing needed. (Use mental math to get this amount: 3 + 6 = 9.)

− 9 = 87

Step 4 Compare your calculator answer with your estimated answer. Based on your estimate, does your calculator answer make sense?

Yes, because 87 is close to 90.

PRACTICE

Find the subtotal of the checks. Then find the amount of the deposit after subtracting the amount for *cash received.*

1.

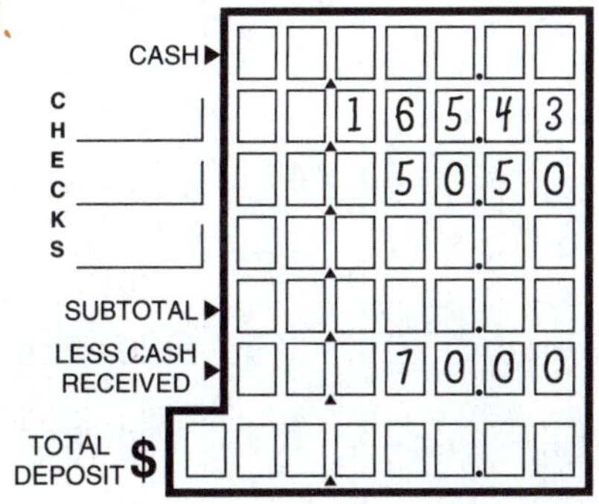

2.

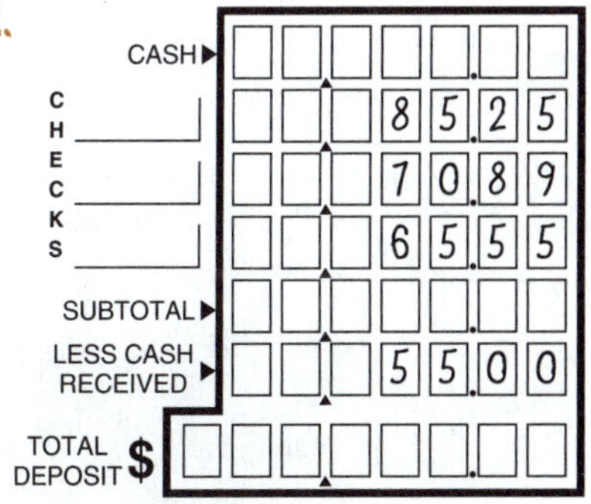

Check your answers on page 322.

ANSWERS AND EXPLANATIONS

PRETEST

PAGES 3–8

1. **three hundred thousand**
2. **seven thousandths**
3. **46,0̲23 < 46,2̲03** Zero is less than 2.
4. **0.76 > 0.456** 0.7̲60 is greater than 0.4̲56.
5. **530,000**
   53̲4,103 The number to the right of the 3 is less than 5. Do not change the underlined digit.
6. **3.73**
   3.72̲5 The number to the right of the 2 is 5; add 1 to the underlined digit.
7. **$2\frac{1}{4}$** Each figure is divided into 4 equal parts. Two figures are completely shaded, and 1 of the 4 parts of the last figure is shaded.
8. **272**
$$\begin{array}{r} \overset{5}{\cancel{6}}\overset{9}{\cancel{0}}\overset{10}{\cancel{0}} \\ -328 \\ \hline 272 \end{array}$$
9. **2.854**
$$\begin{array}{r} \overset{5}{\cancel{6}}.\overset{16}{7}\overset{9}{\cancel{0}}\overset{10}{\cancel{0}} \\ -3.846 \\ \hline 2.854 \end{array}$$
10. **451,744**
$$\begin{array}{r} 743 \\ \times 608 \\ \hline 5\,944 \\ +445\,80 \\ \hline 451{,}744 \end{array}$$
11. **\$3.07**
$$\begin{array}{r} \$3.07 \\ 4\overline{)\$12.28} \\ -\;12 \\ \hline 0\;28 \\ -\;\;28 \\ \hline 0 \end{array}$$
12. **$7\frac{5}{24}$**
$$\begin{array}{rcl} 4\frac{7}{8} & = & 4\frac{21}{24} \\ +2\frac{1}{3} & = & 2\frac{8}{24} \\ \hline & & 6\frac{29}{24} = 6 + 1\frac{5}{24} = 7\frac{5}{24} \end{array}$$
13. **$1\frac{11}{15}$**
$$\begin{array}{rcccl} 6\frac{2}{5} & = & 6\frac{6}{15} & = & \overset{5}{\cancel{6}}\frac{\overset{21}{\cancel{6}}}{15} \\ -4\frac{2}{3} & = & 4\frac{10}{15} & = & 4\frac{10}{15} \\ \hline & & & & 1\frac{11}{15} \end{array}$$
14. **20**
$$\begin{array}{l} 11 + (2 + 1) \times 6 \div 2 \\ 11 + \quad 3 \quad \times 6 \div 2 \\ 11 + \qquad\quad 18 \div 2 \\ 11 + \qquad\qquad 9 \\ \quad 20 \end{array}$$
15. **475**
$$\begin{array}{r} {\scriptstyle 1\,1} \\ 42 \\ 376 \\ +\;57 \\ \hline 475 \end{array}$$
16. **38.201**
$$\begin{array}{r} {\scriptstyle 1\;1\;\;1} \\ 3.270 \\ 26.400 \\ +\;8.531 \\ \hline 38.201 \end{array}$$
17. **\$2.09**
$$\begin{array}{r} \$\overset{4}{\cancel{3}}.\overset{9}{\cancel{0}}\overset{18}{\cancel{8}} \\ -\;2.99 \\ \hline \$2.09 \end{array}$$
18. **0.0162**
$$\begin{array}{r} 0.054 \\ \times\quad 0.3 \\ \hline 0.0162 \end{array}$$
19. **$16\frac{1}{2}$** $4\frac{2}{5} \times 3\frac{3}{4} = \frac{\overset{11}{\cancel{22}}}{\underset{1}{\cancel{5}}} \times \frac{\overset{3}{\cancel{15}}}{\underset{2}{\cancel{4}}} = \frac{33}{2} = 16\frac{1}{2}$
20. **182 r14**
$$\begin{array}{r} 182\text{ r}14 \\ 42\overline{)7{,}658}\phantom{\text{ r}14} \\ -4\,2\phantom{\text{ r}14} \\ \hline 3\,45\phantom{\text{ r}14} \\ -3\,36\phantom{\text{ r}14} \\ \hline 98\phantom{\text{ r}14} \\ -84\phantom{\text{ r}14} \\ \hline 14\phantom{\text{ r}14} \end{array}$$
21. **520**
$$\begin{array}{r} 5\,20 \\ 0.06.\overline{)31.20.} \\ -30 \\ \hline 1\,2 \\ -1\,2 \\ \hline 00 \end{array}$$
22. **$\frac{4}{5}$** $3\frac{1}{2} \div 4\frac{3}{8} = \frac{7}{2} \div \frac{35}{8} = \frac{\overset{1}{\cancel{7}}}{\underset{1}{\cancel{2}}} \times \frac{\overset{4}{\cancel{8}}}{\underset{5}{\cancel{35}}} = \frac{4}{5}$
23. **49** $7^2 = 7 \times 7 = 49$
24. **3.75** $375\% \div 100 = 3.75. = 3.75$
25. **7%** $0.07 \times 100 = 0.07. = 7\%$
26. **$\frac{4}{5}$** $80\% = \frac{80}{100} = \frac{80 \div 20}{100 \div 20} = \frac{4}{5}$

27. **75%**

$$\begin{array}{r} 0.75 \\ 4\overline{)3.00} \\ -2\,8 \\ \hline 20 \\ -20 \\ \hline 0 \end{array}$$

$0.75 = 75\%$

28. **12**

part = base × rate

$8\% = 0.08$

$$\begin{array}{r} 150 \\ \times 0.08 \\ \hline 12.00 \end{array}$$

29. **25%**

rate = part ÷ base

$$\begin{array}{r} 0.25 \\ \$72\overline{)\$18.00} \\ -14\,4 \\ \hline 3\,60 \\ -3\,60 \\ \hline 0 \end{array}$$

$0.25 = 25\%$

30. **36**

base = part ÷ rate

$150\% = 1.5$

$$\begin{array}{r} 36. \\ 1.5.\overline{)54.0.} \\ -45 \\ \hline 9\,0 \\ -9\,0 \\ \hline 0 \end{array}$$

31. **14** $8 \times 35 = 280$; $280 \div 20 = 14$

32. **(3) 144 − 48** Subtract the number of gallons of interior paint sold (48) from the number of gallons of interior paint in stock (144). You do not need the number of gallons of exterior paint.

33. **(2) 36** Two sides measure 12 feet, and two sides measure 6 feet. Add all four measurements to find the perimeter.
$12 + 12 + 6 + 6 = 36$

34. **(3) 20^2** The room is a square because all four sides have the same length. Use the formula $A = s^2$, and substitute 20 for s.

35. **(3) 170** Round the three to the nearest ten, and add to estimate the total.

79	rounds to	80
36	rounds to	40
+52	rounds to	+50
		170

36. **(4) 16 × 8** To find the total amount, multiply the number of packages (16) by the number of buns in each package (8).

37. **(3) $\frac{1}{2}$ or 50%**

$$\frac{\text{number of sections with 2}}{\text{total number of sections}} = \frac{3}{6} = \frac{1}{2} \text{ or } 50\%$$

38. **(2) $\frac{3}{4}$** To make one third of the recipe, divide the amount of broth ($2\frac{1}{4}$ cups) by 3.

$$2\tfrac{1}{4} \div 3 = \tfrac{9}{4} \div \tfrac{3}{1} = \tfrac{\overset{3}{\cancel{9}}}{4} \times \tfrac{1}{\underset{1}{\cancel{3}}} = \tfrac{3}{4}$$

39. **(2) $\frac{3}{\$1.56} = \frac{5}{?}$** Both ratios are in the same order. The first ratio means that 3 pounds cost $1.56. The second ratio means that 5 pounds cost an unknown amount.

40. **(4) 40 × 15%** Multiply the base (40) times the rate (15%) to find the part.

41. **(4) 200** Multiply by 1,000 to convert kilograms to grams. Move decimal point 3 places to the right to multiply by 1,000 (3 zeros). $0.2 \times 1{,}000 = 0.200. = 200$

42. **(4) $2.48 × 4** Find the total cost by multiplying the unit cost ($2.48) by the number of pounds (4).

43. **(2) 6%** Divide the part ($25.20) by the base ($420.00) to find the rate.

$$\begin{array}{r} 0.06 \\ \$420\overline{)\$25.20} \\ -25\,20 \\ \hline 0 \end{array}$$

$0.06 = 6\%$

44. **(3) $18.60 ÷ 30%** Divide the part ($18.60) by the rate (30%) to find the base.

45. **(4) 60 ÷ 12** The area of the rectangle and the length of one side are given. Divide the area by the length to find the width.

46. **(2) 4%** Subtract Abdul's current wage ($5.25) from his new wage ($5.46) to find the amount of change. Then divide by the original amount, his current wage.

$$\begin{array}{r} \$5.46 \\ -\ 5.25 \\ \hline \$0.21 \end{array} \qquad \begin{array}{r} 0.04 \\ \$5.25.\overline{)\$0.21.00} \\ -21.00 \\ \hline 0 \end{array}$$

$0.04 = 4\%$

47. **1,920**

$V = l \times w \times h$
$= 10 \times 12 \times 16$
$= 1{,}920$

48. **18** Round each distance and subtract.

$$\begin{array}{rcr} 36\frac{9}{10} & \text{rounds to} & 37 \\ -19\frac{2}{5} & \text{rounds to} & -19 \\ \hline & & 18 \end{array}$$

49. **345** Multiply the number of gallons of gas (15) by the miles car B can travel on the highway per gallon of gas (23).

$$\begin{array}{r} 23 \\ \times 15 \\ \hline 115 \\ +23 \\ \hline 345 \end{array}$$

50. **$6\frac{1}{4}$ feet** Since 1 foot = 12 inches, divide 75 by 12. Then write the remainder 3 over 12 and reduce to lowest terms.

$$\begin{array}{r} 6 \\ 12\overline{)75} \\ -72 \\ \hline 3 \end{array} \quad 6 \text{ feet } 3 \text{ inches} = 6\tfrac{3}{12} = 6\tfrac{1}{4}$$

51. **$3\frac{3}{4}$** Since 43 is close to 45, think of 8:43 as $8\frac{3}{4}$. Since 36 is close to 30, think of 12:36 as $12\frac{1}{2}$. Then subtract.

$$\begin{array}{rcrcr} 12\frac{1}{2} & = & 12\frac{2}{4} & = & \overset{11}{\cancel{12}}\overset{6}{\cancel{\frac{2}{4}}} \\ -\ 8\frac{3}{4} & = & 8\frac{3}{4} & = & 8\frac{3}{4} \\ \hline & & & & 3\frac{3}{4} \end{array}$$

52. **$\frac{3}{10}$ or 30%**

$$\frac{\text{number of cards with 5 or higher}}{\text{number of cards in hand}} =$$

$$\frac{3}{10} = 30\%$$

53. **$4.04** Add the cost of the statue and the card to find the subtotal. Add the tax to find the total. Then subtract the total from the amount of cash the customer gives Tawanna to find the amount of change.

$$\begin{array}{r} \overset{1}{\$32.50} \\ +\ \ 1.75 \\ \hline \$34.25 \end{array} \qquad \begin{array}{r} \$34.25 \\ +\ \ 1.71 \\ \hline \$35.96 \end{array} \qquad \begin{array}{r} \$\overset{3}{\cancel{4}}\overset{9}{\cancel{0}}.\overset{9}{\cancel{0}}\overset{10}{\cancel{0}} \\ -\ 35.96 \\ \hline \$4.04 \end{array}$$

54. **1,500** The bar for Lincoln County ends between the *3* and *4* marks and represents 3,500 customers. The bar for Hamilton County reaches the *2* mark and represents 2,000 customers. Subtract to find the difference.

$$\begin{array}{r} 3{,}500 \\ -2{,}000 \\ \hline 1{,}500 \end{array}$$

55. **$\frac{9}{16}, \frac{5}{8}, \frac{3}{4}$** Convert all to fractions with denominators of 16: $\frac{5}{8} = \frac{10}{16}$ and $\frac{3}{4} = \frac{12}{16}$. Compare the numerators to order the fractions from least to greatest: $\frac{9}{16}, \frac{10}{16}, \frac{12}{16}$ or $\frac{9}{16}, \frac{5}{8}, \frac{3}{4}$.

56. **25** Find the total of the numbers in the set: 37 + 18 + 21 + 24 = 100
Then divide by the number of items in the set: 100 ÷ 4 = 25

57. **22.5** Arrange the numbers in order. Then find the middle numbers: 18, <u>21</u>, <u>24</u>, 37
Then find the mean of the middle numbers: 21 + 24 = 45; 45 ÷ 2 = 22.5

58. **$\frac{5}{9}$**

$$\frac{\text{20 in factory}}{\text{36 employees}} = \frac{20 \div 4}{36 \div 4} = \frac{5}{9}$$

59. **$750**

$$\begin{array}{lllllll} \textit{interest} & = & \textit{principal} & \times & \textit{rate} & \times & \textit{time} \\ & = & \$3{,}000 & \times & 0.05 & \times & 5 \\ & = & \$750 & & & & \end{array}$$

60. **$^{-}5$ degrees** Start at $^{+}2$ on the number line. Count 7 units to the left.

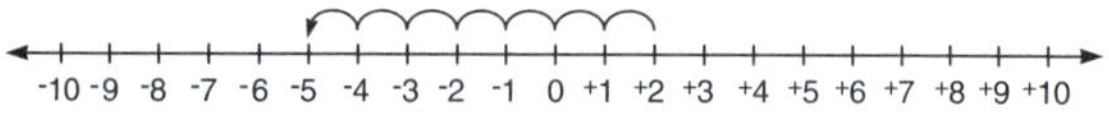

UNIT 1: WHOLE NUMBERS

LESSON 1

PAGE 12

2. **ones**
3. **thousands**
4. **hundreds**
5. **millions**
6. **hundreds**
7. **hundred thousands**
8. **ten millions**
9. **thousands**
10. **hundreds**

PAGE 13

2. **forty-three thousand, eighteen**
3. **one hundred fifteen thousand, two hundred**
4. **five million, four hundred thousand, twelve**
6. **250,911**
7. **12,016**
8. **9,014,560**

PAGE 14

2. **38,000 < 38,500** 0 is less than 5.
3. **179 = 179** The numbers are the same.
4. **210,580 > 210,480** 5 is greater than 4.
5. **1,000,000 < 10,000,000** 10,000,000 has 8 digits; 1,000,000 has only 7 digits.
6. **496 < 4,690** 496 has fewer digits than 4,690.
7. **13,415 = 13,415** The numbers are the same.
8. **802,165 < 803,980** 2 is less than 3.
9. **5,000 < 50,000** 5,000 has fewer digits than 50,000.
10. **1,345 < 1,435** 3 is less than 4.
11. **10,334 = 10,334** The numbers are the same.
12. **479 > 476** 9 is greater than 6.
13. **340,635 < 340,835** 6 is less than 8.
14. **5,010 > 5,001** 1 is greater than 0.
15. **682,489 = 682,489** The numbers are the same.
16. **4,609 < 4,906** 6 is less than 9.

PAGE 15

2. **1,700** 1,723 Since 2 is less than 5, do not change the underlined digit.
3. **7,000** 6,509 The number to the right of 6 is 5; add 1 to the underlined digit.
4. **900** 861 Since 6 is greater than 5, add 1 to the underlined digit.
5. **20,000** 19,580 The number to the right of 9 is 5; add 1 to the underlined digit. The 9 becomes a 10. Write a zero and add 1 to the next place value to the left.
6. **210,000** 209,320 Since 9 is greater than 5, add 1 to the underlined digit.
7. **64,000** 64,299 Since 2 is less than 5, do not change the underlined digit.
8. **5,300,000** 5,256,000 The number to the right of 2 is 5; add 1 to the underlined digit.

PAGE 16

1. **Estimate: $50** $10 + $15 + $25
   Exact: $49
2. **Estimate: 100 feet**
   60 feet + 10 feet + 30 feet
   Exact: 99 feet
3. **$155** $20 + $120 + $15
4. **$600** 18 + 22 = 40 40 × $15 = $600

PAGE 17

2. **$^{-}6$ yards** $0 + (^{+}4) + (^{-}10) = {^{-}6}$
3. **$^{+}$$8** 3 + 5 = 8
4. **$^{-}$$11** $0 + (-8) + (^{-}3) = {^{-}11}$
5. **2 points** 12 − 10 = 2
6. **$^{+}3\%$** $^{-}7 + 10 = 3$
7. **$^{-}1$ degree** $^{-}6 + 5 = {^{-}1}$
8. **$^{+}2$ yards** $^{-}4 + 6 = 2$

PAGE 19

2. **1653**
3. **10241**
4. **18.92**
6. [AC] **73 × 46 = 3358**
7. [AC] **2187 ÷ 3 = 729**
8. [AC] **1406.19 − 94.37 = 1311.82**
9. **$94.85**
10. **$6.30**

PAGES 20–21

1. **hundreds**
2. **ten thousands**
3. **millions**
4. **hundred thousands**
5. **twenty-eight thousand, three hundred two**
6. **one million, seventy-six thousand, five hundred**
7. **42,057**

8. **3,400,590**
9. **5,680 > 856**
   5,680 has more digits than 856.
10. **32,457 = 32,457** The numbers are the same.
11. **82,346 < 82,546** 3 is less than 5.
12. **790,300 > 709,300** 9 is greater than 0.
13. **40** 43 Since 3 is less than 5, do not change the underlined digit.
14. **2,500** 2,453 The number to the right of the underlined digit is 5; add 1 to the underlined digit.
15. **310,000** 307,216 Since 7 is greater than 5, add 1 to the underlined digit.
16. **4,000,000** 4,293,785 Since 2 is less than 5, do not change the underlined digit.
17. **648**
18. **145,800**
19. **1,985**
20. **10,401**
21. **(2) Fewer parts were produced during week 1 than during week 3.**
    12,435 (week 1) > 12,345 (week 3) because 4 > 3. Therefore, more parts were produced during week 1 than during week 3.
22. **(4) week 4**
    14,814 (week 4) > 12,435 (week 1) because 4 > 2.
    14,814 (week 4) > 14,526 (week 2) because 8 > 5.
    14,814 (week 4) > 12,345 (week 3) because 4 > 2.
    14,814 (week 4) > 13,706 (week 5) because 4 > 3.
23. **(5) week 5** 13,706 Since 7 is greater than 5, add 1 to the underlined digit. 13,706 (week 5) rounds to 14,000.
24. **(2) one million, thirty thousand, four hundred two**
25. **(4) $2,300**
    $2,264 Since 6 is greater than 5, add 1 to the underlined digit.
26. **(4) −2 degrees**

LESSON 2

PAGE 23

2.	**494**	1 476 + 18 494	Check	1 18 +476 494
3.	**651**	1 1 387 +264 651	Check	1 1 264 +387 651
4.	**655**	1 473 +182 655	Check	1 182 +473 655
5.	**707**	1 1 148 327 +232 707	Check	1 1 232 327 +148 707
6.	**1,375**	1 1 135 456 +784 1,375	Check	1 1 784 456 +135 1,375
7.	**757**	1 1 564 125 + 68 757	Check	1 1 68 125 +564 757
8.	**1,346**	1 1 817 76 +453 1,346	Check	1 1 453 76 +817 1,346
9.	**195**	1 38 +157 195	Check	1 157 + 38 195
10.	**964**	1 1 450 326 +188 964	Check	1 1 188 326 +450 964
11.	**23,068**	1 2 1 1 3,947 18,889 + 232 23,068	Check	1 2 1 1 232 18,889 + 3,947 23,068

12.	**954**	1 142 7 802 + 3 954	Check	1 3 802 7 +142 954

13. **Calculator answer: $792**
Checked answer: $792 To check addition, add again from bottom to top.

14. **Week 7** After Week 6, the students have read 99 books. Therefore, the students have not read 100 books until Week 7.

PAGE 25

2.	**653**	957 −304 653	Check	653 +304 957
3.	**474**	899 −425 474	Check	474 +425 899
4.	**743**	786 − 43 743	Check	743 + 43 786
5.	**25**	4 13 ~~5~~~~3~~ −28 25	Check	1 25 +28 53
6.	**155**	3 12 ~~4~~~~2~~6 −271 155	Check	1 155 +271 426
7.	**179**	2 12 ~~3~~~~2~~9 −150 179	Check	1 179 +150 329
8.	**467**	13 7 ~~3~~13 ~~8~~~~4~~~~3~~ −376 467	Check	1 1 467 +376 843
9.	**333**	9 4 ~~10~~10 ~~5~~~~0~~~~0~~ −167 333	Check	1 1 333 +167 500
10.	**387**	10 6 ~~0~~15 ~~7~~~~1~~~~5~~ −328 387	Check	1 1 387 +328 715
11.	**877**	9 8 ~~10~~16 ~~9~~~~0~~~~6~~ − 29 877	Check	1 1 877 + 29 906
12.	**38**	4 11 10 ~~5~~~~2~~~~0~~ −482 38	Check	1 1 38 +482 520
13.	**573**	7 12 ~~8~~~~2~~7 −254 573	Check	1 573 +254 827
14.	**318**	7 10 17 ~~8~~~~1~~~~7~~ −499 318	Check	1 1 318 +499 817
15.	**157**	9 6 ~~10~~10 ~~7~~~~0~~~~0~~ −543 157	Check	1 1 157 +543 700
16.	**198**	5 17 14 ~~6~~~~8~~~~4~~ −486 198	Check	1 1 198 +486 684

17. **Calculator answer: $358**
Checked answer: $650 To check subtraction, add your answer and the numbers that were subtracted:
$125 + $89 + $57 + $21 = $292
Check: $358 + $292 = $650

18. **11** To find how many books Gayle needs to order, subtract the amount of books already ordered from the total amount she needs to order:
120 − 43 − 54 − 12 = 11.

PAGE 27

1. **(4) 140** You know that two sides measure 40 feet and two sides measure 30 feet. Add all four sides to find the perimeter.

2. **(5) 20 + 20 + 24 + 24** Since it is a rectangle, two sides of the painting measure 20 inches and two sides measure 24 inches. Add all four measurements to find the perimeter.
3. **(2) 80 − 3** Find the difference in the width of the yard and the opening needed for the sidewalk. Subtract the smaller number from the larger one.
4. **(5) 72** Since the dog run is rectangular in shape, if you know one length and one width, you know the remaining length and width. 30 + 30 + 6 + 6 = 72.
5. **(1) 10 + 10 + 14 + 14** You know that two walls measure 10 feet across and two walls measure 14 feet across. Add all four wall measurements to find the perimeter.
6. **(2) 10 + 10 + 18** The deck is a rectangle, but the railing goes along only three sides. Two of the sides measure 10 feet, and the third side measures 18 feet. Add the three measurements to find the total railing length.
7. **You can find the perimeter of a rectangular room by measuring only one length and one width because in a rectangle you know that the opposite sides are equal in length. Therefore, you really know the lengths of all the sides.**

PAGE 29

1. **(4) $450 − $225** To find how much more Paula pays for rent than for food, subtract what she pays for food from what she pays for rent. You do not need to know how much she earns.
2. **(5) Not enough information is given.** Rasheed delivers water **three** times a week. You know the number of bottles delivered on only two of the days. You do not know the number of bottles delivered on the third day.
3. **(2) 45 × 5** To find how many miles Danielle could drive in 5 hours, multiply how many miles she drives in 1 hour by 5. You do not need to know how many miles per gallon her car gets.
4. **(3) 56 + 17** To find the total number of boxes, add the boxes he collected on Monday. You do not need to know how many boxes Kunio collected on Tuesday.
5. **(5) Not enough information is given.** To find how many pairs of sandals were sold, you must subtract the number of pairs of sandals left in stock (given as 27) from the number of pairs of sandals that were in stock before the sale (not given).
6. **(2) 220** Use addition to find the total. Round the three numbers to the nearest ten and add to estimate the total. 67 is about 70, 73 is about 70, and 81 is about 80. 70 + 70 + 80 = 220
7. Situations and details will vary. Sample answer: I would need to know the price of the item, how many items I was buying, if there was a discount or coupons, and any tax to be added.

PAGE 31

2. **Sports Vehicle** The tallest bar represents the greatest number of cars sold.
3. **225,000** The bar for mini-vans represents 225,000 cars.
4. **200,000** The two smallest bars represent the two least popular types of cars sold. The bar for the 4-door is the smallest and represents about 75,000 cars. The next smallest bar is the convertible and it represents about 125,000 cars. Add to find the total.

$$\begin{array}{r} 75{,}000 \\ +125{,}000 \\ \hline 200{,}000 \end{array}$$

5. **300,000** The bar for sports vehicles represents 450,000 cars. The bar for the 2-door cars represents 150,000. Subtract to find the difference.

```
  450,000
 −150,000
 ────────
  300,000
```

6. **1,025,000** Find values for each bar. Add the values. 150,000 + 75,000 + 450,000 + 225,000 + 125,000 = 1,025,000

7.

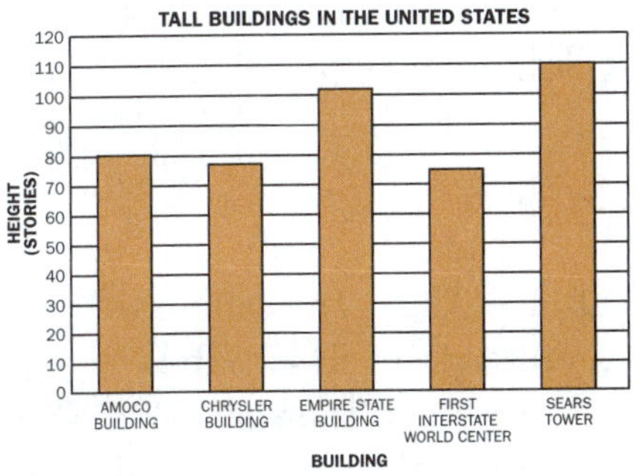

8. **Height or Stories**
9. **Stories, by tens**
10. **Sample answer: Tall Buildings in the United States**
11. **Sears Tower, 110 stories**
12. **First Interstate World Center**
13. **30** The bar for the Sears Tower represents 110 stories. The bar for the Amoco Building represents 80 stories. Subtract to find the difference. 110 − 80 = 30

PAGE 33

1. **(5) laser printer** Look in the column labeled *Item Number* to find the item number listed in the question. Then look in the column labeled *Description* to find the matching item.
2. **(3) monitor, scanner, printer** Look under the column labeled *Item Number* to find the three item numbers that begin with the same three digits—in this case, 414. Look under the column labeled *Description* to find the matching items.
3. **(2) 5 × $6.99** Look in the column labeled *Description* for laser paper. Then look in the column labeled *Qty.* to find the number of items ordered and the column labeled *Item Price* to find the price for each item ordered. Multiply the quantity and the price per item to find the total price for that item.
4. **(4) $55 + $35** Look in the column labeled *Total* to find how much the items ordered will cost.
5. **(4) 1 + 5 + 1 + 1 + 5** Kelly is placing an order for 1 monitor, 5 diskettes, 1 scanner, 1 laser printer, and 5 packages of laser paper.
6. **(5) $699.99 − $459.99** To find the amount of savings, compare prices by subtracting the lower price from the higher price.
7. **The total amount for diskettes is $54.95 because Kelly ordered five diskettes. Multiply $10.99 by 5 to get the total amount.**

PAGES 34–35

1. 733

```
  11
  237
 +496
 ────
  733
```

2. 162

```
  414
  5̸4̸7
 −385
 ────
  162
```

3. $782

```
   11
  $295
 + 487
 ─────
  $782
```

4. 279

```
     9
  5 1̸0̸10
  6̸0̸0̸
 −321
 ────
  279
```

5. $312

```
   5 10
  $6̸0̸7
 − 295
 ─────
  $312
```

6. **928**

```
   1 1
   503
    58
  +367
  ----
   928
```

7. **$689**

```
     1
  $227
   317
 + 145
 -----
  $689
```

8. **$699**

```
  8 18 13
  $993
 - 294
 -----
  $699
```

9. **78**

```
  6 15 12
   762
  -684
  ----
    78
```

10. **1,250**

```
     1
   836
  +414
 -----
 1,250
```

11. **Calculator answer: $3,809**

12. **Calculator answer: 45,205**

13. **89** Subtract the number of deliveries made before Ed's break from the total number of deliveries he needs to make.

```
 1 12 18
   238
  -149
  ----
    89
```

14. **422** Add the number of miles driven before lunch and the number of miles driven after lunch.

```
  1 1
  243
 +179
 ----
  422
```

15. **221** Add the number of each type of customer.

```
  1 1
  153
   26
 + 42
 ----
  221
```

16. **44** Subtract the number of tickets sold for the 5:00 show from the number of tickets sold for the 7:30 show.

```
 127
 -83
 ---
  44
```

17. **(5) 20 + 20 + 14 + 14** You know that two sides measure 20 feet and two sides measure 14 feet. Add all four measurements to find the perimeter.

18. **(4) $92** Add to find the total.

```
    1
  $28
   53
 + 11
 ----
  $92
```

19. **(2) 208 − 185** The difference between Bill's present weight and the weight his doctor advised is the weight Bill should lose. Subtract the smaller amount from the larger one.

20. **(4) $452** Add the cost of tuition and the cost of books and supplies.

```
     1
  $348
 + 104
 -----
  $452
```

21. **(3) comedy** The tallest bar represents the greatest number of rentals. The bar for comedy is the tallest.

22. **(4) 250** The bar for comedy is the tallest. It ends between the 4 and 5 marks and represents 450 rentals. The bar for drama is the smallest. It reaches the 2 mark and represents 200 rentals. Subtract to find the difference.

```
  450
 -200
 ----
  250
```

23. **(2) 1,550** Find values for each bar. Add the values.

350 + 200 + 250 + 450 + 300 = 1,550

LESSON 3

PAGE 37

2. **468**

```
   234
×    2
   468
```

3. **2,304**

```
   5 2
   384
×    6
 2,304
```

4. **1,314**

```
     2
    73
   ×18
   584
  +73
 1,314
```

5. **1,992**

```
     1
    83
   ×24
   332
 +1 66
 1,992
```

6. **12,173**

```
   ~~4~~ [2] ~~6~~ [3]
    259
  ×  47
  1 813
+10 36
 12,173
```

7. **$4,275**

```
   6 4
  $475
×    9
$4,275
```

8. **19,200**

```
     1
    64
  ×300
19,200
```

9. **336,474**

```
    1 ~~1~~ [2]
     837
    ×402
   1 674
+334 80
 336,474
```

10. **35,280**

```
   2 ~~4~~ [1]
    126
   ×280
 10 080
 25 2
 35,280
```

11. **$2,723**

```
   6 6
  $389
×    7
$2,723
```

12. **1,920**

```
     3
    48
   ×40
 1,920
```

13. **33,201**

```
   1 ~~2~~ [4]
    527
  ×  63
  1 581
+31 62
 33,201
```

14. **475,488**

```
   ~~2~~ [1] ~~4~~ [3]
     936
    ×508
   7 488
+468 00
 475,488
```

15. **Calculator answer: 13,170**
$198 \times 55 = 10{,}890$
$57 \times 40 = 2{,}280$
$10{,}890 + 2{,}280 = 13{,}170$
Checked answer: 13,170
When using a calculator, check your multiplication by entering the numbers again.

16. **744** $45 + 48 = 93$ $93 \times 8 = 744$
17. **$1,500** $125 \times \$12 = \$1{,}500$
18. **14,280** $280 \times 51 = 14{,}280$
19. **2,016** $84 \times 24 = 2{,}016$
20. **4,900** $350 \times 14 = 4{,}900$

PAGE 38

2. $215

```
      $215
  7)$1,505
   −14
     10
    − 7
     35
    −35
      0
```

```
   $215
 ×    7
 $1,505
```

3. 913

```
     913
  6)5,478
   −5 4
      7
     −6
     18
    −18
      0
```

```
   913
 ×   6
 5,478
```

4. 165

```
      165
  36)5,940
    −3 6
     2 34
    −2 16
      180
     −180
        0
```

```
   165
 ×  36
   990
  4 95
 5,940
```

PAGE 39

2. 87 r11

```
       87 r11
  24)2,099
    −1 92
      179
     −168
       11
```

```
     87
    ×24
    348
   1 74
  2 088
 +   11
  2,099
```

3. 93 r6

```
     93 r6
  9)843
   −81
     33
    −27
      6
```

```
   93
 ×  9
  837
 +  6
  843
```

4. 246 r4

```
      246 r4
  5)1,234
   −1 0
      23
     −20
      34
     −30
       4
```

```
   246
 ×   5
 1 230
 +   4
 1,234
```

5. 96 r11

```
       96 r11
  18)1,739
    −1 62
      119
     −108
       11
```

```
     96
    ×18
    768
    96
  1 728
 +   11
  1,739
```

6. $35

```
        $35
  62)$2,170
    − 1 86
        310
       −310
          0
```

```
    $35
 ×   62
     70
   2 10
 $2,170
```

7. 46 r7

```
     46 r7
  8)375
  − 32
     55
    −48
      7
```

```
   46
 ×  8
  368
 +  7
  375
```

8. $47

```
        $47
  27)$1,269
     −1 08
        189
       −189
          0
```

```
    $47
    ×27
    329
    94
 $1,269
```

PAGES 40–41

2. 101

```
      101
  26)2,626
    −2 6
       26
      −26
        0
```

```
   101
 ×  26
   606
  2 02
 2,626
```

3. $40

```
     $40
  8)$320
    −32
     00
     −0
      0
```

```
  $40
 ×  8
 $320
```

4. 304

```
     304
  6)1,824
   −1 8
      24
     −24
       0
```

```
   304
 ×   6
 1,824
```

6. 189

```
     189          189
27)5,103        × 27
  −2 7         1 323
   2 40        3 78
  −2 16        5,103
    243
   −243
      0
```

7. 308

```
     308          308
19)5,852        × 19
  −5 7         2 772
     152       3 08
    −152       5,852
       0
```

8. $276

```
    $276         $276
8)$2,208        ×   8
  −1 6         $2,208
     60
    −56
     48
    −48
      0
```

9. 800

```
     800          800
 6)4,800        ×   6
  −4 8          4,800
     00
    − 0
     00
    − 0
      0
```

10. 515

```
     515          515
13)6,695        × 13
  −6 5         1 545
     19        5 15
    −13        6,695
     65
    −65
      0
```

11. $580

```
    $580         $580
4)$2,320        ×   4
  −2 0         $2,320
     32
    −32
     00
    − 0
      0
```

12. 455

```
     455          455
10)4,550        × 10
  −4 0          4,550
     55
    −50
     50
    −50
      0
```

14. 489 r6

```
    489 r6        489
 7)3,429        ×   7
  −2 8          3,423
    62          +   6
   −56          3,429
    69
   −63
     6
```

15. 632

```
     632          632
 9)5,688        ×   9
  −5 4          5,688
     28
    −27
     18
    −18
      0
```

16. 93

```
      93           93
21)1,953          ×21
  −1 89            93
     63           186
    −63          1,953
      0
```

17. 156 r4

```
    156 r4        156
 8)1,252        ×   8
  −  8          1,248
     45         +   4
    −40         1,252
     52
    −48
      4
```

18. 107 r33

```
    107 r33       107
46)4,955        × 46
  −4 6            642
    355          4 28
   −322          4,922
     33          +  33
                 4,955
```

19. **$917**

$$\begin{array}{r} \$917 \\ 4\overline{)\$3{,}668} \\ -3\,6 \\ \hline 6 \\ -4 \\ \hline 28 \\ -28 \\ \hline 0 \end{array} \qquad \begin{array}{r} \$917 \\ \times \quad 4 \\ \hline \$3{,}668 \end{array}$$

20. **346 r4**

$$\begin{array}{r} 346 \text{ r}4 \\ 5\overline{)1{,}734} \\ -1\,5 \\ \hline 23 \\ -20 \\ \hline 34 \\ -30 \\ \hline 4 \end{array} \qquad \begin{array}{r} 346 \\ \times \quad 5 \\ \hline 1{,}730 \\ + \quad 4 \\ \hline 1{,}734 \end{array}$$

21. **Calculator Answer: 68** $1{,}632 \div 24$
Checked Answer: 1,632 68×24
22. **$1,455** $\$17{,}460 \div 12 = \$1{,}455$
23. **231 minutes, or 3 hours and 51 minutes**
$18{,}480 \div 80 = 231$
24. **14** $840 \div 60 = 14$

PAGE 43

2. $\mathbf{n + 16 = 20}$
3. $\mathbf{5n = 50}$
4. $\mathbf{27 - n = 12}$
5. $\mathbf{\frac{n}{6} = 5}$ **or** $\mathbf{n \div 6 = 5}$
6. $$\begin{aligned} a + 12 &= 32 \\ -12 &= -12 \\ \hline \mathbf{a} &= \mathbf{20} \end{aligned} \qquad \begin{aligned} a + 12 &= 32 \\ 20 + 12 &= 32 \end{aligned}$$
7. $$\begin{aligned} d \div 7 &= 8 \\ \times 7 &= \times 7 \\ \hline \mathbf{d} &= \mathbf{56} \end{aligned} \qquad \begin{aligned} d \div 7 &= 8 \\ 56 \div 7 &= 8 \end{aligned}$$
8. $$\begin{aligned} n - 18 &= 56 \\ +18 &= +18 \\ \hline \mathbf{n} &= \mathbf{74} \end{aligned} \qquad \begin{aligned} n - 18 &= 56 \\ 74 - 18 &= 56 \end{aligned}$$
9. $$\begin{aligned} d - 47 &= 24 \\ +47 &= +47 \\ \hline \mathbf{d} &= \mathbf{71} \end{aligned} \qquad \begin{aligned} d - 47 &= 24 \\ 71 - 47 &= 24 \end{aligned}$$
10. $$\begin{aligned} n \div 4 &= 15 \\ \times 4 &= \times 4 \\ \hline \mathbf{n} &= \mathbf{60} \end{aligned} \qquad \begin{aligned} n \div 4 &= 15 \\ 60 \div 4 &= 15 \end{aligned}$$
11. $9n$ is the same as $n \times 9$.
$$\begin{aligned} n \times 9 &= 108 \\ n \times 9 \div 9 &= 108 \div 9 \\ \mathbf{n} &= \mathbf{12} \end{aligned} \qquad \begin{aligned} n \times 9 &= 108 \\ 12 \times 9 &= 108 \end{aligned}$$
12. $3z$ is the same as $z \times 3$.
$$\begin{aligned} z \times 3 &= 54 \\ z \times 3 \div 3 &= 54 \div 3 \\ \mathbf{z} &= \mathbf{18} \end{aligned} \qquad \begin{aligned} z \times 3 &= 54 \\ 18 \times 3 &= 54 \end{aligned}$$
13. $$\begin{aligned} c + 13 &= 13 \\ -13 &= -13 \\ \hline \mathbf{c} &= \mathbf{0} \end{aligned} \qquad \begin{aligned} c + 13 &= 13 \\ 0 + 13 &= 13 \end{aligned}$$
14. $$\begin{aligned} y + 19 &= 45 \\ -19 &= -19 \\ \hline \mathbf{y} &= \mathbf{26} \end{aligned} \qquad \begin{aligned} y + 19 &= 45 \\ 26 + 19 &= 45 \end{aligned}$$
15. $10c$ is the same as $c \times 10$.
$$\begin{aligned} c \times 10 &= 100 \\ c \times 10 \div 10 &= 100 \div 10 \\ \mathbf{c} &= \mathbf{10} \end{aligned} \qquad \begin{aligned} c \times 10 &= 100 \\ 10 \times 10 &= 100 \end{aligned}$$
16. $45d = \$540$ and $45d$ is the same as $d \times 45$.
$$\begin{aligned} d \times 45 &= \$540 \\ d \times 45 \div 45 &= \$540 \div 45 \\ \mathbf{d} &= \mathbf{\$12} \end{aligned}$$
17. $$\begin{aligned} p - 15 &= 125 \\ +15 &= +15 \\ \hline \mathbf{p} &= \mathbf{140} \end{aligned}$$ **pounds**
18. $$\begin{aligned} p + 6 &= 18 \\ -6 &= -6 \\ \hline \mathbf{p} &= \mathbf{12} \end{aligned}$$
19. $$\begin{aligned} n \div 26 &= 3 \\ \times 26 &= \times 26 \\ \hline \mathbf{n} &= \mathbf{78} \end{aligned}$$
20. $$\begin{aligned} a + \$25 &= \$560 \\ -25 &= -25 \\ \hline \mathbf{a} &= \mathbf{\$535} \end{aligned}$$

PAGE 45

2. **Division** When given a total (here, total pay earned in a 40-hour work week) and asked *How much for each unit* (here, each hour) use division.
3. **Multiplication** When given an amount for each (here, admission per person) and the total (here, total number of people who paid), and asked for the total amount (here, money paid) use multiplication.
4. **Subtraction** When given two amounts (here, the budget and the cost of the computer) and asked How much more (here, money needed) use subtraction.
5. **Addition; 18** 12 + 6 = 18
   Estimate to check: 10 + 5 = 15. The answer 18 is reasonable.
6. **Subtraction; 126** 198 − 72 = 126
   Estimate to check: 200 − 70 = 130. The answer 126 is reasonable.
7. **2** Two videos total $34. Three videos total $51. The customer can buy only two videos with $50. Estimate to check: 2 × $20 = $40; 3 × $20 = $60. The answer 2 videos is reasonable.
8. **$112** Add the amount of the bills:
   $53 + $16 + $27 + $5 + $11 = $112.
   Estimate to check:
   $50 + $15 + $30 + $5 + $10 = $110.
   The answer $112 is reasonable.
9. **$346** Subtract $25 from $371.
   Estimate to check: $375 − $25 = $350.
   The answer $346 is reasonable.
10. **221** Add the three amounts:
    67 + 73 + 81 = 221
    Estimate to check: 70 + 70 + 80 = 220.
    The answer 221 is reasonable.

PAGE 47

2. **Mean: 376; Median: 368** To find the mean, first find the total of the numbers in the set. 410 + 350 + 368 = 1,128 Then divide the total by the number of items.
   1,128 ÷ 3 = 376
   To find the median, arrange the numbers in order. Then find the middle number.
   350, 368, 410
3. **Mean: 213; Median: 254** To find the mean, first find the total of the numbers in the set:
   261 + 254 + 105 + 280 + 165 = 1,065.
   Then divide the total by the number of items: 1,065 ÷ 5 = 213
   To find the median, arrange the numbers in order. Then find the middle number.
   105, 165, 254, 261, 280
4. **Mean: 2,350; Median: 2,430** To find the mean, first find the total of the numbers in the set. 2,450 + 2,100 + 1,970 + 2,430 + 2,840 + 1,800 + 2,860 = 16,450
   Then divide the total by the number of items. 16,450 ÷ 7 = 2,350
   To find the median, arrange the numbers in order. Then find the middle number.
   1,800; 1,970; 2,100; 2,430; 2,450; 2,840; 2,860
6. **Median: 202** To find the median, arrange the numbers in order. Then find the average of the two middle numbers.
   146, 162, 184, 198, 206, 212, 234, 523
   198 + 206 = 404 404 ÷ 2 = 202
7. **Median: 50** To find the median, arrange the numbers in order. Then find the average of the two middle numbers.
   26, 32, 41, 46, 54, 73, 81, 81
   46 + 54 = 100 100 ÷ 2 = 50
8. **Median: 3,881** To find the median, arrange the numbers in order. Then find the average of the two middle numbers.
   1,903; 2,950; 3,420; 4,342; 5,651; 7,418
   3,420 + 4,342 = 7,762
   7,762 ÷ 2 = 3,881

9. **Mean: 50** To find the mean, first find the total of the numbers in the set. 25 + 42 + 45 + 37 + 102 + 86 + 46 + 38 + 29 = 450 Then divide the total by the number of items. 450 ÷ 9 = 50.

10. **Mean: 425** To find the mean, first find the total of the numbers in the set. 502 + 147 + 425 + 454 + 518 + 504 = 2,550 Then divide the total by the number of items. 2,550 ÷ 6 = 425

PAGE 49

1. **(1) 40 × 30** To find the area of the exhibit, multiply the length (40) by the width (30).
2. **(4) 13 × 27** To find the amount of carpeting, find the area of the family room by multiplying the length (27) by the width (13).
3. **(3) 6 × 4 ÷ 3** To find the total number of feet, multiply the number of feet needed for each panel (6) by the number of panels (4). Then divide by the number of feet in a yard (3).
4. **(3) 42 × 32** To find the amount of tile flooring, find the area of the basement by multiplying the length (42) by the width (32).
5. **(5) Not enough information is given.** To find the amount of lime needed, Nancy needs to find the area of the garden. Area = length × width, but Nancy only knows the length of the garden; she does not know the width of the garden.
6. **(4) 800** To find the area of a rectangle, multiply the length by the width. The tabletop is 40 inches long and 20 inches wide, so the area of the tabletop is 40 × 20 = 800 square inches.
7. **Area = length × width, or $A = l \times w$. Since, in a square, the length and width are equal, using s for this measurement, $l = s$ and $w = s$. Replacing l and w in the area formula with s, the formula for area of a square becomes $A = s \times s$, or s^2.**

PAGES 50–51

1. **1,368**

$$\begin{array}{r} {}^{1\,1}\\ 456 \\ \times\ \ 3 \\ \hline 1{,}368 \end{array}$$

2. **4,067**

$$\begin{array}{r} 83 \\ \times 49 \\ \hline 747 \\ +3\,32\ \\ \hline 4{,}067 \end{array}$$

3. **$1,782**

$$\begin{array}{r} \$297 \\ \times\ \ \ 6 \\ \hline \$1{,}782 \end{array}$$

4. **124,956**

$$\begin{array}{r} 178 \\ \times 702 \\ \hline 356 \\ +124\,60\ \\ \hline 124{,}956 \end{array}$$

5. **199,165**

$$\begin{array}{r} 653 \\ \times 305 \\ \hline 3\,265 \\ +195\,90\ \\ \hline 199{,}165 \end{array}$$

6. **823 r2**

$$\begin{array}{r} 823\text{ r2} \\ 4\overline{)3{,}294} \\ -3\,2\ \ \\ \hline 09\ \\ -\ \ 8\ \\ \hline 14 \\ -12 \\ \hline 2 \end{array}$$

7. **$920**

$$\begin{array}{r} \$920 \\ 7\overline{)\$6{,}440} \\ -6\,3\ \ \\ \hline 14\ \\ -14\ \\ \hline 00 \end{array}$$

8. **321 r16**

$$\begin{array}{r} 321\text{ r16} \\ 18\overline{)5{,}794} \\ -5\,4\ \ \\ \hline 39\ \\ -36\ \\ \hline 34 \\ -18 \\ \hline 16 \end{array}$$

9. **201 r14**

$$\begin{array}{r} 201\text{ r}14 \\ 43\overline{)8{,}657} \\ -8\,6 \\ \hline 057 \\ -\ 43 \\ \hline 14 \end{array}$$

10. **153**

$$\begin{array}{r} 153 \\ 27\overline{)4{,}131} \\ -2\,7 \\ \hline 1\,43 \\ -1\,35 \\ \hline 81 \\ -81 \\ \hline 0 \end{array}$$

11. **31,200**

$$\begin{array}{r} 800 \\ \times 39 \\ \hline 7200 \\ 2400\ \\ \hline 31{,}200 \end{array}$$

12. **$4,210**

$$\begin{array}{r} \$842 \\ \times \quad 5 \\ \hline \$4{,}210 \end{array}$$

13. **$205**

$$\begin{array}{r} \$205 \\ 8\overline{)\$1{,}640} \\ -1\,6 \\ \hline 040 \\ -\ 40 \\ \hline 0 \end{array}$$

14. **143 r27**

$$\begin{array}{r} 143\text{ r}27 \\ 52\overline{)7{,}463} \\ -5\,2 \\ \hline 2\,26 \\ -2\,08 \\ \hline 183 \\ -156 \\ \hline 27 \end{array}$$

15. **55 miles** To find the mean, first find the total of the numbers in the set.
$62 + 28 + 57 + 65 + 63 = 275$
Then divide the total by the number of items in the set. $275 \div 5 = 55$

16. **62 miles** To find the median, arrange the numbers in order. Find the middle value.
28, 57, <u>62</u>, 63, 65

17. $y = 42$ Use addition to solve for y.

$$\begin{array}{rcr} y - 13 & = & 29 \\ +\ 13 & = & 13 \\ \hline y & = & 42 \end{array}$$

18. **15** Divide the total collected for oil changes by the cost of an oil change.

$$\begin{array}{r} 15 \\ \$22\overline{)\$330} \\ -22 \\ \hline 110 \\ -110 \\ \hline 0 \end{array}$$

19. **18,075**
20. **12,000**
21. **460**
22. **629**
23. **(2) 94 × 50** To find the area, multiply the length times the width.
24. **(1) $56** Multiply Lena's hourly wage by the number of hours she worked. The hours Lena worked on Saturday and Sunday are extra information not needed to find the answer.
$\$8 \times 7 = \$56$
25. **(3) $12 ÷ 6** To find the price of one quart, divide the total cost by the number of quarts bought.
26. **(5) Not enough information is given.** To find the area, multiply the length times the width. The problem does not give the width of the garden.
27. **(1) $5*p* = $20** To find how many pairs of shoes Martin bought, multiply the price for each pair of shoes ($5) by the number of pairs of shoes (p) to get the total ($20).
28. **(5) 42,000** To find how many miles, divide the total miles (126,000) by the number of years (3).
29. **(4) 2,299** To find the total number of miles, add the numbers:
$244 + 996 + 1{,}059 = 2{,}299$

LESSON 4

PAGE 53

2. $5 \times 5 = 25$
3. $1 \times 1 = 1$
4. $7 \times 7 \times 7 = 49 \times 7 = 343$
5. 64
6. 121
7. 324
8. 256
9. 125 $5^3 = 5 \times 5 \times 5 = 25 \times 5 = 125$
10. 512 $8^3 = 8 \times 8 \times 8 = 64 \times 8 = 512$
11. 64 $4^3 = 4 \times 4 \times 4 = 16 \times 4 = 64$
12. 1,000 $10^3 = 10 \times 10 \times 10 = 100 \times 10 = 1{,}000$
13. 11 $11^2 = 121$ Therefore $\sqrt{121} = 11$
14. 4 $4^2 = 16$ Therefore $\sqrt{16} = 4$
15. 17 $17^2 = 289$ Therefore $\sqrt{289} = 17$
16. 14 $14^2 = 196$ Therefore $\sqrt{196} = 14$
17. $21^2 = 441$ $31^2 = 961$
$22^2 = 484$ $32^2 = 1{,}024$
$23^2 = 529$ $33^2 = 1{,}089$
$24^2 = 576$ $34^2 = 1{,}156$
$25^2 = 625$ $35^2 = 1{,}225$
18. $\sqrt{441} = 21$ $\sqrt{3{,}481} = 59$
$\sqrt{625} = 25$ $\sqrt{3{,}721} = 61$
$\sqrt{841} = 29$ $\sqrt{4{,}225} = 65$
$\sqrt{1{,}225} = 35$ $\sqrt{4{,}761} = 69$
$\sqrt{1{,}521} = 39$ $\sqrt{5{,}041} = 71$

PAGE 55

1. **(3) $2 \times 15 + 2 \times 12$** Substitute the numbers into the formula:
$P = 2 \times \underline{15} + 2 \times \underline{12}$
2. **(1) 45×5** The distance formula is $d = r \times t$. Substitute 45 for r and 5 for t.
3. **(2) 4×20** The formula for perimeter of a square is $P = 4 \times s$. Substitute 20 for s.
4. **(3) $150 \div 12$** Use the distance formula $r = d \div t$. Replace 150 for d and 12 (6 hours $\times$ 2 days) for t.
5. **(5) Not enough information is given.** To find how many miles Kathy drove, you need to use the distance formula. You know that $t = 7$ hours. The problem does not give r, the rate at which she drove. The time she left and the number of gallons of gas she put in the car are extra information not needed to solve the problem asked.
6. **(4) length 11 feet, width 10 feet.** Multiply the dimensions in each option. Only (4) is greater than 100 square feet.
Answer (1): $14 \times 5 = 70$
Answer (2): $13 \times 7 = 91$
Answer (3): $12 \times 8 = 96$
Answer (4): $11 \times 10 = 110$
Answer (5): $11 \times 9 = 99$
7. **To rewrite the distance formula in terms of t, divide both sides by r.**

$$d = rt$$
$$d \div r = rt \div r$$
$$d \div r = t$$

PAGE 57

1. **(4) 15^2** The room is a square because all four sides have the same length. Use the formula $A = s^2$ and substitute 15 for s.
2. **(5) $225 \div 9$** Divide the number of square feet by the number of square feet in a square yard (9).
3. **(1) 18^2** The room is a square because all four sides have the same length. Use the formula $A = s^2$ and substitute 18 for s.
4. **(4) 12^2** The floor is a square because all four sides have the same length. Use the formula $A = s^2$ and substitute 12 for s.
5. **(2) $36 \div 9$** Divide the number of square feet by the number of square feet in a square yard (9).
6. **(4) 4×18** Multiply the number of tiles in a square foot by the number of square feet.
7. **You can find the area of a square room by measuring only one side because you can take that one measurement and multiply it by itself ($A = s^2 = s \times s$).**

PAGE 59

1. **(3) 3 × 2 × 5** To find the volume of a rectangular solid, multiply the length (3 feet) times the width (2 feet) times the height (5 feet).
2. **(2) 6 × 4 × 1** The sandbox is a rectangular solid. Multiply the length (6 feet) times the width (4 feet) times the height (1 feet) to find the volume.
3. **(3) 7^3** The box is a cube because each side has the same length. Use the formula $A = s^3$ and substitute 7 for s.
4. **(4) 12 × 10 × 8** To find the volume of the shipping carton, multiply the length (12 inches) times the width (10 inches) times the height (8 inches).
5. **(5) Not enough information is given.** To find the volume of a carton, multiply the length by the width by the height. The problem does not give the height of the carton.
6. **(2) 736** You are asked to find the space left in the truck. You are given the volume of the cartons loaded in the truck. To find the volume of a truck (a rectangular container), multiply the length by the width by the height. $16 \times 8 \times 14 = 1{,}792$ cubic feet. To find the space left in the truck, subtract the volume of the cartons from the volume of the truck. $1{,}792 - 1{,}056 = 736$
7. **$V = s^3 = s \times s \times s$ You can find the volume of a cube by measuring only one side because each side has the same length. Therefore, you already know the measures of the other sides without actually measuring them.**

PAGE 61

1. **450** $\$25 \times 18 = \$450$
   Do not write the dollar sign in the grid. Labels are not needed because the label is stated in the question.

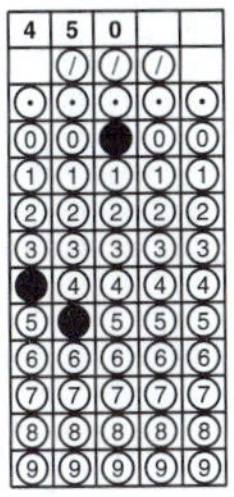

2. **569** $322 + 247 = 569$ miles

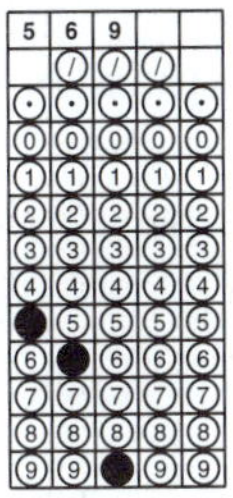

3. **1392** $\$348 \times 4 = \$1{,}392$
   Do not enter commas in the grid.

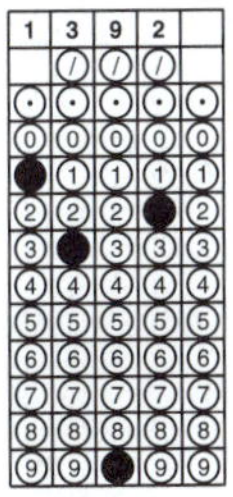

4. **154** $32 \times 3 = 96$ fluid ounces
   Subtract: $250 - 96 = 154$ fluid ounces

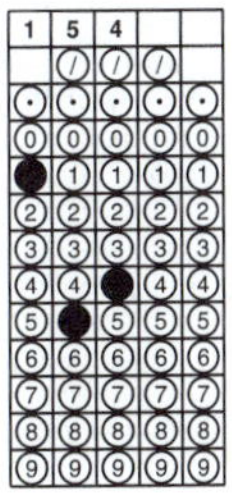

PAGES 62–63

1. $8 \times 8 = 64$
2. $1 \times 1 \times 1 = 1$
3. $3 \times 3 \times 3 = 27$
4. $9 \times 9 = 81$
5. 289
6. 225
7. 5
8. 7
9. 46,225
10. 6,859
11. 18
12. 33
13. **400 sq. ft.** The room is a square because all four sides have the same length. Use the formula $A = s^2$ and substitute 20 for s. $20 \times 20 = 400$
14. **900 cu. ft.** $V = l \times w \times h = 15 \times 12 \times 5 = 180 \times 5 = 900$
15. **720** $V = l \times w \times h = 30 \times 6 \times 4 = 180 \times 4 = 720$
16. **\$126** Multiply the cost per yard by the number of yards bought. $\$14 \times 9 = \$126$
17. **64** Divide the number of square feet by the number of square feet in a square yard (9). $576 \div 9 = 64$
18. **576** $V = l \times w \times h = 12 \times 8 \times 6 = 96 \times 6 = 576$

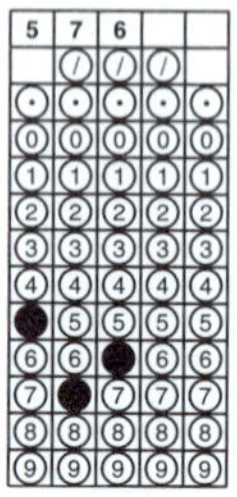

19. **(2) $6 \times 6 \times 6$** To find the volume of a cube, multiply the length (6 inches) times the width (6 inches) times the height (6 inches).
20. **(5) 15^2** The room is a square because all 4 sides have the same length. Use the formula $A = s^2$ and substitute 15 for s.
21. **(5) 160** $V = l \times w \times h = 8 \times 5 \times 4 = 40 \times 4 = 160$ cu. ft.
22. **(3) $42^2 \div 9$** To find the area in square feet, use the formula $A = s^2$ and substitute 42 for s. Then divide the number of square feet by the number of square feet in a square yard (9).
23. **(3) 100** The patio will be a square because all 4 sides are to be the same length. Use the formula $A = s^2$ and substitute 10 for s.
24. **(1) $18 \times 12 \times 3$** To find the volume of a rectangular space, multiply the length (18 inches) times the width (12 inches) times the height (3 inches).
25. **(2) 12** Divide the number of square feet by the number of square feet in a square yard (9). $108 \div 9 = 12$

MATH AT WORK

PAGES 65

1. **(4) 198** You find this by looking at the mileage on Route 78 between Harrisburg, PA and New York, NY.
2. **a.** The northern route is
   $126 + 50 + 416 =$ **592 miles.**
   The southern route is
   $129 + 50 + 169 + 198 =$ **546 miles.**

   b. The **southern route** is the shorter: 592 miles > 546 miles.

   c. Answers will vary. Sample answer: **Even though the southern route is 46 miles shorter (592 − 546), I would suggest that he take the northern route. The one hour delay on the Route 76 will probably make travel on the southern route take longer.**

UNIT 1 REVIEW

PAGES 66–67

1. **ten thousands**
2. **millions**
3. **417 < 1,740** 417 has fewer digits than 1,740.
4. **54,9<u>7</u>2 > 54,9<u>2</u>7** 7 is greater than 2.
5. **4,380,000** 4,3<u>7</u>5,429 7 is in the ten thousands place. The number to the right of 7 is 5; add 1 to the underlined digit.
6. **twenty-three thousand, twelve**
7. **$x = 11$**
   $$\begin{array}{rcr} x + 8 & = & 19 \\ -\ 8 & = & -8 \\ \hline x & = & 11 \end{array}$$
8. **$x = 5$**
   $$\begin{array}{rcr} 4x & = & 20 \\ \div\ 4 & = & \div 4 \\ \hline x & = & 5 \end{array}$$
9. **763**
   $$\begin{array}{r} {}^{1\,1} \\ 426 \\ 84 \\ +253 \\ \hline 763 \end{array}$$
10. **\$992**
    $$\begin{array}{r} {}^{1\,3} \\ \$248 \\ \times\quad 4 \\ \hline \$992 \end{array}$$
11. **\$576**
    $$\begin{array}{r} {}^{8\,14\,14} \\ \$\not{9}\not{5}\not{4} \\ -\ 378 \\ \hline \$576 \end{array}$$
12. **29 r2**
    $$\begin{array}{r} 29\text{ r}2 \\ 27\overline{)785} \\ -54 \\ \hline 245 \\ -243 \\ \hline 2 \end{array}$$
13. **223,608**
    $$\begin{array}{r} {}^{1} \\ {}^{\not{2}\,\not{4}} \\ 726 \\ \times 308 \\ \hline 5\,808 \\ +217\,80 \\ \hline 223{,}608 \end{array}$$
14. **64** $4 \times 4 \times 4 = 16 \times 4 = 64$
15. **257**
    $$\begin{array}{r} {}^{9} \\ {}^{4\ \not{10}\ 10} \\ \not{5}\not{0}\not{0} \\ -243 \\ \hline 257 \end{array}$$
16. **8** $8 \times 8 = 64$
17. **\$313**
    $$\begin{array}{r} \$313 \\ 8\overline{)\$2{,}504} \\ -2\,4 \\ \hline 10 \\ -\ 8 \\ \hline 24 \\ -24 \\ \hline 0 \end{array}$$
18. **305 r11**
    $$\begin{array}{r} 305\text{ r}11 \\ 13\overline{)3{,}976} \\ -3\,9 \\ \hline 076 \\ -\ 65 \\ \hline 11 \end{array}$$

19. **78 inches**
$80 + 76 + 75 + 80 + 79 = 390$

$$\begin{array}{r} 78 \\ 5\overline{)390} \\ -35 \\ \hline 40 \\ -40 \\ \hline 0 \end{array}$$

20. **31 degrees** Arrange the numbers in order from least to greatest. Find the middle number.
18, 21, 24, <u>31</u>, 35, 37, 42
21. **12,844,368**
22. **68**
23. **6,889**
24. **235**
25. **(5) 48 + 48 + 20 + 20** Two sides measure 48 feet and two sides measure 20 feet. Add all four measurements to find the perimeter.
26. **(3) $90** Add to find a total. Round the three numbers to the nearest ten and estimate the total.

$57	rounds to	60
13	rounds to	10
+ 21	rounds to	+ 20
		$90

27. **(2) $14 × 43** To find the amount Ramon earned this week, multiply the amount he earns per hour ($14) times the number of hours he worked this week (43). The number of hours he worked last week is extra information.
28. **(4) 350** The tallest bar is for December. It nearly reaches the 5 mark and represents about 500 sweat shirts. The bar for September is the smallest. It ends between the 1 and 2 marks and represents about 150 sweat shirts. Subtract to find the difference.

$$\begin{array}{r} \overset{4\,10}{\cancel{5}\cancel{0}0} \\ -150 \\ \hline 350 \end{array}$$

29. **196** The room is a square because all four sides have the same length. Use the formula $A = s^2$ and substitute 14 for s.

UNIT 1: MINI-TEST

PAGES 68–69

1. **(4) $1,536** Multiply the number of hours Tony worked (128) by his hourly wage ($12). 128 × $12 = $1,536. The fact that he works 30 to 40 hours per week is extra information.
2. **(2) 29** Find the mean to find the average. Add: 29 + 34 + 47 + 18 + 22 + 24 = 174. Divide by the number of seasons: 174 ÷ 6 = 29 home runs per season.
3. **(1) $1,988** The deposit adds to her savings. Add: $1,362 + $250 = $1,612. Subtract from her goal:
$3,600 − $1,612 = $1,988.
4. **(3) 84** The frame goes around the outside edge of the mirror, so you need to find the perimeter. 24 + 24 + 18 + 18 = 84 inches.
5. **26** Divide: 312 ÷ 12 = 26.

6. **(4) 20** There were 35 sales on Thursday. The smallest bar (Monday) represents 15 sales. 35 − 15 = 20 sales
7. **(5) 84 × 40** Find the area of a rectangle by multiplying length by width.

8. **(3) 6 × 6 × 6** The exponent 3 shows the number of times the base 6 should appear in the multiplication problem.
9. **(2) $32** Arrange the five prices in order and find the middle value:
   $24 $26 <u>$32</u> $46 $52
10. **(1) $420 ÷ (4 + 8)** You need to divide the entry fee amount ($420) by the total number of players (4 + 8).
11. **204** Multiply: 34 × 6 = 204 cookies.

UNIT 2: FRACTIONS

LESSON 5

PAGE 73

2. $\frac{2}{3}$ There are 3 circles in the group, and 2 of the 3 circles (or $\frac{2}{3}$) are shaded.
3. $\frac{1}{2}$ The figure is divided into 2 parts, and 1 of the 2 parts (or $\frac{1}{2}$) is shaded.
4. $\frac{3}{5}$ The figure is divided into 5 parts, and 3 of the 5 parts (or $\frac{3}{5}$) are shaded.
5. $\frac{3}{4}$ There are 4 figures in the group, and 3 of the 4 figures (or $\frac{3}{4}$) are shaded.
6. $\frac{5}{8}$ The figure is divided into 8 parts, and 5 of the 8 parts (or $\frac{5}{8}$) are shaded.

7. 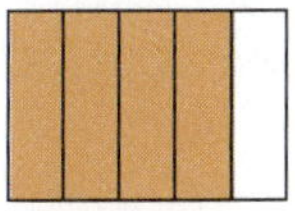

8. 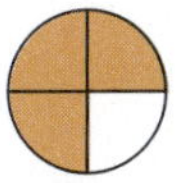

9. 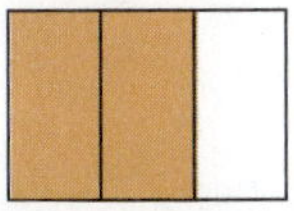

10. 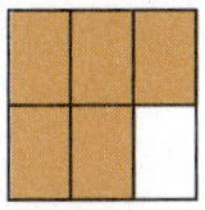

11. 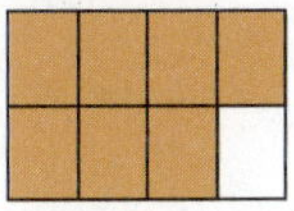

12. 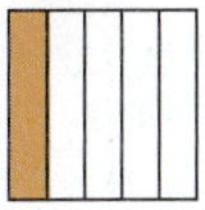

13. $\frac{5}{14}$
14. $\frac{8}{25}$
15. $\frac{17}{24}$
16. $\frac{8}{15}$

PAGE 75

2. $\frac{5}{2}$, $2\frac{1}{2}$ Look at the figures. Two whole figures and one-half of the third figure are shaded.
3. $\frac{19}{8}$, $2\frac{3}{8}$ Look at the figures. Two whole figures and three-eighths of the third figure are shaded.

5. 3 $\quad \begin{array}{r} 3 \\ 6\overline{)18} \\ -18 \\ \hline 0 \end{array} \rightarrow 3$

6. $2\frac{4}{5}$ $\quad \begin{array}{r} 2 \\ 5\overline{)14} \\ -10 \\ \hline 4 \end{array} \rightarrow 2\frac{4}{5}$

7. $3\frac{1}{3}$ $\quad \begin{array}{r} 3 \\ 3\overline{)10} \\ -\ 9 \\ \hline 1 \end{array} \rightarrow 3\frac{1}{3}$

8. 5 $\quad \begin{array}{r} 5 \\ 4\overline{)20} \\ -20 \\ \hline 0 \end{array} \rightarrow 5$

9. $1\frac{7}{8}$ $\quad \begin{array}{r} 1 \\ 8\overline{)15} \\ -\ 8 \\ \hline 7 \end{array} \rightarrow 1\frac{7}{8}$

10. $\frac{11}{6}$ $\quad 1\frac{5}{6} = \frac{(1 \times 6)}{6} + \frac{5}{6} = \frac{6}{6} + \frac{5}{6} = \frac{11}{6}$

11. $\frac{5}{2}$ $\quad 2\frac{1}{2} = \frac{(2 \times 2)}{2} + \frac{1}{2} = \frac{4+1}{2} = \frac{5}{2}$

12. $\frac{7}{1}$ $\quad 7 = \frac{7}{1}$

13. $\frac{19}{5}$ $\quad 3\frac{4}{5} = \frac{(5 \times 3)}{5} + \frac{4}{5} = \frac{15+4}{5} = \frac{19}{5}$

14. $\frac{12}{1}$ $\quad 12 = \frac{12}{1}$

15. $\frac{43}{4}$ $\quad 10\frac{3}{4} = \frac{(4 \times 10)}{4} + \frac{3}{4}$
$= \frac{40 + 3}{4}$
$= \frac{43}{4}$

16. $\frac{7}{2}$ $\quad 3\frac{1}{2} = \frac{(3 \times 2)}{2} + \frac{1}{2}$
$= \frac{6 + 1}{2}$
$= \frac{7}{2}$

17. $6\frac{1}{3}$

PAGE 76

2. $\frac{2}{3} \times \frac{12}{18}$ $\quad 2 \times 18 = 36$, $3 \times 12 = 36$
$\frac{2}{3}$ and $\frac{12}{18}$ are equivalent

3. $\frac{1}{2} \times \frac{5}{10}$ $\quad 1 \times 10 = 10$, $2 \times 5 = 10$
$\frac{1}{2}$ and $\frac{5}{10}$ are equivalent

4. $\frac{3}{6} \times \frac{5}{12}$ $\quad 3 \times 12 = 36$, $6 \times 5 = 30$
$\frac{3}{6}$ and $\frac{5}{12}$ are <u>not</u> equivalent

5. $\frac{4}{6} \times \frac{8}{12}$ $\quad 4 \times 12 = 48$, $6 \times 8 = 48$
$\frac{4}{6}$ and $\frac{8}{12}$ are equivalent

6. $\frac{4}{7} \times \frac{8}{14}$ $\quad 4 \times 14 = 56$, $7 \times 8 = 56$
$\frac{4}{7}$ and $\frac{8}{14}$ are equivalent

7. $\frac{5}{6} \times \frac{25}{30}$ $\quad 5 \times 30 = 150$, $6 \times 25 = 150$
$\frac{5}{6}$ and $\frac{25}{30}$ are equivalent

8. $\frac{3}{5} \times \frac{6}{15}$ $\quad 3 \times 15 = 45$, $5 \times 6 = 30$
$\frac{3}{5}$ and $\frac{6}{15}$ are <u>not</u> equivalent

9. $\frac{6}{10} \times \frac{24}{40}$ $\quad 6 \times 40 = 240$, $10 \times 24 = 240$
$\frac{6}{10}$ and $\frac{24}{40}$ are equivalent

10. $\frac{3}{5} \times \frac{20}{25}$ $\quad 3 \times 25 = 75$, $5 \times 20 = 100$
$\frac{3}{5}$ and $\frac{20}{25}$ are <u>not</u> equivalent

11. $\frac{1}{2} \times \frac{4}{8}$ $\quad 1 \times 8 = 8$, $2 \times 4 = 8$
$\frac{1}{2}$ and $\frac{4}{8}$ are equivalent

12. $\frac{2}{3} \times \frac{6}{9}$ $\quad 2 \times 9 = 18$, $3 \times 6 = 18$
$\frac{2}{3}$ and $\frac{6}{9}$ are equivalent

13. $\frac{7}{8} \times \frac{21}{24}$ $\quad 7 \times 24 = 168$, $8 \times 21 = 168$
$\frac{7}{8}$ and $\frac{21}{24}$ are equivalent

14. $\frac{3}{8} \times \frac{6}{16}$ $\quad 3 \times 16 = 48$, $8 \times 6 = 48$
$\frac{3}{8}$ and $\frac{6}{16}$ are equivalent

15. $\frac{8}{9} \times \frac{16}{18}$ $\quad 8 \times 18 = 144$, $9 \times 16 = 144$
$\frac{8}{9}$ and $\frac{16}{18}$ are equivalent

16. $\frac{1}{4} \times \frac{8}{16}$ $\quad 1 \times 16 = 16$, $4 \times 8 = 32$
$\frac{1}{4}$ and $\frac{8}{16}$ are <u>not</u> equivalent

PAGE 77

2. $\frac{2}{3}$ $\quad \frac{8}{12} = \frac{8 \div 4}{12 \div 4} = \frac{2}{3}$
3. $\frac{2}{5}$ $\quad \frac{4}{10} = \frac{4 \div 2}{10 \div 2} = \frac{2}{5}$
4. $\frac{3}{4}$ $\quad \frac{6}{8} = \frac{6 \div 2}{8 \div 2} = \frac{3}{4}$
5. $\frac{4}{5}$ $\quad \frac{20}{25} = \frac{20 \div 5}{25 \div 5} = \frac{4}{5}$
6. $\frac{1}{3}$ $\quad \frac{12}{36} = \frac{12 \div 12}{36 \div 12} = \frac{1}{3}$
7. $\frac{2}{3}$ $\quad \frac{6}{9} = \frac{6 \div 3}{9 \div 3} = \frac{2}{3}$
8. $\frac{1}{2}$ $\quad \frac{8}{16} = \frac{8 \div 8}{16 \div 8} = \frac{1}{2}$
9. $\frac{4}{5}$ $\quad \frac{16}{20} = \frac{16 \div 4}{20 \div 4} = \frac{4}{5}$
10. $\frac{5}{7}$ $\quad \frac{10}{14} = \frac{10 \div 2}{14 \div 2} = \frac{5}{7}$
11. $\frac{1}{3}$ $\quad \frac{5}{15} = \frac{5 \div 5}{15 \div 5} = \frac{1}{3}$
12. $\frac{5}{6}$ $\quad \frac{20}{24} = \frac{20 \div 4}{24 \div 4} = \frac{5}{6}$
13. $\frac{2}{3}$ $\quad \frac{18}{27} = \frac{18 \div 9}{27 \div 9} = \frac{2}{3}$
14. $\frac{3}{4}$ $\quad \frac{18}{24} = \frac{18 \div 6}{24 \div 6} = \frac{3}{4}$
15. $\frac{2}{3}$ $\quad \frac{22}{33} = \frac{22 \div 11}{33 \div 11} = \frac{2}{3}$
16. $\frac{3}{4}$ $\quad \frac{21}{28} = \frac{21 \div 7}{28 \div 7} = \frac{3}{4}$

PAGE 78

2. $\frac{8}{12}$ $\quad \frac{2}{3} = \frac{2 \times 4}{3 \times 4} = \frac{8}{12}$
3. $\frac{3}{18}$ $\quad \frac{1}{6} = \frac{1 \times 3}{6 \times 3} = \frac{3}{18}$
4. $\frac{14}{20}$ $\quad \frac{7}{10} = \frac{7 \times 2}{10 \times 2} = \frac{14}{20}$

5. $\frac{15}{24}$ $\quad \frac{5}{8} = \frac{5 \times 3}{8 \times 3} = \frac{15}{24}$

6. $\frac{12}{30}$ $\quad \frac{2}{5} = \frac{2 \times 6}{5 \times 6} = \frac{12}{30}$

7. $\frac{8}{18}$ $\quad \frac{4}{9} = \frac{4 \times 2}{9 \times 2} = \frac{8}{18}$

8. $\frac{3}{12}$ $\quad \frac{1}{4} = \frac{1 \times 3}{4 \times 3} = \frac{3}{12}$

9. $\frac{28}{32}$ $\quad \frac{7}{8} = \frac{7 \times 4}{8 \times 4} = \frac{28}{32}$

10. $\frac{15}{25}$ $\quad \frac{3}{5} = \frac{3 \times 5}{5 \times 5} = \frac{15}{25}$

11. $\frac{21}{36}$ $\quad \frac{7}{12} = \frac{7 \times 3}{12 \times 3} = \frac{21}{36}$

12. $\frac{12}{40}$ $\quad \frac{3}{10} = \frac{3 \times 4}{10 \times 4} = \frac{12}{40}$

13. $\frac{20}{45}$ $\quad \frac{4}{9} = \frac{4 \times 5}{9 \times 5} = \frac{20}{45}$

14. $\frac{18}{32}$ $\quad \frac{9}{16} = \frac{9 \times 2}{16 \times 2} = \frac{18}{32}$

15. $\frac{24}{56}$ $\quad \frac{3}{7} = \frac{3 \times 8}{7 \times 8} = \frac{24}{56}$

16. $\frac{72}{100}$ $\quad \frac{18}{25} = \frac{18 \times 4}{25 \times 4} = \frac{72}{100}$

PAGE 79

2. $>$ $\quad \frac{3}{4}$ X $\frac{3}{8}$ $\quad 3 \times 8 = 24$, $4 \times 3 = 12$ $\quad 24 > 12$

3. $<$ Since both fractions have the same denominator (16), compare the numerators: $9 < 12$, therefore $\frac{9}{16} < \frac{12}{16}$.

4. $<$ $\quad \frac{2}{4}$ X $\frac{2}{3}$ $\quad 2 \times 3 = 6$, $4 \times 2 = 8$ $\quad 6 < 8$

5. $>$ $\quad \frac{2}{3}$ X $\frac{3}{5}$ $\quad 2 \times 5 = 10$, $3 \times 3 = 9$ $\quad 10 > 9$

6. $=$ $\quad \frac{8}{10}$ X $\frac{4}{5}$ $\quad 8 \times 5 = 40$, $10 \times 4 = 40$ $\quad 40 = 40$

7. $>$ $\quad \frac{7}{8}$ X $\frac{3}{4}$ $\quad 7 \times 4 = 28$, $8 \times 3 = 24$ $\quad 28 > 24$

8. $<$ Since both fractions have the same denominator (9), compare the numerators: $5 < 8$, therefore $\frac{5}{9} < \frac{8}{9}$.

9. $>$ Since both fractions have the same denominator (12), compare the numerators: $11 > 8$, therefore $\frac{11}{12} > \frac{8}{12}$.

10. $<$ $\quad \frac{5}{6}$ X $\frac{7}{8}$ $\quad 5 \times 8 = 40$, $6 \times 7 = 42$ $\quad 40 < 42$

11. $<$ $\quad \frac{3}{8}$ X $\frac{4}{5}$ $\quad 3 \times 5 = 15$, $8 \times 4 = 32$ $\quad 15 < 32$

12. $>$ $\quad \frac{3}{4}$ X $\frac{2}{3}$ $\quad 3 \times 3 = 9$, $4 \times 2 = 8$ $\quad 9 > 8$

13. $=$ $\quad \frac{1}{3}$ X $\frac{3}{9}$ $\quad 1 \times 9 = 9$, $3 \times 3 = 9$ $\quad 9 = 9$

14. $>$ Since both fractions have the same denominator (21), compare the numerators: $15 > 13$, therefore $\frac{15}{21} > \frac{13}{21}$.

15. $<$ Since both fractions have the same denominator (24), compare the numerators: $15 < 18$, therefore $\frac{15}{24} < \frac{18}{24}$.

16. $=$ $\quad \frac{3}{4}$ X $\frac{18}{24}$ $\quad 3 \times 24 = 72$, $4 \times 18 = 72$ $\quad 72 = 72$

PAGE 81

1. **(4) D** Look at the ruler. The $3\frac{1}{16}$-inch mark is one small mark greater than (to the right of) the 3-inch mark.
2. **(2) $\frac{3}{4}$ is greater than $\frac{1}{2}$** Look at the measuring cup. The $\frac{3}{4}$-cup mark is above (greater than) the $\frac{1}{3}$-, $\frac{1}{2}$-, and $\frac{2}{3}$-cup marks and below (less than) the 1- and $1\frac{1}{4}$-cup marks.
3. **(2) $2\frac{1}{4}$ is equal to $2\frac{2}{8}$** The $\frac{1}{4}$ mark is the same as two $\frac{1}{8}$ marks. Thus, $\frac{1}{4}$ and $\frac{2}{8}$ are the same, or equal.
4. **(1) below the $\frac{1}{4}$ mark** On the ruler, $\frac{1}{8}$ is smaller (less) than $\frac{1}{4}$, so, on the measuring cup, the $\frac{1}{8}$ mark would go below the $\frac{1}{4}$ mark.
5. **(4) $3\frac{1}{4}$ is greater than $3\frac{1}{8}$** $3\frac{1}{4} = 3\frac{2}{8}$ which is more than $3\frac{1}{8}$.
6. **(5) Not enough information is given.** To find the weight of all the fruit, Dean needs to know the weight of the bananas as well as the pears and grapes.
7. **Each fraction has a whole and a fraction amount. The fractions are equal because $\frac{2}{4}$ reduces to $\frac{1}{2}$. Draw two figures. Divide each figure into two equal parts. Shade $1\frac{1}{2}$ parts. Draw another set of two figures the same size as the first two figures. Divide each figure into four equal parts. Shade $1\frac{2}{4}$ parts. The shaded parts are the same, or equal.**

PAGES 82–83

1. $\frac{5}{6}$ The figure is divided into 6 parts, and 5 parts are shaded.
2. $\frac{3}{5}$ There are 5 circles in the group, and 3 are shaded.
3. $\frac{7}{2}$ Each figure is divided into 2 parts, and 7 parts are shaded.
4. $\frac{14}{9}$ Each figure is divided into 9 parts, and 14 parts are shaded.
5. $1\frac{1}{3}$ Each figure is divided into 3 parts. One figure is completely shaded, and 1 of the 3 parts of the last figure is shaded.
6. $2\frac{3}{4}$ Each figure is divided into 4 parts. Two figures are completely shaded, and 3 of the 4 parts of the last figure are shaded.
7. $1\frac{5}{6}$ $\quad 6\overline{)11}$ = 1; $11 - 6 = 5$ $\rightarrow$ $1\frac{5}{6}$
8. 5 $\quad 5\overline{)25}$ = 5; $25 - 25 = 0$ $\rightarrow$ 5
9. $2\frac{3}{8}$ $\quad 8\overline{)19}$ = 2; $19 - 16 = 3$ $\rightarrow$ $2\frac{3}{8}$
10. $4\frac{4}{7}$ $\quad 7\overline{)32}$ = 4; $32 - 28 = 4$ $\rightarrow$ $4\frac{4}{7}$
11. $\frac{7}{1}$ $\quad 7 = \frac{7}{1}$
12. $\frac{15}{8}$ $\quad 1\frac{7}{8} = \frac{(8 \times 1)}{8} + \frac{7}{8} = \frac{8+7}{8} = \frac{15}{8}$
13. $\frac{23}{5}$ $\quad 4\frac{3}{5} = \frac{(5 \times 4)}{5} + \frac{3}{5} = \frac{20+3}{5} = \frac{23}{5}$
14. $\frac{20}{3}$ $\quad 6\frac{2}{3} = \frac{(6 \times 3)}{3} + \frac{2}{3} = \frac{18+2}{3} = \frac{20}{3}$
15. $\frac{1}{2}$ $\quad \frac{5}{10} = \frac{5 \div 5}{10 \div 5} = \frac{1}{2}$
16. $\frac{3}{4}$ $\quad \frac{9}{12} = \frac{9 \div 3}{12 \div 3} = \frac{3}{4}$
17. $\frac{2}{3}$ $\quad \frac{16}{24} = \frac{16 \div 8}{24 \div 8} = \frac{2}{3}$
18. $\frac{3}{4}$ $\quad \frac{24}{32} = \frac{24 \div 8}{32 \div 8} = \frac{3}{4}$
19. $\frac{9}{18}$ $\quad \frac{1}{2} = \frac{1 \times 9}{2 \times 9} = \frac{9}{18}$
20. $\frac{20}{24}$ $\quad \frac{5}{6} = \frac{5 \times 4}{6 \times 4} = \frac{20}{24}$
21. $\frac{24}{32}$ $\quad \frac{3}{4} = \frac{3 \times 8}{4 \times 8} = \frac{24}{32}$
22. $\frac{16}{28}$ $\quad \frac{4}{7} = \frac{4 \times 4}{7 \times 4} = \frac{16}{28}$
23. $>$ $\quad \frac{4}{5} \times \frac{3}{4}$ $\quad 4 \times 4 = 16$, $5 \times 3 = 15$ $\quad 16 > 15$
24. $>$ $\quad \frac{7}{5} \times \frac{5}{6}$ $\quad 7 \times 6 = 42$, $5 \times 5 = 25$ $\quad 42 > 25$
25. $=$ $\quad \frac{4}{8} \times \frac{12}{24}$ $\quad 4 \times 24 = 96$, $8 \times 12 = 96$ $\quad 96 = 96$
26. $>$ $\quad \frac{8}{9} \times \frac{9}{12}$ $\quad 8 \times 12 = 96$, $9 \times 9 = 81$ $\quad 96 > 81$
27. $\frac{13}{40}$ $\quad$ 13 out of 40
28. $\frac{5}{36}$ $\quad$ 5 out of 36
29. (3) $\frac{3}{4}$ $\quad \frac{6}{8} = \frac{6 \div 2}{8 \div 2} = \frac{3}{4}$
30. (3) $3\frac{3}{4}$ $\quad 4\overline{)15}$ = 3 r3; $15 - 12 = 3$ $\rightarrow$ $3\frac{3}{4}$
31. (1) $\frac{9}{1}$ $\quad 9 = \frac{9}{1}$
32. (4) $\frac{3}{4}$ $\quad$ 6 ounces $= \frac{3}{4}$ cup
33. **(2) Stock 1 dropped more.**
    $\frac{3}{4} \times \frac{5}{8}$ $\quad 3 \times 8 = 24$, $4 \times 5 = 20$ $\quad \frac{3}{4} > \frac{5}{8}$
34. (2) $\frac{1}{4}$ $\quad \frac{8}{32} = \frac{8 \div 8}{32 \div 8} = \frac{1}{4}$

LESSON 6

PAGE 85

2. 36
   multiples of 4: 4, 8, 12, 16, 20, 24, 28, 32, (36)
   multiples of 9: 9, 18, 27, (36)
3. 18
   multiples of 2: 2, 4, 6, 8, 10, 12, 14, 16, (18)
   multiples of 9: 9, (18)
4. 21
   multiples of 3: 3, 6, 9, 12, 15, 18, (21)
   multiples of 7: 7, 14, (21)
5. 12
   multiples of 4: 4, 8, (12)
   multiples of 12: (12), 24
6. 15
   multiples of 3: 3, 6, 9, 12, (15)
   multiples of 5: 5, 10, (15)
7. 10
   multiples of 2: 2, 4, 6, 8, (10)
   multiples of 5: 5, (10)
   multiples of 10: (10), 20
8. 24
   multiples of 3: 3, 6, 9, 12, 15, 18, 21, (24)
   multiples of 4: 4, 8, 12, 16, 20, (24)
   multiples of 8: 8, 16, (24)
9. 20
   multiples of 4: 4, 8, 12, 16, (20)
   multiples of 5: 5, 10, 15, (20)
   multiples of 10: 10, (20)
10. $\frac{18}{30}, \frac{25}{30}$
11. $\frac{7}{35}, \frac{30}{35}$
12. $\frac{5}{10}, \frac{2}{10}$
13. $\frac{32}{40}, \frac{35}{40}$
14. $\frac{3}{4}, \frac{1}{4}$
15. $\frac{16}{24}, \frac{18}{24}$
16. $\frac{2}{3}, \frac{2}{3}$
17. $\frac{16}{40}, \frac{35}{40}$
18. $\frac{10}{15}, \frac{9}{15}$
19. $\frac{21}{28}, \frac{22}{28}$
20. $\frac{40}{48}, \frac{9}{48}$
21. $\frac{20}{24}, \frac{21}{24}$
22. $\frac{5}{8}, \frac{2}{8}, \frac{4}{8}$
    multiples of 8: (8), 16
    multiples of 4: 4, (8), 12
    multiples of 2: 2, 4, 6, (8)
    $\frac{5}{8} = \frac{5}{8}$
    $\frac{1}{4} = \frac{1 \times 2}{4 \times 2} = \frac{2}{8}$
    $\frac{1}{2} = \frac{1 \times 4}{2 \times 4} = \frac{4}{8}$
23. $\frac{20}{30}, \frac{18}{30}, \frac{15}{30}$
    multiples of 3: 3, 6, 9, 12, 15, 18, 21, 24, 27, (30)
    multiples of 5: 5, 10, 15, 20, 25, (30)
    multiples of 2: 2, 4, 6, ..., 24, 26, 28, (30)
    $\frac{2}{3} = \frac{2 \times 10}{3 \times 10} = \frac{20}{30}$
    $\frac{3}{5} = \frac{3 \times 6}{5 \times 6} = \frac{18}{30}$
    $\frac{1}{2} = \frac{1 \times 15}{2 \times 15} = \frac{15}{30}$
24. $\frac{10}{18}, \frac{6}{18}, \frac{15}{18}$
    multiples of 9: 9, (18)
    multiples of 3: 3, 6, 9, 12, 15, (18)
    multiples of 6: 6, 12, (18)
    $\frac{5}{9} = \frac{5 \times 2}{9 \times 2} = \frac{10}{18}$
    $\frac{1}{3} = \frac{1 \times 6}{3 \times 6} = \frac{6}{18}$
    $\frac{5}{6} = \frac{5 \times 3}{6 \times 3} = \frac{15}{18}$

PAGE 87

2. $1\frac{3}{5}$ $\quad \frac{7}{10} + \frac{9}{10} = \frac{16}{10} = 1\frac{6}{10} = 1\frac{3}{5}$
3. $\frac{17}{20}$ $\quad \frac{1}{4} + \frac{3}{5} = \frac{5}{20} + \frac{12}{20} = \frac{17}{20}$
4. $1\frac{1}{2}$ $\quad \frac{5}{6} + \frac{2}{3} = \frac{5}{6} + \frac{4}{6} = \frac{9}{6} = 1\frac{3}{6} = 1\frac{1}{2}$
5. $1\frac{1}{24}$ $\quad \frac{3}{8} + \frac{2}{3} = \frac{9}{24} + \frac{16}{24} = \frac{25}{24} = 1\frac{1}{24}$
6. $1\frac{3}{35}$ $\quad \frac{2}{7} + \frac{4}{5} = \frac{10}{35} + \frac{28}{35} = \frac{38}{35} = 1\frac{3}{35}$
7. $5\frac{3}{5}$
   $$\begin{array}{r} 2\frac{1}{5} \\ +3\frac{2}{5} \\ \hline 5\frac{3}{5} \end{array}$$

8. 7
$$\begin{array}{r} 4\frac{5}{12} \\ +2\frac{7}{12} \\ \hline 6\frac{12}{12} \end{array} = 6 + 1 = 7$$

9. $4\frac{11}{12}$
$$\begin{array}{rcl} 1\frac{1}{4} & = & 1\frac{3}{12} \\ +3\frac{2}{3} & = & 3\frac{8}{12} \\ \hline & & 4\frac{11}{12} \end{array}$$

10. $8\frac{1}{15}$
$$\begin{array}{rcl} 5\frac{2}{3} & = & 5\frac{10}{15} \\ +2\frac{2}{5} & = & 2\frac{6}{15} \\ \hline & & 7\frac{16}{15} \end{array} = 7 + 1\frac{1}{15} = 8\frac{1}{15}$$

11. $6\frac{19}{24}$
$$\begin{array}{rcl} 6\frac{5}{8} & = & 6\frac{15}{24} \\ +\ \frac{1}{6} & = & \frac{4}{24} \\ \hline & & 6\frac{19}{24} \end{array}$$

12. $6\frac{5}{12}$
$$\begin{array}{rcl} 3\frac{2}{3} & = & 3\frac{8}{12} \\ +2\frac{3}{4} & = & 2\frac{9}{12} \\ \hline & & 5\frac{17}{12} \end{array} = 5 + 1\frac{5}{12} = 6\frac{5}{12}$$

13. $6\frac{11}{16}$
$$\begin{array}{rcl} 4\frac{3}{8} & = & 4\frac{6}{16} \\ +2\frac{5}{16} & = & 2\frac{5}{16} \\ \hline & & 6\frac{11}{16} \end{array}$$

14. $21\frac{7}{16}$
$$\begin{array}{rcl} 12\frac{1}{4} & = & 12\frac{4}{16} \\ +\ 9\frac{3}{16} & = & 9\frac{3}{16} \\ \hline & & 21\frac{7}{16} \end{array}$$

15. $9\frac{5}{12}$
$$\begin{array}{rcl} 5\frac{3}{4} & = & 5\frac{9}{12} \\ +3\frac{2}{3} & = & 3\frac{8}{12} \\ \hline & & 8\frac{17}{12} \end{array} = 8 + 1\frac{5}{12} = 9\frac{5}{12}$$

16. $4\frac{1}{6}$
$$\begin{array}{rcl} 1\frac{3}{4} & = & 1\frac{9}{12} \\ 1\frac{1}{4} & = & 1\frac{3}{12} \\ \frac{1}{2} & = & \frac{6}{12} \\ +\ \frac{2}{3} & = & \frac{8}{12} \\ \hline & & 2\frac{26}{12} \end{array} = 2 + 2\frac{2}{12} = 4\frac{1}{6}$$

PAGE 89

2. $\frac{11}{24}$ $\quad \frac{5}{6} - \frac{3}{8} =$
$\frac{20}{24} - \frac{9}{24} = \frac{11}{24}$

3. $\frac{5}{16}$ $\quad \frac{9}{16} - \frac{1}{4} =$
$\frac{9}{16} - \frac{4}{16} = \frac{5}{16}$

4. $3\frac{1}{3}$ $\quad 6 - 2\frac{2}{3} =$
$5\frac{3}{3} - 2\frac{2}{3} = 3\frac{1}{3}$

5. $\frac{5}{24}$ $\quad \frac{7}{8} - \frac{2}{3} =$
$\frac{21}{24} - \frac{16}{24} = \frac{5}{24}$

6. $\frac{1}{20}$ $\quad \frac{4}{5} - \frac{3}{4} =$
$\frac{16}{20} - \frac{15}{20} = \frac{1}{20}$

7. $3\frac{1}{2}$
$$\begin{array}{r} 4\frac{7}{8} \\ -1\frac{3}{8} \\ \hline 3\frac{4}{8} \end{array} = 3\frac{1}{2}$$

8. $5\frac{1}{12}$
$$\begin{array}{rcl} 7\frac{3}{4} & = & 7\frac{9}{12} \\ -2\frac{2}{3} & = & 2\frac{8}{12} \\ \hline & & 5\frac{1}{12} \end{array}$$

9. $2\frac{7}{9}$
$$\begin{array}{rcl} 8 & = & \overset{7}{\not 8}\frac{9}{9} \\ -5\frac{2}{9} & = & 5\frac{2}{9} \\ \hline & & 2\frac{7}{9} \end{array}$$

10. $2\frac{1}{3}$
$$\begin{array}{rcl} 6\frac{1}{6} & = & \overset{5}{\not 6}\frac{\overset{7}{\not 1}}{6} \\ -3\frac{5}{6} & = & 3\frac{5}{6} \\ \hline & & 2\frac{2}{6} \end{array} = 2\frac{1}{3}$$

11. $\frac{5}{8}$
$$\begin{array}{rcccl} 9\frac{1}{4} & = & 9\frac{2}{8} & = & \overset{8}{\not 9}\frac{\overset{10}{\not 2}}{8} \\ -8\frac{5}{8} & = & 8\frac{5}{8} & = & 8\frac{5}{8} \\ \hline & & & & \frac{5}{8} \end{array}$$

12. $5\frac{11}{15}$
$$\begin{array}{rcccl} 7\frac{2}{5} & = & 7\frac{6}{15} & = & \overset{6}{\not 7}\frac{\overset{21}{\not 6}}{15} \\ -1\frac{2}{3} & = & 1\frac{10}{15} & = & 1\frac{10}{15} \\ \hline & & & & 5\frac{11}{15} \end{array}$$

13. $6\frac{1}{2}$
$$\begin{array}{r} 9\frac{3}{4} \\ -3\frac{1}{4} \\ \hline 6\frac{2}{4} \end{array} = 6\frac{1}{2}$$

14. $26\frac{7}{16}$
$$\begin{array}{rcl} 35 & = & 34\frac{16}{16} \\ -\ 8\frac{9}{16} & = & 8\frac{9}{16} \\ \hline & & 26\frac{7}{16} \end{array}$$

15. $2\frac{1}{4}$
$$\begin{array}{rcl} 4\frac{3}{4} & = & 4\frac{3}{4} \\ -2\frac{1}{2} & = & 2\frac{2}{4} \\ \hline & & 2\frac{1}{4} \end{array}$$

16. $3\frac{5}{6}$

$$\begin{array}{rcl} 8\frac{1}{2} & = & \overset{7}{\not 8}\frac{\overset{9}{\not 3}}{6} \\ -4\frac{2}{3} & = & 4\frac{4}{6} \\ \hline & & 3\frac{5}{6} \end{array}$$

PAGE 91

1. **(3) $1\frac{3}{4} + 2$** Use the row for jumper. Add the number of yards needed for size 4 ($1\frac{3}{4}$) and for size 6 (2).
2. **(2) $1\frac{7}{8} + 1\frac{1}{8} + \frac{3}{4}$** Use the column for size 5. Add the number of yards needed for the jumper ($1\frac{7}{8}$); blouse B ($1\frac{1}{8}$); and facing, blouse B ($\frac{3}{4}$).
3. **(5) $5 - 1\frac{3}{4}$** Subtract the yards needed for the jumper ($1\frac{3}{4}$) from the number of yards of fabric Russ has (5).
4. **(1) $1\frac{3}{4} + 1\frac{3}{8}$** Add the number of yards needed for the jumper in size 4 ($1\frac{3}{4}$) and blouse A in size 6 ($1\frac{3}{8}$) to find the total number of yards of fabric.
5. **(2) blouse A, size 6** $1\frac{1}{2} = 1\frac{4}{8}$, $1\frac{4}{8} > 1\frac{3}{8}$. Yvonne has enough fabric to make blouse A in size 6. Options (1), (4), and (5), are all patterns smaller than size 6. Option (3) is a pattern in size 6, but requires more fabric than Yvonne has.

PAGE 93

1. **(5) $17\frac{1}{2}$** Add each distance to find the total

$$\begin{array}{rcl} 6\frac{3}{10} & = & 6\frac{3}{10} \\ 6\frac{4}{5} & = & 6\frac{8}{10} \\ +4\frac{2}{5} & = & 4\frac{4}{10} \\ \hline & & 16\frac{15}{10} \end{array}$$

$= 16 + 1\frac{5}{10} = 17\frac{5}{10} = 17\frac{1}{2}$ miles

2. **(4) $1\frac{1}{10}$** One-Stop Gas and Food is located between the city pool and the hospital. Jose drives $3\frac{3}{10}$ miles from the city pool to One-Stop Gas and Food and $4\frac{2}{5}$ miles from the city pool to the hospital. Subtract to find the distance from One-Stop Gas and Food to the hospital.

$$\begin{array}{rcl} 4\frac{2}{5} & = & 4\frac{4}{10} \\ -3\frac{3}{10} & = & 3\frac{3}{10} \\ \hline & & 1\frac{1}{10} \end{array}$$

3. **(1) $8\frac{7}{10}$** Add each distance to find the total distance.

$$\begin{array}{r} 1\frac{1}{10} \\ 1\frac{9}{10} \\ +5\frac{7}{10} \\ \hline 7\frac{17}{10} \end{array} = 7 + 1\frac{7}{10} = 8\frac{7}{10}$$

4. **(4) $\frac{3}{5}$** Subtract:
$6\frac{3}{10} - 5\frac{7}{10} = 5\frac{13}{10} - 5\frac{7}{10} = \frac{6}{10} = \frac{3}{5}$
5. **(5) Not enough information is given.** The map does not show where José lives. Therefore, you do not know the distance he lives from work.
6. $6\frac{3}{10}$ rounds to 6
$6\frac{4}{5}$ rounds to 7
$4\frac{2}{5}$ rounds to 4
$1\frac{1}{10}$ rounds to 1
$1\frac{9}{10}$ rounds to 2
$5\frac{7}{10}$ rounds to 6
Jose's route is about 26 miles.
(6 + 7 + 4 + 1 + 2 + 6)

PAGE 95

1. **(3) 8** Kareem adds $4\frac{1}{4}$ and $3\frac{3}{4}$ hours to find the total.

$$\begin{array}{r} 4\frac{1}{4} \\ +3\frac{3}{4} \\ \hline 7\frac{4}{4} \end{array} = 7 + 1 = 8$$

2. **(3)** $3\frac{3}{4}$ Since 50 is close to 45, think of 8:50 as $8\frac{3}{4}$. Since 32 is close to 30, think of 12:32 as $12\frac{1}{2}$. Then subtract.

$$\begin{array}{rcrcr} 12\frac{1}{2} & = & 12\frac{2}{4} & & \overset{11}{\cancel{12}}\overset{6}{\cancel{\frac{2}{4}}} \\ -\ 8\frac{3}{4} & = & -\ 8\frac{3}{4} & & -\ 8\frac{3}{4} \\ \hline & & & & 3\frac{3}{4} \text{ hours} \end{array}$$

3. **(1)** $3\frac{3}{4}$ Since 24 is close to 30, think of 1:24 as $1\frac{1}{2}$. Since 10 is close to 15, think of 5:10 as $5\frac{1}{4}$. Then subtract.

$$\begin{array}{rcrcr} 5\frac{1}{4} & = & 5\frac{1}{4} & = & \overset{4}{\cancel{5}}\overset{5}{\cancel{\frac{1}{4}}} \\ -1\frac{1}{2} & = & 1\frac{2}{4} & = & -1\frac{2}{4} \\ \hline & & & & 3\frac{3}{4} \text{ hours} \end{array}$$

4. **(1)** $7\frac{1}{2}$ Add the hours Paul worked in the morning and the hours he worked in the afternoon.

$$\begin{array}{r} 3\frac{3}{4} \\ +3\frac{3}{4} \\ \hline 6\frac{6}{4} \end{array} = 6 + 1\frac{2}{4} = 7\frac{2}{4} = 7\frac{1}{2}$$

5. **(5) Not enough information is given.** You need to know what time Paul went to lunch to answer the question.
6. **(3)** $2\frac{1}{2}$ $40\frac{3}{4} - 38\frac{1}{4} = 2\frac{2}{4} = 2\frac{1}{2}$
7. **Create a fraction with a numerator based on the number of minutes being discussed and a denominator based on the number of minutes in an hour.**

20 minutes = $\frac{1}{3}$ hour
20 minutes $= \frac{20}{60} = \frac{20 \div 20}{60 \div 20} = \frac{1}{3}$

40 minutes = $\frac{2}{3}$ hour
40 minutes $= \frac{40}{60} = \frac{40 \div 20}{60 \div 20} = \frac{2}{3}$

50 minutes = $\frac{5}{6}$ hour
50 minutes $= \frac{50}{60} = \frac{50 \div 10}{60 \div 10} = \frac{5}{6}$

PAGE 97

1. $1\frac{1}{16}$
2. $\frac{1}{15}$
3. $1\frac{11}{40}$
4. $\frac{1}{5}$
5. $1\frac{7}{40}$
6. $4\frac{1}{8}$
7. $2\frac{9}{10}$
8. $9\frac{11}{24}$
9. $2\frac{23}{30}$
10. $7\frac{37}{40}$
11. $4\frac{1}{8}$ **feet**

$$\begin{array}{rcr} 8\frac{1}{4} & = & 8\frac{2}{8} \\ -4\frac{1}{8} & = & -4\frac{1}{8} \\ \hline & & 4\frac{1}{8} \end{array}$$

12. $20\frac{13}{16}$

$$\begin{array}{rcr} 12\frac{1}{2} & = & 12\frac{8}{16} \\ +\ 8\frac{5}{16} & = & +\ 8\frac{5}{16} \\ \hline & & 20\frac{13}{16} \end{array}$$

13. $4\frac{1}{4}$

$$\begin{array}{rcr} 2\frac{3}{4} & = & 2\frac{3}{4} \\ +\ 1\frac{1}{2} & = & +\ 1\frac{2}{4} \\ \hline & & 4\frac{1}{4} \end{array}$$

14. $4\frac{1}{6}$

$$\begin{array}{rcr} 42\frac{1}{2} & = & 42\frac{3}{6} \\ -\ 38\frac{1}{3} & = & -38\frac{2}{6} \\ \hline & & 4\frac{1}{6} \end{array}$$

PAGES 98–99

1. $\frac{3}{8}$ $\frac{15}{16} - \frac{9}{16} = \frac{6}{16} = \frac{3}{8}$
2. $\frac{9}{10}$ $\frac{1}{5} + \frac{7}{10} =$
$\frac{2}{10} + \frac{7}{10} = \frac{9}{10}$
3. $5\frac{5}{6}$

$$\begin{array}{rcr} 8 & = & \overset{7}{\cancel{8}}\frac{6}{6} \\ -2\frac{1}{6} & = & 2\frac{1}{6} \\ \hline & & 5\frac{5}{6} \end{array}$$

4. $1\frac{5}{24}$ $\quad \frac{5}{6} + \frac{3}{8} =$

$$\frac{20}{24} + \frac{9}{24} = \frac{29}{24} = 1\frac{5}{24}$$

5. 8

$$\begin{array}{r} 3\frac{7}{9} \\ +4\frac{2}{9} \\ \hline 7\frac{9}{9} = 7 + 1 = 8 \end{array}$$

6. $2\frac{3}{5}$

$$\begin{array}{r} 4\frac{9}{10} \\ -2\frac{3}{10} \\ \hline 2\frac{6}{10} = 2\frac{3}{5} \end{array}$$

7. $5\frac{3}{8}$

$$\begin{array}{rl} 2\frac{5}{8} = & 2\frac{5}{8} \\ +2\frac{3}{4} = & 2\frac{6}{8} \\ \hline & 4\frac{11}{8} = 4 + 1\frac{3}{8} = 5\frac{3}{8} \end{array}$$

8. $\frac{3}{5}$

9. $2\frac{1}{8}$

10. $9\frac{1}{12}$

11. **(3)** $7\frac{1}{8}$ Use the column for size 12. Add the number of yards of fabric in the main color ($5\frac{5}{8}$) and the number of yards of contrasting fabric ($1\frac{1}{2}$) needed for dress B.

$$\begin{array}{rl} 5\frac{5}{8} = & 5\frac{5}{8} \\ +1\frac{1}{2} = & 1\frac{4}{8} \\ \hline & 6\frac{9}{8} = 6 + 1\frac{1}{8} = 7\frac{1}{8} \end{array}$$

12. **(3)** $\frac{3}{8}$ Use the column for size 8. Subtract the number of yards of lace needed for dress B ($2\frac{1}{8}$) from the number of yards of lace needed for dress A ($2\frac{1}{2}$).

$$\begin{array}{rl} 2\frac{1}{2} = & 2\frac{4}{8} \\ -2\frac{1}{8} = & 2\frac{1}{8} \\ \hline & \frac{3}{8} \end{array}$$

13. **(3) $5\frac{1}{8}$ yards** Add the amounts of lace needed for dress A in size 8 ($2\frac{1}{2}$) and size 10 ($2\frac{5}{8}$).

$$\begin{array}{rl} 2\frac{1}{2} = & 2\frac{4}{8} \\ +\ 2\frac{5}{8} = & 2\frac{5}{8} \\ \hline & 4\frac{9}{8} = 4 + 1\frac{1}{8} = 5\frac{1}{8} \text{ yards} \end{array}$$

14. **6** Sundra drives $3\frac{2}{5}$ miles from the courthouse to the art museum, $1\frac{1}{5}$ miles from the art museum to the science museum, and $1\frac{9}{10}$ miles from the science museum to the shopping mall. To estimate the total distance, round each distance and add.

$$\begin{array}{rr} 3\frac{2}{5} \text{ rounds to} & 3 \\ 1\frac{1}{5} \text{ rounds to} & 1 \\ +1\frac{9}{10} \text{ rounds to} & 2 \\ \hline & 6 \end{array}$$

15. $6\frac{1}{2}$ Add each distance to find the total distance. See the explanation for Item 14.

$$\begin{array}{rl} 3\frac{2}{5} = & 3\frac{4}{10} \\ 1\frac{1}{5} = & 1\frac{2}{10} \\ +1\frac{9}{10} = & 1\frac{9}{10} \\ \hline & 5\frac{15}{10} = 5 + 1\frac{5}{10} = 6\frac{5}{10} = 6\frac{1}{2} \end{array}$$

16. **$8\frac{3}{5}$ miles** Use your answer from Question 15 ($6\frac{1}{2}$). Add the distance from the mall to the bus depot.

$$\begin{array}{rl} 6\frac{1}{2} = & 6\frac{5}{10} \\ +\ 2\frac{1}{10} = & 2\frac{1}{10} \\ \hline & 8\frac{6}{10} = 8\frac{3}{5} \text{ miles} \end{array}$$

17. **12** Sundra drives $5\frac{3}{10}$ miles from the bus depot to the library. The entire route is $17\frac{1}{5}$ miles. Round each distance and subtract.

$$\begin{array}{rr} 17\frac{1}{5} \text{ rounds to} & 17 \\ -\ 5\frac{3}{10} \text{ rounds to} & 5 \\ \hline & 12 \end{array}$$

18. $3\frac{3}{4}$ Since 24 is close to 30, think of 8:24 as $8\frac{1}{2}$. Since 13 is close to 15, think of 12:13 as $12\frac{1}{4}$. Then subtract.

$$\begin{array}{rlll} 12\frac{1}{4} = & 12\frac{1}{4} = & \overset{11}{\cancel{12}}\frac{\overset{5}{\cancel{1}}}{4} \\ -\ 8\frac{1}{2} = & 8\frac{2}{4} = & 8\frac{2}{4} \\ \hline & & 3\frac{3}{4} \end{array}$$

19. $4\frac{3}{4}$ Since 58 is close to 60, think of 12:58 as 1. Since 43 is close to 45, think of 5:43 as $5\frac{3}{4}$. Then subtract.

$$\begin{array}{r} 5\frac{3}{4} \\ -1 \\ \hline 4\frac{3}{4} \end{array}$$

20. $8\frac{1}{2}$ Add the hours Joy worked in the morning ($3\frac{3}{4}$) and the hours she worked in the afternoon ($4\frac{3}{4}$).

$$\begin{array}{l} 3\frac{3}{4} \\ +4\frac{3}{4} \\ \hline 7\frac{6}{4} = 7 + 1\frac{2}{4} = 8\frac{2}{4} = 8\frac{1}{2} \end{array}$$

LESSON 7

PAGE 101

2. $\frac{3}{10}$ $\quad \frac{3}{8}\times\frac{4}{5} = \frac{3}{\underset{2}{\cancel{8}}}\times\frac{\overset{1}{\cancel{4}}}{5} = \frac{3}{10}$

3. $\frac{2}{21}$ $\quad \frac{\overset{1}{\cancel{3}}}{7}\times\frac{2}{\underset{3}{\cancel{9}}} = \frac{2}{21}$

4. $2\frac{1}{4}$ $\quad \frac{3}{8}\times 6 = \frac{3}{\underset{4}{\cancel{8}}}\times\frac{\overset{3}{\cancel{6}}}{1} = \frac{9}{4} = 2\frac{1}{4}$

5. $11\frac{1}{2}$ $\quad 5\times 2\frac{3}{10} = \frac{\overset{1}{\cancel{5}}}{1}\times\frac{23}{\underset{2}{\cancel{10}}} = \frac{23}{2} = 11\frac{1}{2}$

6. $1\frac{1}{2}$ $\quad \frac{5}{8}\times 2\frac{2}{5} = \frac{\overset{1}{\cancel{5}}}{\underset{2}{\cancel{8}}}\times\frac{\overset{3}{\cancel{12}}}{\underset{1}{\cancel{5}}} = \frac{3}{2} = 1\frac{1}{2}$

7. $3\frac{1}{9}$ $\quad 3\frac{1}{2}\times\frac{8}{9} = \frac{7}{\underset{1}{\cancel{2}}}\times\frac{\overset{4}{\cancel{8}}}{9} = \frac{28}{9} = 3\frac{1}{9}$

8. $3\frac{1}{2}$ $\quad 2\frac{1}{3}\times 1\frac{1}{2} = \frac{7}{\underset{1}{\cancel{3}}}\times\frac{\overset{1}{\cancel{3}}}{2} = \frac{7}{2} = 3\frac{1}{2}$

9. 12 $\quad 1\frac{4}{5}\times 6\frac{2}{3} = \frac{\overset{3}{\cancel{9}}}{\underset{1}{\cancel{5}}}\times\frac{\overset{4}{\cancel{20}}}{\underset{1}{\cancel{3}}} = \frac{12}{1} = 12$

10. 18 $\quad \frac{3}{4}\times 24 = \frac{3}{\underset{1}{\cancel{4}}}\times\frac{\overset{6}{\cancel{24}}}{1} = \frac{18}{1} = 18$

11. $5\frac{1}{4}$ miles $\quad 3\frac{1}{2}\times 1\frac{1}{2} = \frac{7}{2}\times\frac{3}{2} = \frac{21}{4} = 5\frac{1}{4}$

12. $10\frac{1}{2}$ $\quad 15\frac{3}{4}\times\frac{2}{3} = \frac{\overset{21}{\cancel{63}}}{\underset{2}{\cancel{4}}}\times\frac{\overset{1}{\cancel{2}}}{\underset{1}{\cancel{3}}} = \frac{21}{2} = 10\frac{1}{2}$

13. $25\frac{1}{2}$ $\quad 4\frac{1}{4}\times 6 = \frac{17}{\underset{2}{\cancel{4}}}\times\frac{\overset{3}{\cancel{6}}}{1} = \frac{51}{2} = 25\frac{1}{2}$

PAGE 103

2. $1\frac{1}{3}$ $\quad \frac{5}{6}\div\frac{5}{8} = \frac{\overset{1}{\cancel{5}}}{\underset{3}{\cancel{6}}}\times\frac{\overset{4}{\cancel{8}}}{\underset{1}{\cancel{5}}} = \frac{4}{3} = 1\frac{1}{3}$

3. $\frac{1}{3}$ $\quad \frac{4}{15}\div\frac{4}{5} = \frac{\overset{1}{\cancel{4}}}{\underset{3}{\cancel{15}}}\times\frac{\overset{1}{\cancel{5}}}{\underset{1}{\cancel{4}}} = \frac{1}{3}$

4. $\frac{4}{5}$ $\quad \frac{2}{3}\div\frac{5}{6} = \frac{2}{\underset{1}{\cancel{3}}}\times\frac{\overset{2}{\cancel{6}}}{5} = \frac{4}{5}$

5. $\frac{7}{36}$ $\quad \frac{7}{12}\div 3 = \frac{7}{12}\div\frac{3}{1} = \frac{7}{12}\times\frac{1}{3} = \frac{7}{36}$

6. $1\frac{3}{4}$ $\quad 2\div 1\frac{1}{7} = \frac{2}{1}\div\frac{8}{7} = \frac{\overset{1}{\cancel{2}}}{1}\times\frac{7}{\underset{4}{\cancel{8}}} = \frac{7}{4} = 1\frac{3}{4}$

7. $\frac{15}{44}$ $\quad \frac{5}{8}\div 1\frac{5}{6} = \frac{5}{8}\div\frac{11}{6} = \frac{5}{\underset{4}{\cancel{8}}}\times\frac{\overset{3}{\cancel{6}}}{11} = \frac{15}{44}$

8. $12\frac{1}{2}$ $\quad 3\frac{3}{4}\div\frac{3}{10} = \frac{15}{4}\div\frac{3}{10} = \frac{\overset{5}{\cancel{15}}}{\underset{2}{\cancel{4}}}\times\frac{\overset{5}{\cancel{10}}}{\underset{1}{\cancel{3}}} = \frac{25}{2} = 12\frac{1}{2}$

9. $\frac{28}{45}$ $\quad 1\frac{2}{5}\div 2\frac{1}{4} = \frac{7}{5}\div\frac{9}{4} = \frac{7}{5}\times\frac{4}{9} = \frac{28}{45}$

10. $2\frac{5}{6}$ $\quad 4\frac{1}{4}\div 1\frac{1}{2} = \frac{17}{4}\div\frac{3}{2} = \frac{17}{\underset{2}{\cancel{4}}}\times\frac{\overset{1}{\cancel{2}}}{3} = \frac{17}{6} = 2\frac{5}{6}$

11. 24 $\quad 15\div\frac{5}{8} = \frac{\overset{3}{\cancel{15}}}{1}\times\frac{8}{\underset{1}{\cancel{5}}} = 24$

12. 22
$5\frac{1}{2}\div\frac{1}{4} = \frac{11}{2}\div\frac{1}{4} = \frac{11}{\underset{1}{\cancel{2}}}\times\frac{\overset{2}{\cancel{4}}}{1} = \frac{22}{1} = 22$

13. $2\frac{1}{2}$ $\quad 6\frac{1}{4}\div 2\frac{1}{2} = \frac{25}{4}\div\frac{5}{2} = \frac{\overset{5}{\cancel{25}}}{\underset{2}{\cancel{4}}}\times\frac{\overset{1}{\cancel{2}}}{\underset{1}{\cancel{5}}} = \frac{5}{2} = 2\frac{1}{2}$

14. 10
$12\frac{1}{2}\div 1\frac{1}{4} = \frac{25}{2}\div\frac{5}{4} = \frac{\overset{5}{\cancel{25}}}{\underset{1}{\cancel{2}}}\times\frac{\overset{2}{\cancel{4}}}{\underset{1}{\cancel{5}}} = \frac{10}{1} = 10$

PAGE 105

1. (2) 100 Multiply the length of each shelf ($1\frac{2}{3}$ feet) by the number of shelves (5). Then multiply by 12 to change feet to inches.

$1\frac{2}{3}\times 5 = \frac{5}{3}\times\frac{5}{1} = \frac{25}{3} = 8\frac{1}{3}$ feet

$8\frac{1}{3}\times 12 = \frac{25}{\underset{1}{\cancel{3}}}\times\frac{\overset{4}{\cancel{12}}}{1} = \frac{100}{1} = 100$ inches

2. (4) 4

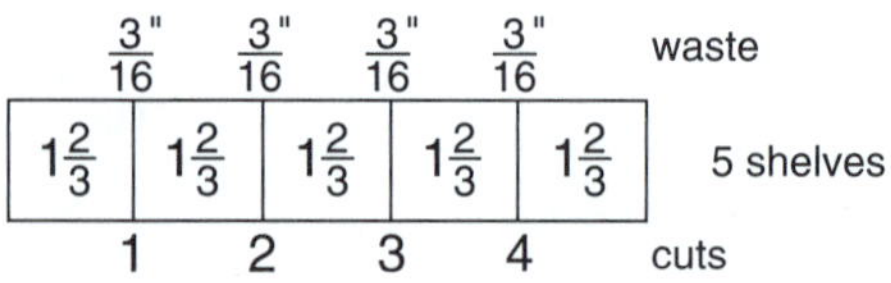

3. **(5)** $\frac{3}{4}$ Multiply the waste per cut ($\frac{3}{16}$ inch) by the number of cuts (4; see the diagram for Question 2).

$\frac{3}{16} \times 4 = \frac{3}{\cancel{16}_4} \times \frac{\cancel{4}^1}{1} = \frac{3}{4}$ inch

4. **(4)** $27\frac{1}{2}$ Multiply the length of each board ($6\frac{7}{8}$ feet) by the number of shelves (4). $6\frac{7}{8} \times 4 = 24\frac{28}{8}$ or $24\frac{7}{2} = 24 + 3\frac{1}{2} = 27\frac{1}{2}$.

5. **(2)** $\frac{3}{4}$ Divide the total waste ($2\frac{1}{4}$ inches) by the number of cuts (3). $2\frac{1}{4} \div 3 = \frac{9}{4} \div 3 = \frac{\cancel{9}^3}{4} \times \frac{1}{\cancel{3}_1} = \frac{3}{4}$.

6. **Your diagram should show a board with five cuts separating the board into six equal pieces.**

PAGE 107

1. $3\frac{1}{2}$
2. $\frac{32}{45}$
3. $\frac{9}{16}$
4. 15
5. $1\frac{5}{16}$
6. 2
7. 335
8. 10
9. $4\frac{1}{20}$
10. 20
11. 8 $20 \div 2\frac{1}{2} = 8$
12. $56\frac{1}{4}$ $3\frac{3}{4} \times 15 = 56\frac{1}{4}$
13. $32\frac{2}{3}$ $2\frac{1}{3} \times 14 = 32\frac{2}{3}$
14. $\frac{3}{4}$ **pound**

$4\frac{1}{2} \div 6 = \frac{3}{4}$

PAGE 109

1. $\frac{41}{12}$

$1\frac{2}{3} + 1\frac{3}{4} = \frac{5}{3} + \frac{7}{4} = \frac{20}{12} + \frac{21}{12} = \frac{41}{12}$

Remember, change mixed numbers to improper fractions to fill in the grid.

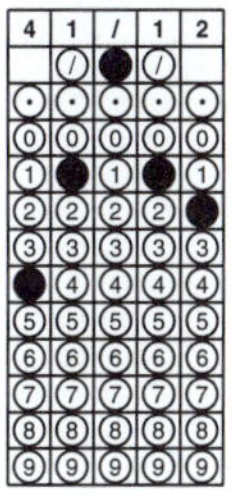

2. $\frac{11}{8}$

$5\frac{1}{2} \div 4 = \frac{11}{2} \div \frac{4}{1} = \frac{11}{2} \times \frac{1}{4} = \frac{11}{8}$

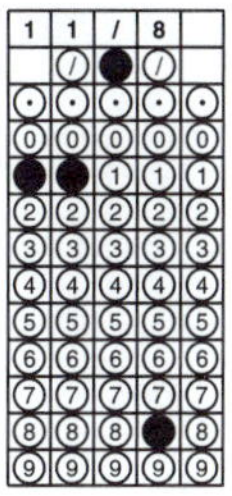

3. $\frac{25}{2}$

$2\frac{1}{2} \times 5 = \frac{5}{2} \times \frac{5}{1} = \frac{25}{2}$

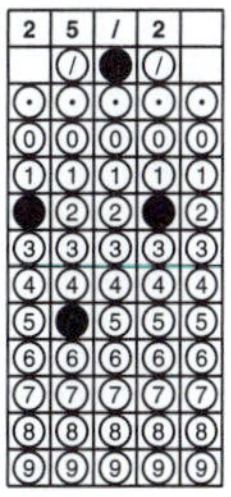

4. $\frac{3}{2}$

$9\frac{1}{4} - 7\frac{3}{4} = 8\frac{5}{4} - 7\frac{3}{4} = 1\frac{1}{2} = \frac{3}{2}$

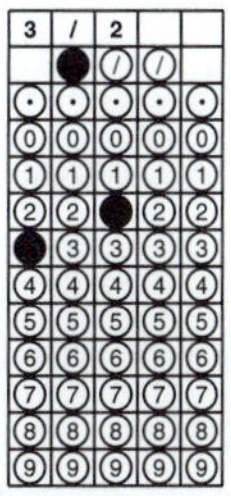

PAGES 110–111

1. $\frac{4}{15}$ $\frac{4}{5}\times\frac{1}{3}=\frac{4}{15}$

2. $2\frac{2}{3}$ $\frac{4}{9}\div\frac{1}{6}=\frac{4}{\underset{3}{\cancel{9}}}\times\frac{\overset{2}{\cancel{6}}}{1}=\frac{8}{3}=2\frac{2}{3}$

3. $1\frac{1}{24}$ $\frac{5}{8}\div\frac{3}{5}=\frac{5}{8}\times\frac{5}{3}=\frac{25}{24}=1\frac{1}{24}$

4. $\frac{5}{42}$ $\frac{\overset{1}{\cancel{2}}}{7}\times\frac{5}{\underset{6}{\cancel{12}}}=\frac{5}{42}$

5. 8 $7\div\frac{7}{8}=\frac{7}{1}\div\frac{7}{8}=\frac{\overset{1}{\cancel{7}}}{1}\times\frac{8}{\underset{1}{\cancel{7}}}=\frac{8}{1}=8$

6. $4\frac{1}{2}$ $6\times\frac{3}{4}=\frac{\overset{3}{\cancel{6}}}{1}\times\frac{3}{\underset{2}{\cancel{4}}}=\frac{9}{2}=4\frac{1}{2}$

7. $\frac{6}{25}$ $\frac{\overset{2}{\cancel{4}}}{\underset{5}{\cancel{15}}}\times\frac{\overset{3}{\cancel{9}}}{\underset{5}{\cancel{10}}}=\frac{6}{25}$

8. $\frac{9}{10}$ $\frac{2}{5}\div\frac{4}{9}=\frac{\overset{1}{\cancel{2}}}{5}\times\frac{9}{\underset{2}{\cancel{4}}}=\frac{9}{10}$

9. $11\frac{1}{3}$ $4\times2\frac{5}{6}=\frac{\overset{2}{\cancel{4}}}{1}\times\frac{17}{\underset{3}{\cancel{6}}}=\frac{34}{3}=11\frac{1}{3}$

10. $4\frac{1}{2}$ $5\div1\frac{1}{9}=\frac{5}{1}\div\frac{10}{9}=\frac{\overset{1}{\cancel{5}}}{1}\times\frac{9}{\underset{2}{\cancel{10}}}=\frac{9}{2}=4\frac{1}{2}$

11. $1\frac{5}{6}$ $2\frac{3}{4}\times\frac{2}{3}=\frac{11}{\underset{2}{\cancel{4}}}\times\frac{\overset{1}{\cancel{2}}}{3}=\frac{11}{6}=1\frac{5}{6}$

12. $3\frac{3}{4}$ $4\frac{1}{2}\div1\frac{1}{5}=\frac{9}{2}\div\frac{6}{5}=\frac{\overset{3}{\cancel{9}}}{2}\times\frac{5}{\underset{2}{\cancel{6}}}=\frac{15}{4}=3\frac{3}{4}$

13. $1\frac{1}{3}$ $3\frac{1}{5}\div2\frac{2}{5}=\frac{16}{5}\div\frac{12}{5}=\frac{\overset{4}{\cancel{16}}}{\underset{1}{\cancel{5}}}\times\frac{\overset{1}{\cancel{5}}}{\underset{3}{\cancel{12}}}=\frac{4}{3}=1\frac{1}{3}$

14. $4\frac{1}{7}$ $2\frac{5}{12}\times1\frac{5}{7}=\frac{29}{\underset{1}{\cancel{12}}}\times\frac{\overset{1}{\cancel{12}}}{7}=\frac{29}{7}=4\frac{1}{7}$

15. $2\frac{1}{10}$

16. $\frac{1}{3}$

17. $3\frac{3}{4}$

18. $\frac{4}{5}$

19. **(3) $\frac{3}{4}\times 2$** To double the recipe, multiply the amount of bread crumbs by 2.

20. **(2) $\frac{5}{8}$** Since 6 is one third of 18, Carla needs to make one third of the recipe. Multiply the amount of mayonnaise by $\frac{1}{3}$.

$$1\frac{7}{8}\times\frac{1}{3}=\frac{\overset{5}{\cancel{15}}}{8}\times\frac{1}{\underset{1}{\cancel{3}}}=\frac{5}{8}$$

21. **(4) 6**

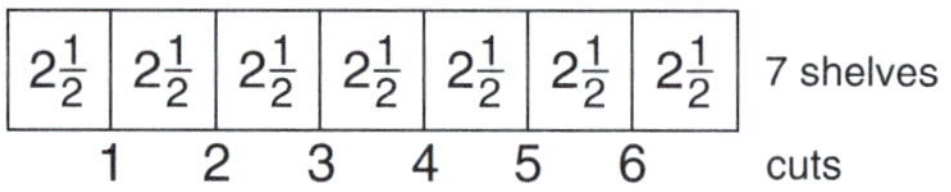

22. **$\frac{5}{2}$** Multiply the hours worked by the part of the day spent answering the phone. Remember to enter a mixed number as an improper fraction on a grid.

$$7\frac{1}{2}\times\frac{1}{3}=\frac{\overset{5}{\cancel{15}}}{2}\times\frac{1}{\underset{1}{\cancel{3}}}=\frac{5}{2}$$

23. **40** Divide the total width by the width of each strip.

$$45\div1\frac{1}{8}=\frac{45}{1}\div\frac{9}{8}=\frac{\overset{5}{\cancel{45}}}{1}\times\frac{8}{\underset{1}{\cancel{9}}}=\frac{40}{1}=40$$

MATH AT WORK

PAGE 113

1. **(3) $6\frac{1}{8}$ feet** The length of the whole wall with the door is $15\frac{1}{4}$ feet, the same as the length of the opposite wall. Subtract the sum of the door (3 ft.) and the part of the wall to the left of the door ($6\frac{1}{8}$ ft.) from the length of the whole wall to get the part of the wall to the right of the door.

   $15\frac{1}{4} - (3 + 6\frac{1}{8}) = 15\frac{1}{4} - 9\frac{1}{8} = 15\frac{2}{8} - 9\frac{1}{8} = 6\frac{1}{8}$

2. **(2) $46\frac{1}{2}$** Jake needs baseboard for the perimeter minus the door. (There is no baseboard on the door.)

   $P = (2l + 2w) - \text{door}$
   $= (2 \times 15\frac{1}{4} + 2 \times 9\frac{1}{2}) - 3$
   $= (\frac{\cancel{2}^1}{1} \times \frac{61}{\cancel{4}_2} + \frac{\cancel{2}^1}{1} \times \frac{19}{\cancel{2}_1}) - 3$
   $= (\frac{61}{2} + 19) - 3$
   $= 30\frac{1}{2} + 19 - 3 = 46\frac{1}{2}$

3. **Jake needs a total of 96 feet of stripping material for this room.**

   $15\frac{1}{4} \times 2 = 30\frac{1}{2}$ feet (two lengths of room)

   $9\frac{1}{2} \times 2 = 19$ feet (two widths of room)

   $30\frac{1}{2} + 19 = 49\frac{1}{2}$ feet (2 lengths + two widths of ceiling molding)

   $49\frac{1}{2} - 3 = 46\frac{1}{2}$ feet (perimeter of ceiling minus the door area = baseboard)

   $49\frac{1}{2} + 46\frac{1}{2} = 96$ feet (baseboard and ceiling molding lengths)

UNIT 2 REVIEW

PAGES 114–115

1. $1\frac{7}{12}$ $\frac{5}{6} + \frac{3}{4} =$
   $\frac{10}{12} + \frac{9}{12} = \frac{19}{12} = 1\frac{7}{12}$

2. $\frac{7}{30}$ $\frac{9}{10} - \frac{2}{3} =$
   $\frac{27}{30} - \frac{20}{30} = \frac{7}{30}$

3. $\frac{19}{20}$ $\frac{2}{5} + \frac{3}{10} + \frac{1}{4} = \frac{8}{20} + \frac{6}{20} + \frac{5}{20} = \frac{19}{20}$

4. $7\frac{5}{24}$
   $4\frac{7}{8} = 4\frac{21}{24}$
   $+2\frac{1}{3} = 2\frac{8}{24}$
   $6\frac{29}{24} = 6 + 1\frac{5}{24} = 7\frac{5}{24}$

5. $1\frac{13}{20}$
   $3\frac{2}{5} = 3\frac{8}{20} = \cancel{3}^2\frac{\cancel{8}^{28}}{20}$
   $-1\frac{3}{4} = 1\frac{15}{20} = 1\frac{15}{20}$
   $1\frac{13}{20}$

6. $\frac{14}{15}$ $\frac{7}{12} \div \frac{5}{8} = \frac{7}{\cancel{12}_3} \times \frac{\cancel{8}^2}{5} = \frac{14}{15}$

7. $\frac{5}{6}$ $\frac{\cancel{15}^5}{\cancel{16}_2} \times \frac{\cancel{8}^1}{\cancel{9}_3} = \frac{5}{6}$

8. $8\frac{1}{3}$ $3\frac{1}{3} \times 2\frac{1}{2} = \frac{\cancel{10}^5}{3} \times \frac{5}{\cancel{2}_1} = \frac{25}{3} = 8\frac{1}{3}$

9. $\frac{15}{16}$ $3\frac{3}{8} \div 3\frac{3}{5} = \frac{27}{8} \div \frac{18}{5} = \frac{\cancel{27}^3}{8} \times \frac{5}{\cancel{18}_2} = \frac{15}{16}$

10. $\frac{11}{12}$

11. $1\frac{5}{12}$

12. $55\frac{11}{16}$

13. 8

14. $5\frac{23}{24}$

15. $1\frac{7}{45}$

16. $\frac{3}{20}$ 3 out of 20

17. **40** Divide the weight of a box by the weight of a packet.

    $30 \div \frac{3}{4} = \frac{30}{1} \div \frac{3}{4} = \frac{\cancel{30}^{10}}{1} \times \frac{4}{\cancel{3}_1} = \frac{40}{1} = 40$

18. $4\frac{5}{8}$ Since there are 12 inches in one foot, divide the number of inches by 12.

    $55\frac{1}{2} \div 12 = \frac{111}{2} \div \frac{12}{1} = \frac{111}{2} \times \frac{1}{12} =$
    $\frac{111}{24} = 4\frac{15}{24} = 4\frac{5}{8}$

19. **25** Since there are 3 feet in one yard, multiply the number of yards by 3.

    $8\frac{1}{3} \times 3 = \frac{25}{\cancel{3}_1} \times \frac{\cancel{3}^1}{1} = \frac{25}{1} = 25$

20. **(3) $5\frac{3}{10}$** Henry drives $1\frac{3}{10}$ miles from point C to point D, $1\frac{1}{5}$ miles from point D to point E, and $2\frac{4}{5}$ miles from point E to the garage. To find the total distance, add each distance.

$$\begin{aligned} 1\tfrac{3}{10} &= 1\tfrac{3}{10} \\ 1\tfrac{1}{5} &= 1\tfrac{2}{10} \\ +2\tfrac{4}{5} &= 2\tfrac{8}{10} \end{aligned}$$

$$4\tfrac{13}{10} = 4 + 1\tfrac{3}{10} = 5\tfrac{3}{10}$$

21. **(2) $2\frac{1}{3} \times 2$** Since 16 is 2 times 8, Clara needs to double the recipe. Multiply the amount of milk ($2\frac{1}{3}$ cups) by 2.

22. **(4) $3\frac{1}{4} - 1\frac{5}{8}$** Use the column for size 14 to find the number of yards needed for blouse B ($3\frac{1}{4}$) and the skirt ($1\frac{5}{8}$). Subtract to find how many more yards of fabric Pearl needs for the blouse than for the skirt.

23. $\frac{33}{8}$ Use the column for size 12. Add the number of yards needed for blouse A ($2\frac{7}{8}$) and the number of yards needed for the skirt ($1\frac{1}{4}$):

$$\begin{aligned} 2\tfrac{7}{8} &= 2\tfrac{7}{8} \\ +1\tfrac{1}{4} &= 1\tfrac{2}{8} \end{aligned}$$

$$3\tfrac{9}{8} = 3 + 1\tfrac{1}{8} = 4\tfrac{1}{8} = \tfrac{33}{8}$$

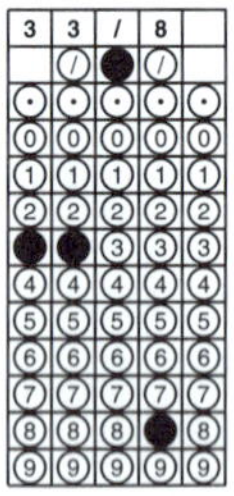

UNIT 2: MINI-TEST

PAGES 116–117

1. **(3) 24** Divide the total amount by the amount per serving. $30 \div 1\frac{1}{4} = \frac{30}{1} \div \frac{5}{4} =$

$$\frac{\overset{6}{\cancel{30}}}{1} \times \frac{4}{\underset{1}{\cancel{5}}} = 24$$

2. **(2) \$52** Multiply the amount paid per hour by the number of hours worked.
$\$8 \times 6\frac{1}{2} = \frac{\$\cancel{8}^4}{1} \times \frac{13}{\cancel{2}^1} = \$52$

3. **(4) $3\frac{5}{12}$** Add to find the total amount of coffee. $1\frac{2}{3} + 1\frac{3}{4} = 1\frac{8}{12} + 1\frac{9}{12} = 2\frac{17}{12} = 2 + 1\frac{5}{12} = 3\frac{5}{12}$

4. **(4) 20** Add the length of each side to find the total distance. $5\frac{2}{4} + 6\frac{3}{4} + 7\frac{3}{4} = 18\frac{8}{4} = 18 + 2 = 20$

5. $\frac{47}{16}$ Subtract to find the length of tubing the welder needs to cut off. $39\frac{5}{16} - 36\frac{3}{8} = 39\frac{5}{16} - 36\frac{6}{16} = 38\frac{21}{16} - 36\frac{6}{16} = 2\frac{15}{16} = \frac{47}{16}$

Remember to enter a mixed number as an improper fraction in the grid.

6. **(5) $32\frac{1}{3} \times 5\frac{1}{4}$** To find the area, multiply the length, $32\frac{1}{2}$, by the width, $5\frac{1}{4}$.

7. **(3) $1\frac{1}{8}$** Subtract to find the difference.
$4\frac{3}{8} - 3\frac{1}{4} = 4\frac{3}{8} - 3\frac{2}{8} = 1\frac{1}{8}$

8. **(2) $5\frac{3}{4}$** Add to find the total amount of rainfall. $2\frac{1}{2} + 3\frac{1}{4} = 2\frac{2}{4} + 3\frac{1}{4} = 5\frac{3}{4}$

9. **(5) 36** Multiply to find the total weight.

$$24 \times 1\frac{1}{2} = \frac{\overset{12}{\cancel{24}}}{1} \times \frac{3}{\cancel{2}_1} = 36$$

10. **(4) $2\frac{1}{2} \div \frac{1}{4}$**
Divide the total amount by the weight in each box.

11. $\frac{5}{2}$ Multiply to find the hours Marta spent on labels. Leave answer as an improper fraction, to enter in the grid.

$$7\frac{1}{2} \times \frac{1}{3} = \frac{\overset{5}{\cancel{15}}}{2} \times \frac{1}{\underset{1}{\cancel{3}}} = \frac{5}{2}$$

UNIT 3: DECIMALS

LESSON 8

PAGE 121

1. **(3) 6 dollars** The 6 is in the dollars column.
2. **(3) 40 cents** The 4 equals 4 dimes, or 40 cents.
3. **(3) 20 dollars** The 2 is in the tens column, so you have 2 tens or 20 dollars.
4. **(3) 49 dollars and 55 cents** \$49.55 means 49 dollars and 55 cents.
5. **(4) six ten thousandths** The 6 is in the ten thousandths column.
6. **(3) two thousandths** The 2 is in the thousandths column.
7. **(1) one tenth** The 1 is in the tenths column.
8. **10 has one zero, so the tenths place is one place to the right of the decimal point; 100 has two zeros, so the hundredths place is two places to the right of the decimal point; 1,000 has three zeros, so the thousandths place is three places to the right of the decimal point; and so on.**

PAGE 122

2. **two hundred fifty-six thousandths**
3. **two and nine hundredths**
4. **six and eight hundred five thousandths**
5. **24.356**
6. **3.78**
7. **0.491**
8. **267.3**
9. **14.036**
10. **5.7084**

PAGE 123

2. $\frac{9}{20}$ $\quad 0.45 = \frac{45}{100} = \frac{45 \div 5}{100 \div 5} = \frac{9}{20}$
3. $\frac{3}{20}$ $\quad 0.15 = \frac{15}{100} = \frac{15 \div 5}{100 \div 5} = \frac{3}{20}$
4. $\frac{16}{25}$ $\quad 0.64 = \frac{64}{100} = \frac{64 \div 4}{100 \div 4} = \frac{16}{25}$
5. $\frac{1}{8}$ $\quad 0.125 = \frac{125 \div 125}{1{,}000 \div 125} = \frac{1}{8}$
6. $1\frac{4}{5}$ $\quad 1.8 = 1\frac{8}{10} = 1\frac{8 \div 2}{10 \div 2} = 1\frac{4}{5}$
7. $\frac{1}{10}$ $\quad 0.10 = \frac{10}{100} = \frac{10 \div 10}{100 \div 10} = \frac{1}{10}$
8. $\frac{1}{200}$ $\quad 0.005 = \frac{5}{1000} = \frac{5 \div 5}{1000 \div 5} = \frac{1}{200}$
9. $\frac{8}{25}$ $\quad 0.32 = \frac{32}{100} = \frac{32 \div 4}{100 \div 4} = \frac{8}{25}$
10. $\frac{3}{8}$ $\quad 0.375 = \frac{375}{1000} = \frac{375 \div 25}{1000 \div 25} = \frac{15 \div 5}{40 \div 5} = \frac{3}{8}$
11. $2\frac{319}{1{,}000}$ $\quad 2.319 = 2\frac{319}{1{,}000}$
12. $\frac{1}{250}$ $\quad 0.004 = \frac{4}{1000} = \frac{4 \div 4}{1000 \div 4} = \frac{1}{250}$
13. $3\frac{7}{250}$ $\quad 3.028 = 3\frac{28}{1000} = 3\frac{28 \div 4}{1000 \div 4} = 3\frac{7}{250}$
14. $\frac{17}{20}$ $\quad 0.85 = \frac{85}{100} = \frac{85 \div 5}{100 \div 5} = \frac{17}{20}$
15. $\frac{5}{8}$ $\quad 0.625 = \frac{625}{1000} = \frac{625 \div 25}{1000 \div 25} = \frac{25 \div 5}{40 \div 5} = \frac{5}{8}$
16. $4\frac{3}{1{,}250}$ $\quad 4.0024 = 4\frac{24}{10{,}000} = 4\frac{24 \div 8}{10{,}000 \div 8} = 4\frac{3}{1{,}250}$

PAGE 125

2. **0.43 > 0.09**
   $0.\underline{4}3 \quad 0.\underline{0}9$ 4 is greater than 0.
3. **0.73 > 0.542**
   Add a zero: $0.\underline{7}30 \quad 0.\underline{5}42$
   Then compare: 7 is greater than 5.
4. **8.058 < 8.58**
   Add a zero: $8.\underline{0}58 \quad 8.\underline{5}80$
   Then compare: 0 is less than 5.
5. **2.58 = 2.580**
   Add a zero: 2.580 2.580
   The numbers are the same.
6. **53.005 > 52.008**
   $5\underline{3}.005 \quad 5\underline{2}.008$ 3 is greater than 2.
7. **0.863 < 0.9**
   Add two zeros: $0.\underline{8}63 \quad 0.\underline{9}00$
   Then compare: 8 is less than 9.

8. **2.001 < 2.01**
   Add a zero: $2.0\underline{0}1$ $2.0\underline{1}0$
   Then compare: 0 is less than 1.
9. **1.32 > 1.319**
   Add a zero: $1.3\underline{2}0$ $1.3\underline{1}9$
   Then compare: 2 is greater than 1.
10. **5 = 5.00**
   Add two zeros: 5.00   5.00
   The numbers are the same.
11. **0.95 < 0.954**
   Add a zero: $0.95\underline{0}$ $0.95\underline{4}$
   Then compare: 0 is less than 4.
12. **0.005 < 0.05**
   Add a zero: $0.0\underline{0}5$ $0.0\underline{5}0$
   Then compare: 0 is less than 5.
13. **0.43 > 0.34**
   $0.\underline{4}3$ $0.\underline{3}4$ 4 is greater than 3.
14. **0.54 > 0.054**
   Add a zero: $0.\underline{5}40$ $0.\underline{0}54$
   Then compare: 5 is greater than 0.
15. **5.2 = 5.200**
   Add two zeros: 5.200   5.200
   The numbers are the same.
17. **8** $\underline{8}.276$ Since 2 is less than 5, do not change the underlined digit.
18. **46.4** $46.\underline{3}518$ The number to the right of 3 is 5; add 1 to the underlined digit.
19. **71.048** $71.04\underline{8}3$ Since 3 is less than 5, do not change the underlined digit.
20. **56.28** $56.2\underline{8}19$ Since 1 is less than 5, do not change the underlined digit.
21. **0.5** $0.\underline{5}462$ Since 4 is less than 5, do not change the underlined digit.
22. **102** $10\underline{1}.667$ Since 6 is greater than 5, add 1 to the underlined digit.
23. **25.024** $25.02\underline{3}5$ The number to the right of 3 is 5, add 1 to the underlined digit.
24. **Adrian**
   $0.21\underline{5} > 0.21\underline{3}$ 5 is greater than 3.
25. **96** $9\underline{5}.7$ Since 7 is greater than 5, add 1 to the underlined digit.
26. **Brand B**
   $\$2.\underline{7}9 < \$2.\underline{8}3$ 7 is less than 8.
27. **0.5, 0.57, 0.69, 0.75**
   0.5 < 0.57 < 0.69 < 0.75

PAGE 127

1. **(4) 5.8 + 5.8 + 2.7 + 2.7** To find the perimeter, add the lengths of all four sides.
2. **(2) 2.7 − 0.7** Find the difference between the length of the side and the opening.
3. **(5) Not enough information is given.** To find the perimeter, you need the length and width of the swimming pool.
4. **(3) 10.2 + 3.1 + 3.1** Add the lengths of the three sides that do not touch the side of the building.
5. **(3) 60.2 − 15.8** Find the difference between the length of the side and the length of the yellow strip.
6. Answers will vary.

PAGES 128–129

1. **seven hundredths**
2. **two ten thousandths**
3. **six thousandths**
4. **five tenths**
5. **eighty-two hundredths**
6. **three and five hundred thirteen thousandths**
7. **eighteen and three thousand four hundred six ten thousandths**
8. **seven hundred ninety and four thousandths**
9. **2.43**
10. **0.2343**
11. **37.621**
12. **429.09**
13. $\frac{1}{20}$ $0.05 = \frac{5}{100} = \frac{1}{20}$
14. $\frac{7}{25}$ $0.28 = \frac{28}{100} = \frac{7}{25}$
15. $\frac{3}{4}$ $0.75 = \frac{75}{100} = \frac{3}{4}$
16. $\frac{1}{40}$ $0.025 = \frac{25}{1000} = \frac{1}{40}$
17. $\frac{24}{25}$ $0.96 = \frac{96}{100} = \frac{24}{25}$
18. $3\frac{41}{200}$ $3.205 = 3\frac{205}{1000} = 3\frac{41}{200}$
19. $5\frac{1}{1,250}$ $5.0008 = 5\frac{8}{10,000} = 5\frac{1}{1,250}$
20. $9\frac{7}{20}$ $9.35 = 9\frac{35}{100} = 9\frac{7}{20}$

21. **0.759 < 0.795**
0.759 0.795
5 is less than 9.

22. **0.326 <0.54**
Add a zero: 0.326 0.540
Then compare: 3 is less than 5.

23. **0.6 = 0.60**
Add a zero: 0.60 0.60
The numbers are the same.

24. **32.574 > 32.547**
32.574 32.547
7 is greater than 4.

25. **3.20** 3.196 Since 6 is greater than 5, add 1 to the underlined digit.

26. **6.5** 6.453 The number to the right of 4 is 5; add 1 to the underlined digit.

27. **1** 0.724 Since 7 is greater than 5, add 1 to the underlined digit.

28. **42.936** 42.9358 Since 8 is greater than 5, add 1 to the underlined digit.

29. **(2) 0.016**

30. **(4) 5.2 + 5.2 + 3.4 + 3.4** To find the perimeter, add the lengths of all four sides.

31. **(1) 4.2 − 2.7** Subtract the number of meters of fabric Calvin used from the number of meters of fabric he had.

32. **(4) Ron finished before Al.**
Ron = 10.06 < 10.25 = Al
0 is less than 2.

33. **(2) three and fifty-nine thousandths**

34. **(4) 24.6350 > 24.6305**
24.6350 > 24.6305
5 is greater than 0.

35. **(4) $1.33** 1.325 The number to the right of 2 is 5; add 1 to the underlined digit.

36. **(2)** $\frac{2}{5}$ $0.4 = \frac{4}{10} = \frac{2}{5}$

LESSON 9

PAGE 131

2. **8.953**
$$\begin{array}{r} \scriptstyle 0\,16\,13 \\ 17.368 \\ -\ 8.415 \\ \hline 8.953 \end{array}$$

3. **10.58**
$$\begin{array}{r} \scriptstyle 3\ 12 \\ 24.28 \\ -13.70 \\ \hline 10.58 \end{array}$$

4. **65.757**
$$\begin{array}{r} \scriptstyle 1\ \ 1 \\ 28.467 \\ +37.290 \\ \hline 65.757 \end{array}$$

5. **5.25**
$$\begin{array}{r} \scriptstyle 0\,11\ 10 \\ 12.05 \\ -\ 6.80 \\ \hline 5.25 \end{array}$$

6. **20.24**
$$\begin{array}{r} \scriptstyle 1\,1 \\ 3.70 \\ 14.24 \\ +\ 2.30 \\ \hline 20.24 \end{array}$$

7. **88.784**
$$\begin{array}{r} \scriptstyle 1\,1 \\ 24.810 \\ 35.700 \\ +28.274 \\ \hline 88.784 \end{array}$$

8. **8.09**
$$\begin{array}{r} \scriptstyle 3\,10 \\ 8.40 \\ -0.31 \\ \hline 8.09 \end{array}$$

9. **8.047**
$$\begin{array}{r} \scriptstyle 0\,14\ 7\ 9\,10 \\ 14.800 \\ -\ 6.753 \\ \hline 8.047 \end{array}$$

10. **17.88**
$$\begin{array}{r} 12.00 \\ +\ 5.88 \\ \hline 17.88 \end{array}$$

11. **6.18**
$$\begin{array}{r} \scriptstyle 1\ \ 1 \\ 2.43 \\ 0.57 \\ +3.18 \\ \hline 6.18 \end{array}$$

12. 17.484

    5 16 12 10 10
    67.310
    −49.826
    17.484

13. 27.93

    4 16 14
    57.43
    −29.50
    27.93

14. 25.208

    1
    0.300
    21.508
    + 3.400
    25.208

15. 17.75

    1
    4.26
    5.10
    +8.39
    17.75

16. 3.703

    0 11 15 9 10
    12.600
    − 8.897
    3.703

17. **$364.20** Subtract for each check written, and add for each deposit.

18. **$116.00** Add the two amounts Roberto was paid for overtime that weekend.

PAGE 133

1. **(1) $39.99 + 58.00** To find the subtotal, add the cost of the two jackets.
2. **(3) subtotal + $6.86** To find the total including tax, add the tax to the subtotal.
3. **(1) total − $45.00** To find how much Mrs. Wilson still owes on the coats, subtract the layaway cost from the total.
4. **(2) $30.00 + $24.00 + $12.00** To find the subtotal, add the cost of the three items.
5. **(5) (3 × $18) + (2 × $5.99)** To find the total of the sale before tax, find the cost of the three shirts, the cost of the two containers, and add their totals.
6. **(2) subtotal − discount + tax** Coupons help customers save money so the amount of the discount is subtracted from the subtotal. The tax is added to the new cost of the purchase.
7. **Step 1: Find the subtotal by adding the costs of all three items. Step 2: Find the total by adding the sales tax to the subtotal. Step 3: Find the change by subtracting the total from the amount the customer gives the sales clerk.**

PAGE 135

1. **(4) $14.50 × (38 + 40)** Work the parentheses first to find the total number of hours worked. Then multiply the total hours by the hourly wage.
2. **(5) (24 × $9.95) + (18 × $11.75)** To find how much Glen makes per week, work the parentheses first to find out how much he earns for each job. Then sum the totals.
3. **(5) (205 + 165 + 104 + 190) ÷ 4 + 40** Find the mean of the bowlers: (205 + 165 + 104 + 190) ÷ 4. Then add 40 to the mean.
4. **(5) ($6.50 + $9.25) × ($9.25 − $6.50)** First, add the two amounts. Then find the difference of the two amounts. Last, multiply the two totals.
5. **(1) (3 × $11.00) + (5.5 × $8.40)** To find the cost of the sale before tax, work the parentheses first to find the cost of 3 yards of fabric and the cost of 5.5 yards of fabric. Then sum the two totals.
6. **(3) $40 − 2 × $8.50 − 2 × $3.50 − $4.50** You don't need parentheses because the multiplication steps will be done first following the order of operations.
7. **13** $(6 \times 2) + 4 - 3 = 12 + 4 - 3 = 13$
   33 $6 \times (2 + 4) - 3 = 6 \times 6 - 3 = 36 - 3 = 33$
   18 $6 \times (2 + 4 - 3) = 6 \times 3 = 18$

PAGE 137

1. **(2) 2.9** The point for week 4 is at the ninth small line above 2 on the scale.
2. **(3) from week 3 to week 4** The steepest part of the graph is between the points for week 3 and week 4.

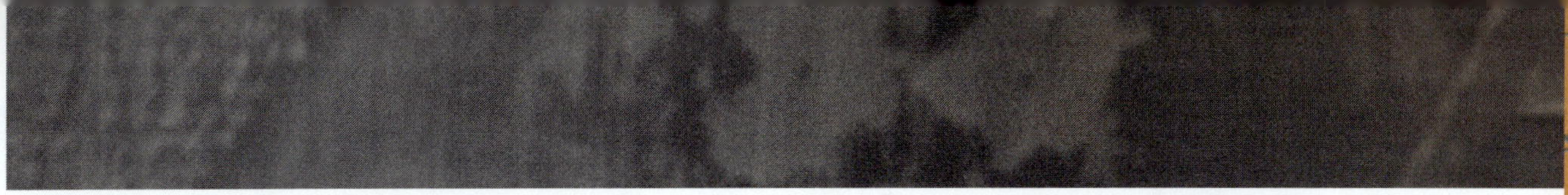

3. **(2) 2.5** Subtract the number of miles Ahmad walked each day during week 1 (1.3) from the number of miles he walked each day during week 6 (3.8).

$$\begin{array}{r} 3.8 \\ -1.3 \\ \hline 2.5 \end{array}$$

4. **(3) \$1.5 million** Sales dropped from Year 4 to Year 5. Subtract the sales for Year 5 (\$2 million) from the sales for Year 4 (\$3.5 million). \$3.5 − \$2 = \$1.5.

5. **(2) \$12 million** Add the amounts for Years 3 through 6: 3 + 3.5 + 2 + 3.5 = 12

6. **The lines show the increases and decreases in the amount of sales from Year 1 to Year 6. The steeper the line, the greater the increase or decrease in sales. The steepest part of the graph is between points for Years 4 and 5, and Years 5 and 6. This means, for example, that the greatest increase in sales was from Year 5 to Year 6. When the line is flatter this means that there was less increase or decrease in sales.**

PAGES 138–139

1. **77.04** $$\begin{array}{r} \scriptstyle 1 \\ 35.20 \\ +41.84 \\ \hline 77.04 \end{array}$$

2. **22.88** $$\begin{array}{r} \scriptstyle 3\ 1210 \\ 74.30 \\ -51.42 \\ \hline 22.88 \end{array}$$

3. **78.924** $$\begin{array}{r} \scriptstyle 812\ 12\,5\,10 \\ 93.260 \\ -14.336 \\ \hline 78.924 \end{array}$$

4. **20.825** $$\begin{array}{r} \scriptstyle 1\ 1\ \ 1 \\ 15.930 \\ +\ 4.895 \\ \hline 20.825 \end{array}$$

5. **23.766** $$\begin{array}{r} \scriptstyle 1 \\ 8.450 \\ +15.316 \\ \hline 23.766 \end{array}$$

6. **8.207** $$\begin{array}{r} \scriptstyle 1\ 1 \\ 4.260 \\ +3.947 \\ \hline 8.207 \end{array}$$

7. **6.899** $$\begin{array}{r} \scriptstyle 6\ 1218 \\ 7.389 \\ -0.490 \\ \hline 6.899 \end{array}$$

8. **1.12** $$\begin{array}{r} \scriptstyle 8\,10 \\ 5.90 \\ -4.78 \\ \hline 1.12 \end{array}$$

9. **21.673** $$\begin{array}{r} \scriptstyle 1\,1 \\ 3.810 \\ 12.463 \\ +\ 5.400 \\ \hline 21.673 \end{array}$$

10. **178.307** $$\begin{array}{r} \scriptstyle 1\ \ 1 \\ 42.180 \\ 53.700 \\ +82.427 \\ \hline 178.307 \end{array}$$

11. **2.899** $$\begin{array}{r} \scriptstyle 0\,11\ 121612 \\ 12.372 \\ -\ 9.473 \\ \hline 2.899 \end{array}$$

12. **1.477** $$\begin{array}{r} \scriptstyle 6\ 121610 \\ 7.370 \\ -5.893 \\ \hline 1.477 \end{array}$$

13. **20.78** $$\begin{array}{r} \scriptstyle 1 \\ 8.25 \\ 9.47 \\ +3.06 \\ \hline 20.78 \end{array}$$

14. **22.046** $$\begin{array}{r} \scriptstyle 1\,2 \\ 15.430 \\ 2.800 \\ +\ 3.816 \\ \hline 22.046 \end{array}$$

15. **9.015** $$\begin{array}{r} \scriptstyle 2\ \ 1 \\ 4.245 \\ 1.970 \\ +2.800 \\ \hline 9.015 \end{array}$$

16. **1.876** $$\begin{array}{r} \scriptstyle 2\ 17\,9\,10 \\ 3.800 \\ -1.924 \\ \hline 1.876 \end{array}$$

17. **148.7734**

18. **1,336.2735**

19. **(4) \$100 – (\$30.00 + \$24.00 + \$4.46)** First, sum the expenses. Then subtract this amount from \$100.

20. **(3) 10** $4 + (2 \times 6) - (12 \div 2)$

$$\begin{array}{lll} 4 + 12 & - & 6 \\ 16 & - & 6 = 10 \end{array}$$

21. **$30** Round the cost of the tape player and the cost of the tape and add.

$$\begin{array}{rcr} \$24.99 & \text{rounds to} & \overset{1}{\$2}5 \\ + \quad 4.99 & \text{rounds to} & + \quad 5 \\ \hline & & \$30 \end{array}$$

22. **$29.98** Add the cost of the tape player and the cost of the tape.

$$\begin{array}{r} \overset{1\,1}{\$24.99} \\ + \quad 4.99 \\ \hline \$29.98 \end{array}$$

23. **$32.08** Add the tax to the subtotal (see the explanation for Question 22.)

$$\begin{array}{r} \overset{1\,1}{\$29.98} \\ + \quad 2.10 \\ \hline \$32.08 \end{array}$$

24. **$7.92** Subtract the total (see the explanation for Question 23) from the amount the customer gave George.

$$\begin{array}{r} \overset{3\;9\;\;9\,10}{\$\not{4}\not{0}.\not{0}\not{0}} \\ - \quad 32.08 \\ \hline \$7.92 \end{array}$$

25. **(3) 2,400** The line represents 0.2 thousand or 200 students. The point above Year 2 on the graph is located at 2.4, which represents 2,400 students.

26. **(4) Year 5** The increase from Year 4 to Year 5 represented a change of 800 students, the greatest change, either increase or decrease, on the graph.

27. **(3) 1,400** Subtract the number of students enrolled in Year 1 from the number in Year 6.

$$\begin{array}{r} \overset{2\;\,12}{\not{3},\not{2}00} \\ -1,800 \\ \hline 1,400 \end{array}$$

LESSON 10

PAGE 140

2. **72.8**

$$\begin{array}{r} 9.1 \\ \times \; 8 \\ \hline 72.8 \end{array}$$

3. **6.15**

$$\begin{array}{r} 12.3 \\ \times \; 0.5 \\ \hline 6.15 \end{array}$$

4. **0.0728**

$$\begin{array}{r} 1.04 \\ \times 0.07 \\ \hline 0.0728 \end{array}$$

5. **0.375**

$$\begin{array}{r} 0.75 \\ \times \; 0.5 \\ \hline 0.375 \end{array}$$

6. **0.816**

$$\begin{array}{r} 136 \\ \times 0.006 \\ \hline 0.816 \end{array}$$

7. **25.6**

$$\begin{array}{r} 128 \\ \times \; 0.2 \\ \hline 25.6 \end{array}$$

8. **27.68**

$$\begin{array}{r} 17.3 \\ \times \; 1.6 \\ \hline 10\;38 \\ + \; 17\;3 \\ \hline 27.68 \end{array}$$

9. **0.0126**

$$\begin{array}{r} 0.42 \\ \times 0.03 \\ \hline 0.0126 \end{array}$$

10. **1.845**

$$\begin{array}{r} 2.05 \\ \times \; 0.9 \\ \hline 1.845 \end{array}$$

11. **18.972**

$$\begin{array}{r} 5.27 \\ \times \; 3.6 \\ \hline 3\;162 \\ +15\;81 \\ \hline 18.972 \end{array}$$

12. **0.496**

$$\begin{array}{r} 6.2 \\ \times 0.08 \\ \hline 0.496 \end{array}$$

13. **152.928**

```
     28.32
  ×   5.4
    11 328
 + 141 60
   152.928
```

PAGE 141

2. **1.5 × $0.79**

3.

4.

PAGE 143

2. **0.63**

```
    0.63
 9)5.67
  −5 4
     27
    −27
      0
```

3. **5.9**

```
     5.9
 7)41.3
  −35
    6 3
   −6 3
      0
```

4. **2.8**

```
      2.8
 12)33.6
   −24
     9 6
    −9 6
       0
```

5. **8.5**

```
       8.5
 13)110.5
   −104
       6 5
      −6 5
         0
```

6. **0.306**

```
   0.306
 4)1.224
  −1 2
     24
    −24
      0
```

7. **0.042**

```
     0.042
 21)0.882
   −  84
       42
      −42
        0
```

8. **0.002**

```
        0.002
 0.5.)0.0.010
        −  10
             0
```

9. **0.31**

```
           0.31
 0.08.)0.02.48
       −   2 4
             8
            −8
             0
```

10. **8.53**

```
           8.53
 0.12.)1.02.36
       −  96
          6 3
         −6 0
            36
           −36
             0
```

11. **.0175**

```
   0.0175
 6)0.1050
   −  6
      45
     −42
       30
      −30
        0
```

12. **42.5**

```
              42.5
 0.012.)0.510.0
         −48
           30
          −24
            6 0
           −6 0
              0
```

13. **3.5**

```
        3.5
 3.6.)12.6.0
     −10 8
       1 8 0
      −1 8 0
           0
```

14. **0.06**

```
        0.06
0.25.)0.01.50
      -  1 50
            0
```

15. **16,000**

```
          16,000.
0.004.)64.000.
       -4
        24
       -24
         0
```

16. **5.151**

```
          5.151
5.1.)26.2.701
    -25 5
        77
       -51
        260
       -255
          51
         -51
           0
```

17. **372**

18. **$16.50**

PAGE 145

1. **(5) 450 cm** Multiply by 100 to convert meters to centimeters.
   $4.5 \times 100 = 4.50. = 450$
2. **(4) 5,500 mg** Multiply by 1,000 to convert grams to milligrams.
   $5.5 \times 1{,}000 = 5.500. = 5{,}500$
3. **(2) 160 cm** First multiply 0.4 by 4 to find the perimeter of the square.
   $0.4 \times 4 = 1.6$ m
   Then multiply by 100 to convert meters to centimeters.
   $1.6 \times 100 = 1.60. = 160$ cm
4. **(3) 0.1 L** Divide by 1,000 to convert milliliters to liters.
   $100 \div 1{,}000 = 0.100. = 0.1$ L
5. **(5) Not enough information is given.** To find how many meters long the race is, you need to know the distance Monica ran.
6. **(1) 298,000 mg** Multiply by 1,000 to convert grams to milligrams.
   $298 \times 1{,}000 = 298.000. = 298{,}000$
7. **Answers will vary. To convert amounts from grams to kilograms, divide by 1,000. To convert amounts from liters to milliliters, multiply by 1,000.**

PAGE 147

1. **(4) $149.97** Find the total amount by multiplying the unit price ($49.99) by the quantity (3).
2. **(2) 4 × $135** Find the total amount by multiplying the unit price ($135) by the quantity (4).
3. **(1) 48** Sum the numbers in the quantity column (12 + 8 + 15 + 3 + 6 + 4) to find the total (48).
4. **(4) $117.80** Find the total amount of the washcloths ($4.49 × 12 = $53.88) and the total amount of the hand towels ($7.99 × 8 = $63.92). Sum the two totals ($53.88 + $63.92 = $117.80).
5. **(5) $1,236.77** Find the TOTAL by finding the total amount of each unit item and finding their sum
   ($53.88 + $63.92 + $375 + $149.97 + $54 + $540 = $1,236.77)
6. **(5) Not enough information is given.** To find the total amount of the blenders, you would need to multiply the unit price (unknown) by the quantity (4).
7. **(3) $189** To find the total amount for the mugs, find the amount for the first order ($9 × 6 = $54) and the amount for the second order ($9 × 15 = $135). Then sum the amounts ($54 + $135 = $189).
8. **$696.77 Subtract the total amount of the tote bags (4 × $135 = $540) from the TOTAL found in Question 5 ($1,236.77).**

PAGE 149

1. **(4) $0.99 × 6** Find the total cost by multiplying the unit cost ($0.99) by the number of calendar refills (6).
2. **(3) $19.95 ÷ 5** Find the unit cost by dividing the total cost ($19.95) by the number of tapes (5).

3. **(1) $47.94 ÷ 6** Find the unit cost by dividing the total cost ($47.94) by the number of reams of paper (6).
4. **(5) $0.19 × 25** Find the total cost by multiplying the unit cost ($0.19) by the number of pencils (25).
5. **(5) Not enough information is given.** Find the total cost by multiplying the unit cost of the paper towels (unknown) by the number of rolls of paper towels (25).
6. **(4) $1.69 ÷ 4** Find the unit cost by dividing the total cost ($1.69) by the number of bulbs (4).
7. **Recall division is the inverse operation of multiplication. You know total cost = number of units × unit cost. To find unit cost, divide both sides by the number of units: total cost ÷ number of units = unit cost.**

PAGE 151

1. **12.5** Mike will make 5 × 2 = 10 trips during the week. 1.25 × 10 = 12.5 miles

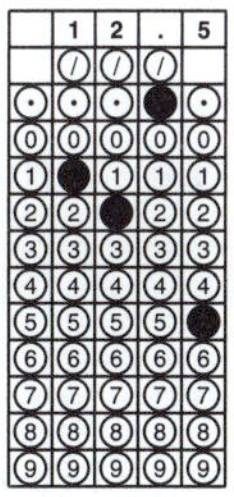

2. **14.8** Divide: 177.6 ÷ 12 = 14.8 ounces.

3. **0.16** Subtract: 1.45 − 1.29 = 0.16 therms. Notice that it is not necessary to grid in the zero to the left of the decimal point. This zero does not affect the value of the number.

4. **1.75** Multiply: 0.035 × 50 = 1.75 grams.

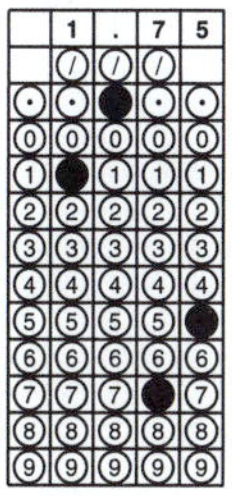

PAGES 152–153

1. **19.2**

$$\begin{array}{r} 3.2 \\ \times\ 6 \\ \hline 19.2 \end{array}$$

2. **0.0824**

$$\begin{array}{r} 2.06 \\ \times 0.04 \\ \hline 0.0824 \end{array}$$

3. **1.251**

$$\begin{array}{r} 4.17 \\ \times\ 0.3 \\ \hline 1.251 \end{array}$$

4. **33.48**

$$\begin{array}{r} 12.4 \\ \times\ 2.7 \\ \hline 8\ 68 \\ +24\ 8\ \ \\ \hline 33.48 \end{array}$$

5. **0.13536**

$$\begin{array}{r} 0.752 \\ \times\ 0.18 \\ \hline 6016 \\ +752\ \\ \hline 0.13536 \end{array}$$

6. **0.76**

```
    0.76
  6)4.56
   -4 2
      36
     -36
       0
```

7. **0.825**

```
     0.825
   4)3.300
    -3 2
       10
     -  8
        20
       -20
         0
```

8. **3.87**

```
          3.87
  0.5.)1.9.35
      -1.5
         4 3
        -4 0
           35
          -35
            0
```

9. **67.75**

```
            67.75
  0.08.)5.42.00
       -4.8
          62
         -56
           6 0
          -5 6
             40
            -40
              0
```

10. **9.3**

```
             9.3
  1.5.)13.9.5
      -13.5
          4 5
         -4 5
            0
```

11. **0.252**
12. **130.19**
13. **12,500** Multiply by 1,000 to convert grams to milligrams.
12.5 × 1,000 = 12.500. = 12,500
14. **14.9 miles** Divide the total distance by the number of days.
74.5 ÷ 5 = 14.9
15.

16. **5.578** Divide: 16.734 ÷ 3 = 5.578 meters.

17. **(2) $197.00 ÷ 4** Find the unit cost by dividing the total cost ($197.00) by the number of tires (4).
18. **(3) $2.40** Multiply the cost of each pound by the number of pounds.
$0.60 × 4 = $2.40
19. **(5) $644.93** Find the total amount for each of the two items. Then find their sum.
(4 × $149) + (7 × $6.99) =
$596 + $48.93 = $644.93
20. **(4) 24.3 miles per gallon** Divide the total miles by the number of gallons of gas.
291.6 ÷ 12 = 24.3
21. **(4) $1.29 × 31** Find the total cost by multiplying the unit cost ($1.29) by the number of notebooks (31).
22. **(1) 5 ÷ 1.2** Divide the total length (5 meters) by the length of each piece (1.2 meters).
23. **3.98** Divide by 100 to convert centimeters to meters.
398 ÷ 100 = 3.98

MATH AT WORK

PAGE 155

1. **(5) $1,512.24** (3 × $345.23) + $476.55
2. **(3) Subtotal**
3. **(4) $1,386.74** Subtotal − cash received= $1,512.24 − $125.50

UNIT 3 REVIEW

PAGES 156–157

1. **six ten-thousandths**
2. **eight hundredths**
3. **0.25 > 0.025** Add a zero: $0.\underline{2}50$ $0.\underline{0}25$ Then compare: 2 is greater than 0.
4. **0.97 = 0.970** Add a zero: 0.970 0.970 Then compare: 0.970 and 0.970 are the same.
5. **5.36** $5.3\underline{6}2$ Since 2 is less than 5, do not change the underlined digit.
6. **7.4** $7.\underline{3}51$ The number to the right of 3 is 5; add 1 to the underlined digit.
7. $\frac{21}{50}$ $\frac{42 \div 2}{100 \div 2} = \frac{21}{50}$
8. **0.76**
$$\begin{array}{r} 0.76 \\ 25\overline{)19.0} \\ -17\,5 \\ \hline 1\,50 \\ -1\,50 \\ \hline 0 \end{array}$$
9. **11.785**
$$\begin{array}{r} {\scriptstyle 1} \\ 3.925 \\ 4.600 \\ +3.260 \\ \hline 11.785 \end{array}$$
10. **7.86**
$$\begin{array}{r} {\scriptstyle 1\,14\,13} \\ \not{2}\not{5}.\not{3}6 \\ -17.50 \\ \hline 7.86 \end{array}$$
11. **5.143**
$$\begin{array}{r} {\scriptstyle 3\,9\,10} \\ 8.\not{4}\not{0}\not{0} \\ -3.257 \\ \hline 5.143 \end{array}$$
12. **9.4**
$$\begin{array}{r} 9.4 \\ 4\overline{)37.6} \\ -36 \\ \hline 1\,6 \\ -1\,6 \\ \hline 0 \end{array}$$
13. **50.4**
$$\begin{array}{r} 6.3 \\ \times\ 8 \\ \hline 50.4 \end{array}$$
14. **2.286**
$$\begin{array}{r} 2.54 \\ \times\ 0.9 \\ \hline 2.286 \end{array}$$
15. **45**
$$\begin{array}{r} 45. \\ 0.06.\overline{)2.70.} \\ -2\,4 \\ \hline 30 \\ -30 \\ \hline 0 \end{array}$$
16. **468**
$$\begin{array}{r} 468 \\ 0.016.\overline{)7.488.} \\ -6\,4 \\ \hline 1\,08 \\ -\ 96 \\ \hline 128 \\ -128 \\ \hline 0 \end{array}$$
17. **15.56**
$$\begin{array}{r} {\scriptstyle 1\ 1} \\ 3.54 \\ 6.83 \\ +5.19 \\ \hline 15.56 \end{array}$$
18. **0.51**
$$\begin{array}{r} 0.51 \\ 0.07.\overline{)0.03.57} \\ -3\,5 \\ \hline 07 \\ -\ 7 \\ \hline 0 \end{array}$$
19. **42.48**
$$\begin{array}{r} {\scriptstyle 4\ 13\,10} \\ 4\not{5}.\not{4}\not{0} \\ -\ 2.92 \\ \hline 42.48 \end{array}$$
20. **0.0285**
$$\begin{array}{r} 0.57 \\ \times 0.05 \\ \hline 0.0285 \end{array}$$

21. **$5.51** Add the cost of the shoes and the socks to find the subtotal. Then add the tax to find the total. Finally, subtract the total from the amount Bryant gave the clerk.

$$\begin{array}{r} \scriptstyle 1\,1\;1 \\ \$64.95 \\ +\;\;5.99 \\ \hline \$70.94 \end{array} \qquad \begin{array}{r} \scriptstyle 1 \\ \$70.94 \\ +\;\;3.55 \\ \hline \$74.49 \end{array} \qquad \begin{array}{r} \scriptstyle 7\,9\;\,9\,10 \\ \$\not{8}\not{0}.\not{0}\not{0} \\ -\;74.49 \\ \hline \$5.51 \end{array}$$

22. **11** $9 \div (4 - 1) + 8$
$9 \div 3 + 8 = 3 + 8 = 11$

23. **$0.26** Find the unit cost.
$\$3.12 \div 12 = \$0.26$

24. **$1.08** $0.40 \times \$2.69 = \$1.076$, which rounds to $1.08

25. **$151.96** Sum the total amounts of the two items. $(4 \times \$12.99) + (5 \times \$20) = \$51.96 + \$100 = \$151.96$

26. **(1) (4 × $12.50) + $15.95 + $14.87** Find the total cost by multiplying the cost of each tire ($42.50) by the number of tires (4), and then adding the amount for balancing ($15.95) and tax ($14.87).

27. **6.1** Add the number of parts produced in November (3.2 thousand) and the number produced in December (2.9 thousand).

$$\begin{array}{r} \scriptstyle 1\phantom{.2 \text{ thousand}} \\ 3.2 \text{ thousand} \\ +2.9 \text{ thousand} \\ \hline 6.1 \text{ thousand} \end{array}$$

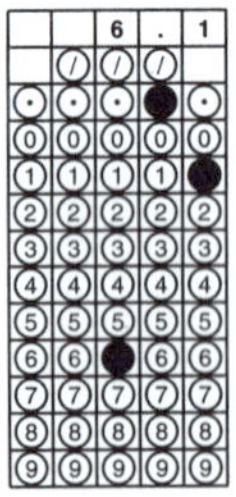

UNIT 3 MINI-TEST

PAGES 158–159

1. **(3) $30.00** Find the cost for twelve months if you pay monthly: $\$16.50 \times 12 = \$198$. Subtract to find the difference: $\$198 - \$168 = \$30$.

2. **(3) 0.1355** Subtract the smaller measurement from the larger one: $2.15 - 2.0145 = 0.1355$ centimeters.

3. **(4) 16,000** Divide: $20{,}000 \div 1.25 = 16{,}000$ bushels.

4. **(1) $683.30** Find the value of Cyber shares ($\$24.50 \times 10 = \$245.00$) and the value of Titan shares ($\$14.61 \times 30 = \$438.30$). Find the sum of these subtotals: $\$245.00 + \$438.30 = \$683.30$.

5. **14.07** Add: $3.655 + 1.85 + 2.54 + 6.025 = 14.07$ pounds.

6. **(4) 7,000 − 5,750** The lines in between the labeled lines on the vertical scale divide the thousands into fourths, so each line represents 250 sales. The dot for Year 5 represents 7.0 thousand, or 7,000 sales. The dot for Year 1 represents 5.75 thousand, or 5,750 sales. Only option (4) shows the difference between these numbers.

7. **(2) Year 3** The steepest line segment showing increase connects Year 2 and Year 3.

8. **(2) 2.7** $1.4 - 0.2 \times 6 + 2.5 =$
$1.4 - 1.2 + 2.5 =$
$0.2 + 2.5 = 2.7$

9. **(5) $40 − ($3.59 × 5 + $14.79)** The multiplication step will be done first and then the $14.79 will be added to the product. The amount in parentheses represents the total cost of the plants, which will then be subtracted from $40.

10. **(3) 3,000** There are 1,000 liters in 1 kiloliter. 3 × 1,000 = 3,000 liters

11. **8.5** Subtract: 12.75 − 4.25 = 8.5 hours.

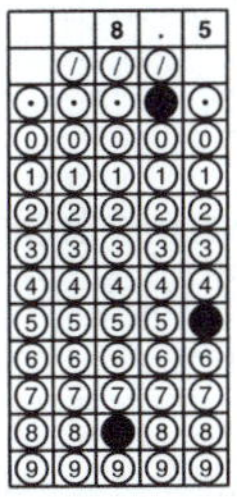

UNIT 4: RATIOS, PROPORTIONS, AND PERCENTS

LESSON 11

PAGE 163

1. **(2)** $\dfrac{\textbf{1 cup sugar}}{\textbf{6 cups flour}}$

2. **(4)** $\dfrac{\textbf{1,267 nonfiction}}{\textbf{1,055 fiction}}$

3. **(1)** $\dfrac{\textbf{\$0.24}}{\textbf{1 oz.}}$

4. **(4)** $\dfrac{\textbf{\$18}}{\textbf{1 hr.}}$

5. **(2)** $\dfrac{\textbf{19 hours}}{\textbf{26 hours}}$
   Subtract the 19 hours Daksha spends baking cakes from the total hours she spends baking cakes and cookies: 45 − 19 = 26 hours baking cookies.

6. **Car 1 gets more miles per gallon. Car 1 gets 15 miles to 1 gallon of gasoline. Car 2 gets 100 miles to 10 gallons. Divide 100 by 10 to find how many miles Car 2 gets to 1 gallon.**
   $\frac{100}{10} = \frac{100 \div 10}{10 \div 10} = \frac{10}{1}$ **or 10 miles to 1 gallon.**

PAGE 165

2. $\frac{16}{21}$

   $$\frac{\text{16 employees}}{\text{21 employees}} = \frac{16}{21}$$

3. $\frac{\text{1 gallon}}{\text{200 square feet}}$

   $$\frac{\text{2 gallons of paint}}{\text{400 square feet}} = \frac{2 \div 2}{400 \div 2} = \frac{1}{200}$$

4. $\frac{\text{1 pound}}{\$0.75}$

   $$\frac{\text{2 pounds of apples}}{\$1.50} = \frac{2 \div 2}{\$1.50 \div 2} = \frac{1}{\$0.75}$$

5. $\frac{2}{5}$

   $$\frac{\text{4 hours}}{\text{10 hours}} = \frac{4 \div 2}{10 \div 2} = \frac{2}{5}$$

6. $\frac{2}{5}$

   $$\frac{\text{8 grams}}{\text{20 grams}} = \frac{8 \div 4}{20 \div 4} = \frac{2}{5}$$

7. $\frac{\text{1 pair}}{\$4}$

   $$\frac{\text{3 pairs of socks}}{\$12} = \frac{3 \div 3}{\$12 \div 3} = \frac{\text{1 pair of socks}}{\$4}$$

8. $\frac{\text{3 women}}{\text{2 men}}$

   $$\frac{\text{12 women}}{\text{8 men}} = \frac{12 \div 4}{8 \div 4} = \frac{3}{2}$$

9. $\frac{\$14}{\text{1 hr.}}$

   $$\frac{\$280}{\text{20 hr.}} = \frac{\$280 \div 20}{20 \div 20} = \frac{\$14}{\text{1 hr.}}$$

10. $\frac{8}{3}$

   $$\frac{\text{40 books}}{\text{15 books}} = \frac{40 \div 5}{15 \div 5} = \frac{8}{3}$$

11. $\frac{9}{31}$

   $$\frac{\text{9 rainy days}}{\text{31 days in the month}} = \frac{9}{31}$$

12. $\frac{2}{3}$

   $$\frac{\text{12 won}}{\text{18 played}} = \frac{12 \div 6}{18 \div 6} = \frac{2}{3}$$

13. $\frac{5}{2}$

$\frac{35 \text{ cars}}{14 \text{ vans}} = \frac{35 \div 7}{14 \div 7} = \frac{5}{2}$

14. $\frac{23}{1}$

$\frac{230 \text{ miles}}{10 \text{ gallons}} = \frac{230 \div 10}{10 \div 10} = \frac{23}{1}$

15. $\frac{3}{5}$

$\frac{15 \text{ cash}}{25 \text{ charge}} = \frac{15 \div 5}{25 \div 5} = \frac{3}{5}$

16. $\frac{\$6}{1}$

$\frac{\$42}{7 \text{ hours}} = \frac{\$42 \div 7}{7 \div 7} = \frac{\$6}{1}$

PAGE 167

2. **6** $4 \times 18 = 72$; $72 \div 12 = 6$
3. **6** $18 \times 10 = 180$; $180 \div 30 = 6$
4. **2** $7 \times 4 = 28$; $28 \div 14 = 2$
5. **55** $20 \times 11 = 220$; $220 \div 4 = 55$
6. **25** $30 \times 10 = 300$; $300 \div 12 = 25$
7. **35** $42 \times 10 = 420$; $420 \div 12 = 35$
8. **19.2** $24 \times 4 = 96$; $96 \div 5 = 19.2$

$$\begin{array}{r} 19.2 \\ 5\overline{)96.0} \\ \underline{5} \\ 46 \\ \underline{45} \\ 1\,0 \\ \underline{1\,0} \end{array}$$

9. **25** $5 \times 40 = 200$; $200 \div 8 = 25$
10. **$210** $\frac{40 \text{ hr.}}{\$240} = \frac{35 \text{ hr.}}{?}$

$240 \times 35 = 8{,}400$; $8{,}400 \div 40 = 210$

11. **21** $\frac{35 \text{ laps}}{25 \text{ min.}} = \frac{?}{15 \text{ min.}}$

$35 \times 15 = 525$; $525 \div 25 = 21$

12. **17** $\frac{5 \text{ rolls}}{15 \text{ rolls}} = \frac{?}{51 \text{ rolls}}$

$5 \times 51 = 255$; $255 \div 15 = 17$

13. **6** $\frac{45 \text{ min.}}{1 \text{ day}} = \frac{?}{8 \text{ days}}$

$45 \times 8 = 360$; $360 \div 1 =$
$360 \text{ min} \div 60 \text{ min./hr.} = 6 \text{ hr.}$

14. **98.425** $\frac{1 \text{ meter}}{39.37 \text{ inches}} = \frac{2.5}{?}$

$39.37 \times 2.5 = 98.425$; $98.425 \div 1 = 98.425$

15. **1,250** $\frac{2{,}500 \text{ cars}}{4{,}000 \text{ spaces}} = \frac{?}{2{,}000 \text{ spaces}}$

$2{,}500 \times 2{,}000 = 5{,}000{,}000 \div 4{,}000 = 1{,}250$

PAGE 169

1. **(3) 20** The distance on the map from Riverton to Plainview is 1 inch.

$\frac{2 \text{ in.}}{40 \text{ mi.}} = \frac{1 \text{ in.}}{? \text{ mi.}}$ $40 \times 1 = 40$; $40 \div 2 = 20$

2. **(2) 55** The distance on the map from Plainview to Rock Falls is $2\frac{3}{4}$ inches.

$\frac{2 \text{ in.}}{40 \text{ mi.}} = \frac{2\frac{3}{4} \text{ in.}}{? \text{ mi.}}$ $40 \times 2\frac{3}{4} = 110$; $110 \div 2 = 55$

3. **(3) 3**

$\frac{2 \text{ in.}}{40 \text{ mi.}} = \frac{? \text{ in.}}{60 \text{ mi.}}$ $2 \times 60 = 120$; $120 \div 40 = 3$

4. **(4) $7\frac{1}{2}$** The distance on the map from Mesa to Canyon City is $1\frac{1}{2}$ inches.

$\frac{3 \text{ in.}}{15 \text{ mi.}} = \frac{1\frac{1}{2} \text{ in.}}{? \text{ mi.}}$ $15 \times 1\frac{1}{2} = 22\frac{1}{2}$; $22\frac{1}{2} \div 3 = 7\frac{1}{2}$

5. **(4) $11\frac{1}{4}$** The distance on the map from Bluffton to Canyon City is $2\frac{1}{4}$ inches.

$\frac{3 \text{ in.}}{15 \text{ mi.}} = \frac{2\frac{1}{4} \text{ in.}}{? \text{ mi.}}$ $15 \times 2\frac{1}{4} = 33\frac{3}{4}$; $33\frac{3}{4} \div 3 = 11\frac{1}{4}$

6. **The distance from Mesa to Canyon City is $7\frac{1}{2}$ miles. The distance from Canyon City to Bluffton is $11\frac{1}{4}$ miles. The total miles traveled is $7\frac{1}{2} + 11\frac{1}{4}$, or $18\frac{3}{4}$ miles. Round $18\frac{3}{4}$ miles to 20 miles. If Jerome drives 40 miles in 1 hour, then he will drive 20 miles in $\frac{1}{2}$ hour. Therefore, it takes Jerome about 30 minutes to drive from Mesa to Bluffton.**

PAGES 170–171

1. $\frac{5}{7}$

$\frac{20 \text{ in office}}{28 \text{ in warehouse}} = \frac{20 \div 4}{28 \div 4} = \frac{5}{7}$

2. $\frac{5}{12}$

$\frac{15 \text{ with field goals}}{36 \text{ total points}} = \frac{15 \div 3}{36 \div 3} = \frac{5}{12}$

3. **36** $6 \times 42 = 252; 252 \div 7 = 36$

4. **20** $8 \times 30 = 240; 240 \div 12 = 20$

5. **265 miles**

$$\frac{106 \text{ miles}}{2 \text{ hours}} = \frac{? \text{ miles}}{5 \text{ hours}} \quad \begin{aligned} 106 \times 5 &= 530 \\ 530 \div 2 &= 265 \end{aligned}$$

6. **12**

$$\frac{400 \text{ total parts}}{3 \text{ defective parts}} = \frac{1{,}600 \text{ total parts}}{? \text{ defective parts}}$$

$3 \times 1{,}600 = 4{,}800$
$4{,}800 \div 400 = 12$

7. **$1.65**

$$\frac{3}{\$0.99} = \frac{5}{?} \quad \begin{aligned} \$0.99 \times 5 &= \$4.95 \\ \$4.95 \div 3 &= \$1.65 \end{aligned}$$

8. **$6\frac{2}{5}$ hours**

$$\frac{2 \text{ hours}}{5 \text{ pictures}} = \frac{? \text{ hours}}{16 \text{ pictures}} \quad \begin{aligned} 2 \times 16 &= 32 \\ 32 \div 5 &= 6\tfrac{2}{5} \end{aligned}$$

9. **$7\frac{1}{2}$**

$$\frac{3 \text{ parts blue}}{4 \text{ parts gray}} = \frac{? \text{ gallons blue}}{10 \text{ gallons gray}}$$

$3 \times 10 = 30 \quad 30 \div 4 = 7\frac{1}{2}$

10. **400 miles**

$$\frac{300 \text{ miles}}{12 \text{ gallons}} = \frac{? \text{ miles}}{16 \text{ gallons}}$$

$300 \times 16 = 4{,}800 \quad 4{,}800 \div 12 = 400$

11. **240**

$$\frac{5 \text{ women}}{9 \text{ employees}} = \frac{? \text{ women}}{432 \text{ employees}}$$

$5 \times 432 = 2{,}160 \quad 2{,}160 \div 9 = 240$

12. **$10\frac{1}{2}$** The distance on the map from Mountainview to Somerset is $1\frac{3}{4}$ inches.

$$\frac{2 \text{ in.}}{12 \text{ mi.}} = \frac{1\frac{3}{4} \text{ in.}}{? \text{ mi.}} \quad \begin{aligned} 12 \times 1\tfrac{3}{4} &= 21 \\ 21 \div 2 &= 10\tfrac{1}{2} \end{aligned}$$

13. **21** The distance on the map from Somerset to Princeton is $3\frac{1}{2}$ inches.

$$\frac{2 \text{ in.}}{12 \text{ mi.}} = \frac{3\frac{1}{2} \text{ in.}}{? \text{ mi.}} \quad \begin{aligned} 12 \times 3\tfrac{1}{2} &= 42 \\ 42 \div 2 &= 21 \end{aligned}$$

14. **5**

$$\frac{2 \text{ in.}}{12 \text{ mi.}} = \frac{? \text{ in.}}{30 \text{ mi.}} \quad \begin{aligned} 2 \times 30 &= 60 \\ 60 \div 12 &= 5 \end{aligned}$$

LESSON 12

PAGE 173

1. **(4) 30** 30% means "30 out of 100," so there are 30 pennies.

2. **(3) $\frac{3}{10}$** $30\% = \frac{30}{100} = \frac{3}{10}$

3. **(1) $0.30** $30\% = \frac{30}{100} = \$0.30$

4. **(2) $\frac{90}{100}$** 90% means "90 out of 100," so the score is $\frac{90}{100}$.

5. **(3) 0.85** $\frac{85}{100} = 0.85$

6. **Count the number of shaded squares. Nine squares are shaded. Nine squares out of 100 are shaded, or $\frac{9}{100}$. The fraction $\frac{9}{100}$ = 0.09. Percent means "out of 100," so 9% of the squares are shaded.**

PAGE 174

2. **$\frac{3}{4}$, 0.75, 75%** 75 of 100 parts are colored. $\frac{75}{100}$ reduces to $\frac{3}{4}$.
$\frac{75 \div 25}{100 \div 25} = \frac{3}{4}$; $\frac{75}{100} = 0.75$;
$\frac{75}{100}$ is the same as 75%.

3. **$\frac{1}{5}$, 0.2, 20%** 20 of 100 parts are colored. $\frac{20}{100}$ reduces to $\frac{1}{5}$.
$\frac{20 \div 20}{100 \div 20} = \frac{1}{5}$; $\frac{20}{100} = 0.2$;
$\frac{20}{100}$ is the same as 20%.

4. **$\frac{1}{100}$, 0.01, 1%** 1 of 100 parts is colored.
$\frac{1}{100}$ is in lowest terms; $\frac{1}{100} = 0.01$;
$\frac{1}{100}$ is the same as 1%.

PAGE 175

2. **0.04** .04. = 0.04

3. **2.5** 2.50. = 2.5

4. **0.065** $6\frac{1}{2}\% = 6.5\%$ = .06.5 = 0.065

5. **0.0525** .05.25 = 0.0525

6. **0.0014** .00.14 = 0.0014

7. **0.9725** .97.25 = 0.9725

8. **1** 1.00. = 1.00

9. **260%** 2.60. = 260%

10. **3%** 0.03. = 3%

11. **62.5%** 0.62.5 = 62.5%

12. **0.08%** 0.00.08 = 0.08%

13. **463.5%** 4.63.5 = 463.5%

14. **20%** 0.20. = 20%

15. **586%** 5.86. = 586%

16. **34.5%** $0.34.\frac{1}{2} = 34\frac{1}{2}\%$ or 34.5%

PAGE 176

1. $\frac{3}{5}$ $\frac{60}{100} = \frac{60 \div 20}{100 \div 20} = \frac{3}{5}$

2. $\frac{2}{25}$ $8\% = \frac{8}{100} = \frac{8 \div 4}{100 \div 4} = \frac{2}{25}$

3. $\frac{13}{25}$ $52\% = \frac{52}{100} = \frac{52 \div 4}{100 \div 4} = \frac{13}{25}$

4. $\frac{1}{100}$

5. $\frac{5}{2}$ $\frac{250}{100} = \frac{250 \div 50}{100 \div 50} = \frac{5}{2}$

6. $\frac{4}{25}$ $\frac{16}{100} = \frac{16 \div 4}{100 \div 4} = \frac{4}{25}$

7. **1** $\frac{100}{100} = \frac{100 \div 100}{100 \div 100} = \frac{1}{1} = 1$

8. $\frac{21}{50}$ $\frac{42}{100} = \frac{42 \div 2}{100 \div 2} = \frac{21}{50}$

9. **50%** $\frac{1}{2} =$ 2)1.0 = 0.5 = 50%
   −1 0
   0

10. **60%** $\frac{3}{5} =$ 5)3.0 = 0.6 = 60%
   −3 0
   0

11. **70%** 10)7.0 = 0.7 = 70%
   −7 0
   0

12. **37.5%** 8)3.0 = 0.375 = 37.5%
   −2 4
   60
   −56
   40
   −40
   0

13. **25%** 4)1.0 = 0.25 = 25%
   − 8
   20
   −20
   0

14. **62.5%** 8)5.0 = 0.625 = 62.5%
   −4 8
   20
   −16
   40
   −40
   0

15. **650%** $6\frac{5}{10} = \frac{65}{10} =$ 10)65 = 6.5 = 650%
   −60
   5 0
   −5 0
   0

16. **300%** $\frac{15}{5} = 3 = 300\%$

PAGE 177

1. **0.37** To convert a percent to a decimal, divide by 100 (move decimal point 2 places to the left).
   37% = .37. = 0.37

2. **0.04** 4% = .04. = 0.04

3. **2.25** 225% = 2.25. = 2.25

4. **0.065** $6\frac{1}{2}\% = 6.5\% = .06.5 = 0.065$

5. **46%** To convert a decimal to a percent, multiply by 100 (move decimal point 2 places to the right).
   0.46 = 0.46. = 46%

6. **8%** $0.08 = 0.08. = 8\%$

7. **250%** $2.5 = 2.50. = 250\%$

8. **37.5%** $0.375 = 0.37.5 = 37.5\%$

9. $\frac{1}{2}$ $50\% = \frac{50}{100} = \frac{50 \div 50}{100 \div 50} = \frac{1}{2}$

10. $\frac{2}{5}$ $40\% = \frac{40}{100} = \frac{40 \div 20}{100 \div 20} = \frac{2}{5}$

11. $\frac{1}{4}$ $25\% = \frac{25}{100} = \frac{25 \div 25}{100 \div 25} = \frac{1}{4}$

12. $\frac{9}{10}$ $90\% = \frac{90}{100} = \frac{90 \div 10}{100 \div 10} = \frac{9}{10}$

13. **70%** $10\overline{)70}$ = 0.7 $\quad 0.7 = 0.70. = 70\%$

14. **75%** $4\overline{)3.00}$ = 0.75 $\quad 0.75 = 0.75. = 75\%$

15. **80%** $5\overline{)4.0}$ = 0.8 $\quad 0.8 = 0.80. = 80\%$

16. **87.5%** $8\overline{)7.000}$ = 0.875 $\quad 0.875 = 0.87.5 = 87.5\%$

LESSON 13

PAGE 179

1. **(3) the part** base = 8 hours, rate = 20%, part = the hours it takes Hector to do the tasks
2. **(2) 8%** The rate is followed by the percent sign.
3. **(2) the rate** base = 64 people, part = 16 are women, rate = the percent of the employees that are women
4. **(1) 6** rate = 50%, part = 3 hours, base = the whole = 6 hours
5. **(3) the part** base = $24, rate = 20%, part = amount saved
6. **(1) the base** rate = 25%, part = $10, base = regular price of the sweater
7. **(4) 12** base = 16 games, rate = 75%, part = 12 games
8. **Problems will vary. Possible answer: Fashions Plus has a 20% discount on goose down jackets. The original price is $500. The customer saves $100. Which number is the part?**
   Part = base × rate
   $100 = $500 × 20%

PAGE 181

2. **$18** $40\% = 0.40$

$$\begin{array}{r} \$45 \\ \times\ .40 \\ \hline 00 \\ 180 \\ \hline \$18.00 \end{array}$$

3. **1.28** $8\% = 0.08$

$$\begin{array}{r} 16 \\ \times .08 \\ \hline 1.28 \end{array}$$

4. **120** $250\% = 2.5$

$$\begin{array}{r} 48 \\ \times 2.5 \\ \hline 240 \\ 96 \\ \hline 120.0 \end{array}$$

5. **$2.25** $4\frac{1}{2}\% = 0.045$

$$\begin{array}{r} \$50 \\ \times .045 \\ \hline 250 \\ 200 \\ \hline \$2.250 \end{array}$$

6. **10.44** $7\frac{1}{4}\% = 0.0725$

$$\begin{array}{r} 144 \\ \times .0725 \\ \hline 720 \\ 288 \\ 1008 \\ \hline 10.4400 \end{array}$$

8. **12** $80\% = 0.80$

$$\begin{array}{r} 15 \\ \times .80 \\ \hline 00 \\ 120 \\ \hline 12.00 \end{array}$$

9\. **400** 5% = 0.05

$$\begin{array}{r} 8.000 \\ \times\ \ .05 \\ \hline 400.00 \end{array}$$

10\. **108** 45% = 0.45

$$\begin{array}{r} 240 \\ \times\ .45 \\ \hline 1200 \\ 960 \\ \hline 108.00 \end{array}$$

11\. **$5,200** 16% = 0.16
$32,500 × 0.16 = $5,200

12\. **$0.36** 27% = 0.27
$1.32 × 0.27 = $0.3564; $0.3564 rounded to the nearest cent is $0.36.

PAGE 183

1. **(2) $650 × 0.12 × 1**
interest = *principal* × *rate* × *time*
= $650 × 0.12 × 1
2. **(1) $75.60** $i = p \times r \times t$
= $420 × 0.18 × 1
= $75.60
3. **(4) $210.80** $i = p \times r \times t$
= $1,240 × 0.085 × 2
= $210.80
4. **(2) $252** $i = p \times r \times t$
= $1,600 × 0.0525 × 3
= $252
5. **(2) $410 × 0.16 × 2** Subtract the down payment from the cost to find the principal.
$560 − $150 = $410
Use the formula to find the interest.
$i = p \times r \times t$ = $410 × 0.16 × 2
6. **(3) $541.20** Add the principal and interest to find the amount to be paid back.
$410 + $131.20 = $541.20
7. **(4) $2,352** Use the formula to find the interest. $i = p \times r \times t$
= $2,100 × 0.12 × 1
= $252
Add the principal and interest to find the amount to be paid back.
$2,100 + $252 = $2,352
8. Estimate: Use the formula to find the interest. $i = p \times r \times t$ (Round .15 to .20)
= $10,000 × 0.20 × 2
= $4,000
Divide the amount to be paid back ($10,000 + $4,000) by the number of monthly payments. Round 24 months (in 2 years) to 25.
$14,000 ÷ 25 is 560 so the monthly payment is about $560.
Actual Calculations:
$i = p \times r \times t$
= $10,000 × 0.15 × 2
= $3,000
$13,000 ÷ 24 = $541.67

PAGE 185

1. **(3) 38%** Add the percents for maintenance, advertising, and operations.
10% + 8% + 20% = 38%
2. **(4) $7,200** Multiply the amount of business expenses by the percent for operations. $36,000 × 20% = $36,000 × 0.2 = $7,200
3. **(5) $20,520** Multiply the business expenses by the percent for salaries.
$36,000 × 57% = $36,000 × 0.57 = $20,520
4. **(2) 56.25%** Add the percents for carpet and lighting costs. 37.5% + 18.75% = 56.25%
5. **(1) $900** Multiply the amount spent on the improvements by the total percent for carpet and lighting. $1,600 × 56.25% = $1,600 × 0.5625 = $900
6. **(3) $300** Add the percents for wallpaper and paint: 12.5% + 6.25% = 18.75%.
Multiply the amount spent on the improvements by the combined percent:
$1,600 × 18.75% = $1,600 × 0.1875 = $300

PAGES 186–187

1. **192** 64% = 0.64
$$\begin{array}{r} 300 \\ \times 0.64 \\ \hline 192.00 \end{array}$$

2. **4.32** 6% = 0.06
$$\begin{array}{r} 72 \\ \times 0.06 \\ \hline 4.32 \end{array}$$

3. **$86.40** 90% = 0.9
$$\begin{array}{r} \$96 \\ \times\ 0.9 \\ \hline \$86.40 \end{array}$$

4. **$224** 400% = 4
$$\begin{array}{r} \$56 \\ \times\ \ 4 \\ \hline \$224 \end{array}$$

5. **45** 125% = 1.25
$$\begin{array}{r} 1.25 \\ \times\ 36 \\ \hline 45.00 \end{array}$$

6. **5.25** $3\frac{1}{2}\% = 0.035$
$$\begin{array}{r} 150 \\ \times 0.035 \\ \hline 5.250 \end{array}$$

7. **$116** Multiply the base ($145) times the rate (80%).
80% = 0.8
$$\begin{array}{r} \$145 \\ \times\ \ 0.8 \\ \hline \$116.0 \end{array}$$

8. **$41.25** Multiply the base ($125) times the rate (33%).
33% = 0.33
$$\begin{array}{r} \$125 \\ \times 0.33 \\ \hline 3\ 75 \\ +37\ 5\ \\ \hline \$41.25 \end{array}$$

9. **$2,040** Subtract the down payment from the cost to find the principal.
$1,750 − $250 = $1,500
Use the formula to find the interest.
$i = p \times r \times t$
$= \$1,500 \times 0.18 \times 2$
$= \$540$
Add the principal and the interest to find the amount to be paid back.
$1,500 + $540 = $2,040

10. **1,650** Multiply the base (3,000) times the rate (55%).
55% = 0.55
$$\begin{array}{r} 3,000 \\ \times\ 0.55 \\ \hline 1,650.00 \end{array}$$

11. **$37.20** Multiply the base ($310) times the rate (12%).
12% = 0.12
$$\begin{array}{r} \$310 \\ \times 0.12 \\ \hline \$37.20 \end{array}$$

12. **$97.15** Add the cost of both appliances.
$475 + $680 = $1,155
Subtract the down payment from the cost to find the principal.
$1,155 − $150 = $1,005
Use the formula to find the interest.
$i = p \times r \times t$
$= \$1,005 \times 0.16 \times 1$
$= \$160.80$
Add the principal and interest to find the amount to be paid back.
$1,005 + $160.80 = $1,165.80
Divide the amount to be paid back by the number of monthly payments.
$1,165.80 ÷ 12 = $97.15

13. **(4) 40 × 15%** Multiply the base (40) times the rate (15%) to find the part.

14. **(2) 40% + 25%** Add the percentages for supervising (40%) and planning (25%).

15. **(1) 8** Multiply the base (40) times the rate (20%).
20% = 0.2 40 × 0.2 = 8

16. **(5) Other** The smallest section of the graph represents *other.*

17. **(5) $770** Multiply the base ($2,200) times the rate (35%).
$2,200 × 0.35 = $770

18. **(3) $880** Add the percents for food and clothes: 25% + 15% = 40%. Multiply the Wagners' total monthly income by the sum. $2,200 × 40% = $2,200 × 0.4 = $880

LESSON 14

PAGE 189

1. **(3) $299 ÷ $1,150** To solve for the rate, divide the part ($299) by the base ($1,150).
2. **(1) $500 ÷ $4,500** Divide the part ($500) by the base ($4,500) to find the rate.
3. **(3) $30 ÷ $200** Divide the part ($30) by the base ($200) to find the rate.
4. **(3) 65 ÷ 110** To find the rate, divide the part (65) by the base (110).
5. **(4) $45 ÷ $380** Divide the part ($45) by the base ($380) to find the rate.
6. **(5) Not enough information is given.** You are given the base ($420) but you do not have the part (how much Flavia spends a week).
7. **(4) 117 ÷ 300** Divide the part (117) by the base (300) to find the rate.
8. **The two equations are the inverse of each other. To find the rate, you divide both sides by the base.**
   part = base × rate
   part ÷ base = base × rate ÷ base
   part ÷ base = rate

PAGE 191

2. **10%**
   $0.1 = 36.0 \div 360$ 0.1 = 10%
3. **400%**
   $4 = 180 \div 45$ 4 = 400%
4. **60%**
   $0.6 = 57.0 \div 95$ 0.6 = 60%
5. **2%**
   $0.02 = 4.00 \div 200$ 0.02 = 2%
6. **250%**
   $2.5 = 15.0 \div 6$ 2.5 = 250%
7. **80%** Divide the part (32) by the base (40) to find the rate.
   $0.8 = 32.0 \div 40$ 0.8 = 80%
8. **60%** Divide the part ($570) by the base ($950) to find the rate.
   $0.6 = 570.0 \div 950$ 0.6 = 60%
9. **25%** Divide the part ($9) by the base ($36) to find the rate.
   $0.25 = \$9.00 \div \$36$ 0.25 = 25%
10. **16%** Divide the part ($73.60) by the base ($460) to find the rate.
    $0.16 = \$73.60 \div \$460$ 0.16 = 16%
11. **2%** Divide the part ($1,100) by the base ($46,000) to find the rate.
    $1,100 ÷ $46,000 = 0.0239, rounded to the nearest percent is 2%.
12. **57%** Divide the part (42) by the base (74) to find the rate.
    42 ÷ 74 = 0.567   0.567 rounds to 0.57
    0.57 = 57%

PAGE 193

1. **(2) grams of fat and number of calories in a gram of fat**
   grams of fat $\times \frac{9 \text{ calories}}{\text{g fat}}$ = calories from fat
2. **(3)** $\frac{9}{200} = \frac{?}{100}$
   $\frac{\text{part}}{\text{base}} = \frac{\text{rate}}{100}$
   $\frac{9}{200} = \frac{?}{100}$
3. **(2) 4.5%** Solve the proportion.
   $\frac{9}{200} = \frac{?}{100}$ 9 × 100 = 900
   900 ÷ 200 = 4.5
4. **(2)** $\frac{18}{120} = \frac{?}{100}$

5. **(1) 15%** Solve the proportion.

$\frac{18}{120} = \frac{?}{100}$ $18 \times 100 = 1{,}800$ $1{,}800 \div 120 = 15$

6. **Andrew doesn't need to avoid either food. For spaghetti, only 0.5% of the calories come from fat, and in oat cereal 1.7% of the calories come from fat. Each of these percents is less than 30%.**

PAGE 195

1. **(4) $1,600** Add the amounts on the graph to find the total.

$$\begin{array}{r} \$600 \\ 300 \\ 250 \\ 350 \\ +\ 100 \\ \hline \$1{,}600 \end{array}$$

2. **(3) 38%** Divide the part ($600) by the base ($1,600; see the explanation for Question 1) to find the rate.

$\$1{,}600\overline{)\$600.000}$ = 0.375 0.375 rounds to 0.38. 0.38 = 38%

3. **(3) 28%** Add the amounts budgeted for personal expenses ($350) and other ($100). $350 + $100 = $450. Then divide the part ($450) by the base ($1,600; see the explanation for Question 1) to find the rate.

$\$1{,}600\overline{)\$450.000}$ = 0.281 0.281 rounds to 0.28. 0.28 = 28%

4. **(5) $200,000** Add the amounts on the graph to find the total. Then multiply by 1,000 because the key says *$ in thousands.*

$$\begin{array}{r} \$50 \\ 85 \\ 30 \\ 20 \\ +\ 15 \\ \hline \$200 \end{array} \qquad \begin{array}{r} \$200 \\ \times\ 1{,}000 \\ \hline \$200{,}000 \end{array}$$

5. **(4) 68%** Add the sales from T-shirts ($50,000) and from sweatshirts ($85,000).
$50,000 + $85,000 = $135,000
Then divide the part ($135,000) by the base ($200,000; see the explanation for Question 4) to find the rate.

$\$200\overline{)\$135.000}$ = 0.675 0.675 rounds to 0.68. 0.68 = 68%

6. **Subtract the sales from jackets ($20,000) from the sales from caps ($30,000).**
$30,000 – $20,000 = $10,000
Then divide the part ($10,000) by the base ($200,000; see the explanation for Question 4) to find the rate.

$\$200\overline{)\$10.00}$ = 0.05 0.05 = 5%

PAGES 196–197

1. **160%**

$10\overline{)16.0}$ = 1.6 1.6 = 160%

2. **75%**

$72\overline{)54.00}$ = 0.75 0.75 = 75%

3. **4%**

$300\overline{)12.00}$ = 0.04 0.04 = 4%

4. **500%**

$15\overline{)75}$ = 5 5 = 500%

5. **40%** Divide the part (240) by the base (600) to find the rate.

$600\overline{)240.0}$ = 0.4 0.4 = 40%

6. **83%** Divide the part ($80) by the base ($96) to find the rate.

$\$96\overline{)\$80.000}$ = 0.833 0.833 rounds to 0.83. 0.83 = 83%

7. **57%** Divide the part (125) by the base (220) to find the rate.

$220\overline{)125.000}$ = 0.568 — 0.568 rounds to 0.57. 0.57 = 57%

8. **30%** Divide the part (12) by the base (40) to find the rate.

$40\overline{)12.0}$ = 0.3 — 0.3 = 30%

9. **20,000** Add the amounts on the graph to find the total. Then multiply by 1,000.
7 + 2 + 5 + 4 + 2 = 20
20 × 1,000 = 20,000

10. **25%** Divide the part (5,000) by the base (20,000; see the explanation for Item 9) to find the rate.

$20{,}000\overline{)5{,}000.00}$ = 0.25 — 0.25 = 25%

11. **45%** Add the amounts for coffee (7,000) and tea (2,000) to find the part.
7,000 + 2,000 = 9,000
Then divide the part by the base (20,000; see the explanation for Item 9) to find the rate.

$20{,}000\overline{)9{,}000.00}$ = 0.45 — 0.45 = 45%

12. **(3) number of calories in a gram of fat**

13. **(4)** $\frac{45}{180} = \frac{?}{100}$

14. **(2) 25%** Solve the proportion.

$\frac{45}{180} = \frac{?}{100}$ 45 × 100 = 4,500

4,500 ÷ 180 = 25

15. **(2) $25.00 ÷ $500.00** Divide the part deducted for the credit union account ($25.00) by the base ($500.00).

16. **(3) 14%** Add the deductions for federal ($63.40) and state ($6.60) income taxes to find the part.

$63.40
+ 6.60
$70.00

Then divide the part by the base ($500.00) to find the rate.

$\$500\overline{)\$70.00}$ = 0.14 — 0.14 = 14%

17. **(2) $20.00** Multiply the base ($500.00) times the rate (4%).

4% = 0.04

$500.00
× 0.04
$20.0000

LESSON 15

PAGE 199

1. **(2) $17.52 ÷ 0.06** Divide the part ($17.52) by the rate (0.06) to find the base (gross pay).
2. **(2) 270 ÷ 0.75** Divide the part (270) by the rate (0.75) to find the base (number of people who came into the store).
3. **(3) 22 ÷ 0.88** Divide the part (22) by the rate (0.88) to find the base (number of questions on the test).
4. **(2) 18 ÷ 0.45** Divide the part (18) by the rate (0.45) to find the base (hours Curtis worked).
5. **(4) 85 ÷ 0.20** Divide the part (85) by the rate (0.20) to find the base (number of sales).
6. **(5) Not enough information is given.** You do not know the rate to answer the question.
7. **(4) $13.60 ÷ 0.40** Divide the part ($13.60) by the rate (0.40) to find the base (regular price).
8. **part = base × rate, divide both sides by the rate**
   part ÷ rate = base × rate ÷ rate
   part ÷ rate = base

PAGE 201

1. **(2) $7.20 ÷ 30%** Divide the part ($7.20) by the rate (30%) to find the base (regular price).
2. **(3) $54.00 ÷ 45%** Divide the part ($54.00) by the rate (45%) to find the base (regular price).
3. **(5) $36.00 − $12.60** Subtract the amount you save ($12.60) from the regular price ($36.00) to find the sale price.
4. **(2) $5.60 ÷ 40%** Divide the part ($5.60) by the rate (40%) to find the base (regular price).
5. **(2) $64.00 × 25%** Multiply the base ($64.00) times the rate (25%) to find the part (savings).
6. **Divide the part ($6.50) by the base ($32.50) to find the discount rate (20%). To find the sale price of the skirt, subtract the amount saved ($6.50) from the regular price of the skirt ($32.50).**

PAGE 203

2. **90** 40% = 0.4 $\quad 0.4\overline{)36.0}$ = 90.
3. **$12** 75% = 0.75 $\quad 0.75\overline{)\$9.00}$ = $12.
4. **19** 300% = 3 $\quad 3\overline{)57}$ = 19
5. **500** 8% = 0.08 $\quad 0.08\overline{)40.00}$ = 500.
6. **20** 175% = 1.75 $\quad 1.75\overline{)35.00}$ = 20.
7. **$75** Divide the part ($22.50) by the rate (30%) to find the base.

   30% = 0.3 $\quad 0.3\overline{)\$22.5.0}$ = $75.
8. **20** Divide the part (14) by the rate (70%) to find the base.

   70% = 0.7 $\quad 0.7\overline{)14.0}$ = 20.
9. **$1,400** Divide the part ($490) by the rate (35%) to find the base.

   35% = 0.35 $\quad 0.35\overline{)\$490.00}$ = $1400.
10. **$685** Divide the part ($34.25) by the rate (5%) to find the base.

    5% = 0.05 $\quad 0.05\overline{)\$34.25}$ = $685.
11. **14,830** 16,313 ÷ 1.1 = 14,830
12. **$7,600** $2,356 ÷ 0.31 = $7,600

PAGE 205

2. **25%** $8.80 − 6.60 = $2.20 $\quad \$8.80\overline{)\$2.20.00}$ = 0.25 $\quad$ 0.25 = 25%
3. **40%** 25 − 15 = 10 $\quad 25\overline{)10.0}$ = 0.4 $\quad$ 0.4 = 40%
4. **12%** $\frac{576}{4,800} = \frac{?}{100}$

   576 × 100 = 57,600
   57,600 ÷ 4,800 = 12
5. **10%** $\frac{7}{70} = \frac{?}{100}$

   7 × 100 = 700
   700 ÷ 70 = 10
6. **6%** $\frac{\$1.44}{\$24.00} = \frac{?}{100}$

   $1.44 × 100 = $144.00
   $144.00 ÷ 24.00 = 6
7. **50%** Subtract the original premium ($750.00) from the new premium ($1,125.00). Then divide by the original premium.

   $1,125.00 − $750.00 = $375.00
   $375.00 ÷ $750.00 = 0.50
   0.50 = 50%
8. **15%** Subtract the new fare ($23.80) from the original fare ($28.00). Then divide by the original fare.

   $28.00 − $23.80 = $4.20
   $4.20 ÷ $28.00 = 0.15
   0.15 = 15%

PAGE 207

1. **(3) 30** The orange bar for dinner in September reaches halfway between the marks for *140* and *160*. So, the number of customers in September was 150. The number of dinner customers in August was 120. Subtract to find the difference. $150 - 120 = 30$

2. **(3) 25%** Divide the change in the number of dinner customers (30; see the explanation for Question 1) by the number of dinner customers in August (120).

   $$120\overline{)30.00} = 0.25$$

   $0.25 = 25\%$

3. **(2) 18%** Add the number of customers for breakfast (60), lunch (130), and dinner (150) in September. Then divide the part (number of breakfast customers, 60) by the base (total customers) to find the rate.

   $60 + 130 + 150 = 340$

   $$340\overline{)60.000} = 0.176$$

   0.176 rounds to 0.18.
   $0.18 = 18\%$

4. **(3) 20%** Subtract the number of customers for breakfast in August (50) from the number of customers for breakfast in September (60). Then divide by the number of customers in August.

   $60 - 50 = 10$

   $$50\overline{)10.0} = 0.2$$

   $0.2 = 20\%$

5. **(1) 7%** Subtract the number of customers for lunch in September (130) from the number of customers for lunch in August (140). Then divide by the number in August.

   $140 - 130 = 10$

   $$140\overline{)10.000} = 0.071$$

   0.071 rounds to 0.07.
   $0.07 = 7\%$

6. **(4) 650** Find the number of customers for September: $60 + 130 + 150 = 340$ (see the explanation for Question 3). Find the number of customers for August: $50 + 140 + 120 = 310$. Add the two totals: $340 + 310 = 650$.

7. **Add the number of customers for breakfast (50), lunch (140), and dinner (120) in August. Then add the number of customers for breakfast (60), lunch (130), and dinner (150) in September. Subtract the total for August (310) from the total for September (340). Then divide by the total for August.**

PAGE 209

1. **330** To find the amount of commission (part), multiply his sales (base) by the commission percent (rate). $\$1,320 \times 25\% = \$1,320 \times 0.25 = \$330$ (Do not grid the dollar sign.)

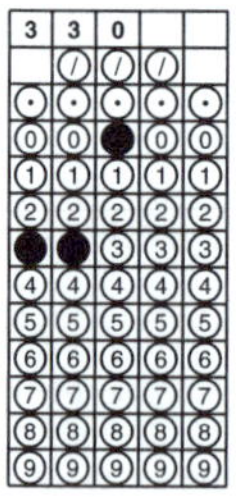

2. **1700** To find the amount of sales (base), divide the commission (part) by the commission percent (rate). $\$425 \div 25\% = \$425 \div 0.25 = \$1,700$

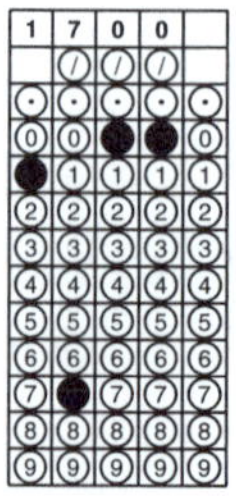

3. **175** To find the amount of interest (part) for one year, multiply the amount of the loan (base) by the interest percent (rate).
$3,500 × 5% = $3,500 × 0.05 = $175

4. **30** To find the number of games played (base), divide the number of games won (part) by the percent (rate).
27 ÷ 90% = 27 ÷ 0.9 = 30

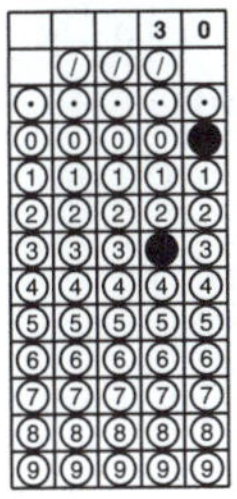

PAGES 210–211

1. **90** 30% = 0.3 $0.3\overline{)27.0}$ = 90.

2. **2,000** 3% = 0.03 $0.03\overline{)60.00}$ = 2000.

3. **$60** 68% = 0.68 $0.68\overline{)\$40.80}$ = $60.

4. **20** 20% = 0.2 $0.2\overline{)4.0}$ = 20.

5. **$28** 125% = 1.25 $1.25\overline{)\$35.00}$ = $28.

6. **48** 75% = 0.75 $0.75\overline{)36.00}$ = 48.

7. **(5) $120 ÷ 80%** Divide the part ($120) by the rate (80%) to find the base.
8. **(3) 9,610 ÷ 62%** Divide the part (9,610) by the rate (62%) to find the base.

9. **(2) 8%** Subtract the original premium ($216.00) from the new premium ($233.28). Then divide by the original premium.
10. **(3) 10%** Subtract Stan's current wage ($6.50) from his new wage ($7.15). Then divide by his current wage.

$7.15 − 6.50 = $0.65 $\$6.50\overline{)\$0.65.0}$ = 0.1   0.1 = 10%

11. **$14.40** Divide the part ($3.60) by the rate (25%) to find the base.

25% = 0.25 $0.25\overline{)3.60.00}$ = $14.40

12. **$10.80** Subtract the amount you save ($3.60) from the regular price (see the explanation for Question 11).

$14.40 − 3.60 = $10.80

13. **89%** Subtract the amount of July's gas bill ($10) from January's gas bill ($95). Then divide by the amount of January's gas bill.

$95 − 10 = $85 $\$95\overline{)\$85.000}$ = 0.894   0.894 rounds to 0.89. 0.89 = 89%

14. **88%** Subtract the amount of January's electric bill ($40) from July's electric bill ($75). Then divide by the amount of January's electric bill.

$75 − 40 = $35 $\$40\overline{)\$35.000}$ = 0.875   0.875 rounds to 0.88. 0.88 = 88%

15. **(3) $525** To find the amount Jim pays for fuel (part), multiply his car expenses (base) by the fuel percent (rate).
$3,500 × 15% = $3,500 × 0.15 = $525

16. **(4) $204.17** To find his monthly car payment, first find the total amount Jim spends on car payments. Multiply the total car expenses (base) by the car payment percent (rate): $3,500 × 70% = $3,500 × 0.7 = $2,450. To find the monthly car payment, divide the answer by 12. $2,450 ÷ 12 = $204.166 = $204.17, rounded to the nearest cent.

17. **2160** To find the amount of commission (part), multiply Trudy's total sales (base) by her commission percent (rate). $12,000 × 18% = $12,000 × 0.18 = $2,160.

LESSON 16

PAGE 213

1. **(4) 135** Multiply the base (500) times the rate (27%) to find the part.
   500 × 0.27 = 135
2. **(2) $\frac{2}{5}$ or 40%** The probability is the sum of the probabilities of running or jogging and walking.
   27% + 13% = 40% which equals $\frac{2}{5}$.
3. **(2) 20%** Divide the part (12) by the base (60) to find the rate.
   12 ÷ 60 = 0.2 = 20%
4. **(4) 67%** Divide the part (28 + 12 = 40) by the base (60) to find the rate.
   40 ÷ 60 = 0.666, which rounds to 67%
5. **0.2% of the population** 60 ÷ 30,000 = 0.002; 0.002 = 0.2%; **Answers may vary: The size of the sample surveyed is probably too small to predict probability accurately. A larger sample of the population would make the survey a better predictor of probability.**

PAGE 214

2. **1 or 100%** A box contains 15 blue marbles. Each of the 15 outcomes is equally likely to occur. The probability is 1 or 100% that a blue marble will be picked from the box. There is no other possibility.
3. **0 or 0%** None of the pairs of socks in the drawer is brown, so there is no possibility that the pair of socks picked will be brown.
4. **0 or 0%** There is 100% chance of rain tomorrow. Since there is 100% chance of rain, then there is no other possibility. Therefore, the probability that it will *not* rain is 0%.

PAGE 215

2. **0, 0%** The numbers on the number cube are from 1 to 6, so you cannot roll a 9. The probability of rolling a 9 is 0 or 0%.
3. **1, 100%** All of the possible rolls of the number cube are numbers less than 7, so the probability of rolling a number less than 7 is 1 or 100%.
4. **$\frac{1}{2}$, 50%**

$$\frac{\text{number of ways to roll an even number}}{\text{number of possible outcomes}} = \frac{3}{6} = \frac{1}{2} \text{ or } 50\%$$

PAGE 217

2. **$\frac{1}{2}$ or 50%** There are 5 even numbers on the wheel.

$$\frac{\text{number of sections with even numbers}}{\text{number of sections}} = \frac{5}{10} = \frac{1}{2} \text{ or } 50\%$$

3. **$\frac{3}{5}$ or 60%** The numbers on the wheel that are greater than 4 are 5, 6, 7, 8, 9, and 10.

$$\frac{\text{number of sections greater than 4}}{\text{number of sections}} = \frac{6}{10} = \frac{3}{5} \text{ or } 60\%$$

4. **1 or 100%** All of the numbers on the wheel are less than 12. So the probability of the wheel stopping on a number less than 12 is 1 or 100%.

5. **$\frac{3}{8}$ or 37.5%**

$$\frac{\text{number of gray marbles}}{\text{number of marbles}} = \frac{3}{8} \text{ or } 37.5\%$$

6. **$\frac{3}{4}$ or 75%**

$$\frac{\text{number of marbles that are black or gray}}{\text{number of marbles}} = \frac{6}{8} = \frac{3}{4} \text{ or } 75\%$$

7. **$\frac{5}{8}$ or 62.5%**

$$\frac{\text{number of marbles that are not black}}{\text{number of marbles}} = \frac{5}{8} \text{ or } 62.5\%$$

8. **$\frac{1}{5}$ or 20%**

$$\frac{\text{number of white marbles}}{\text{number of marbles}} = \frac{4}{20} = \frac{1}{5} \text{ or } 20\%$$

9. **$\frac{7}{100}$ or 7%**

$$\frac{\text{number of defective jeans}}{\text{number of jeans}} = \frac{35}{500} = \frac{35 \div 5}{500 \div 5} = \frac{7}{100} = 7\%$$

PAGE 219

2. **$\frac{1}{4}$ or 25%** After a blank card is drawn, there are 4 cards left, *Drive* and 3 blanks.

$$\frac{\text{number of cards with } \textit{Drive}}{\text{number of cards left}} = \frac{1}{4} \text{ or } 25\%$$

3. **$\frac{2}{3}$ or $66\frac{2}{3}$%** After two blank cards are drawn, there are 3 cards left, *Drive* and 2 blanks.

$$\frac{\text{number of blank cards}}{\text{number of cards left}} = \frac{2}{3} \text{ or } 66\frac{2}{3}\%$$

4. **$\frac{1}{3}$ or $33\frac{1}{3}$%**

$$\frac{\text{number of \$100 bills}}{\text{number of bills}} = \frac{3}{9} = \frac{1}{3} \text{ or } 33\frac{1}{3}\%$$

5. **$\frac{1}{4}$ or 25%** After a $100 bill is drawn, there are 8 bills left, 6 $20 bills and 2 $100 bills.

$$\frac{\text{number of \$100 bills}}{\text{number of bills}} = \frac{2}{8} = \frac{1}{4} \text{ or } 25\%$$

6. **$\frac{2}{7}$ or 28.6%** After a $20 bill and $100 bill are drawn, there are 7 bills left, 5 $20 bills and 2 $100 bills.

$$\frac{\text{number of \$100 bills}}{\text{number of bills}} = \frac{2}{7} \text{ or } 28.6\%$$

$2 \div 7 = 0.2857 = 28.6\%$

7. **$\frac{1}{25}$ or 4%**

$$\frac{\text{Amy's ticket}}{\text{number of tickets}} = \frac{1}{25} \text{ or } 4\%$$

8. **$\frac{8}{22}$ or 36%** After three tickets are drawn, there are 22 tickets left.

$$\frac{\text{number of men}}{\text{number of tickets}} = \frac{8}{22} \text{ or } 36\%$$

PAGES 220–221

1. **100** Add the numbers from each section: $34 + 20 + 10 + 12 + 24 = 100$
2. **20%** Divide the part (20) by the base (100) to find the rate.
3. **$\frac{1}{10}$ or 10%** Divide the part (10) by the base (100) to find the rate.
4. **1 or 100%** The drawer contains only multi-colored scarves. The probability is 1 or 100% that a multi-colored scarf will be taken from the drawer. There is no other possibility.
5. **0 or 0%** There is 100% chance of snow. Since there is 100% chance of snow, there is no other possibility. Therefore, the probability that it will not snow is 0%.
6. **$\frac{1}{5}$, 20%**

$$\frac{\text{number of ways you can choose a green marble}}{\text{number of marbles}} = \frac{1}{5} \text{ or } 20\%$$

7. **$\frac{4}{5}$, 80%**

$$\frac{\text{number of ways you can choose a marble that is not blue}}{\text{number of marbles}} = \frac{4}{5} \text{ or } 80\%$$

8. **$\frac{2}{5}$, 40%**

$$\frac{\text{number of ways you can choose an orange or yellow marble}}{\text{number of marbles}} = \frac{2}{5} \text{ or } 40\%$$

9. **0, 0%**

$$\frac{\text{number of ways you can choose a black marble}}{\text{number of marbles}} = \frac{0}{5} = 0 \text{ or } 0\%$$

10. **(2) $\frac{3}{5}$ or 60%**

$$\frac{\text{number of black marbles}}{\text{number of marbles}} = \frac{6}{10} = \frac{3}{5} \text{ or } 60\%$$

11. **(1) 0 or 0%** There are no green marbles in the box, so there is no possibility of choosing one.

12. **(2) $\frac{1}{10}$ or 10%**

$$\frac{\text{number of gray marbles}}{\text{number of marbles}} = \frac{1}{10} \text{ or } 10\%$$

13. **(5) $\frac{7}{10}$ or 70%**

$$\frac{\text{number of marbles that are not white}}{\text{number of marbles}} = \frac{7}{10} \text{ or } 70\%$$

14. **$\frac{1}{3}$ or $33\frac{1}{3}$%** After the gray marble is drawn, there are 9 marbles left: 6 black and 3 white.

$$\frac{\text{number of white marbles}}{\text{number of marbles}} = \frac{3}{9} = \frac{1}{3} \text{ or } 33\frac{1}{3}\%$$

15. **$\frac{1}{4}$ or 25%** After the gray and white marbles are drawn, there are 8 marbles left: 6 black and 2 white.

$$\frac{\text{number of white marbles}}{\text{number of marbles}} = \frac{2}{8} = \frac{1}{4} \text{ or } 25\%$$

16. **$\frac{5}{7}$ or 71%** After 1 gray, 1 white, and 1 black marble are drawn, there are 7 marbles left: 5 black and 2 white.

$$\frac{\text{number of black marbles}}{\text{number of marbles}} = \frac{5}{7} \text{ or } 71\%$$

17. **$\frac{1}{3}$ or $33\frac{1}{3}$%** After 1 gray, 1 white, and 2 black marbles are drawn, there are 6 marbles left: 4 black and 2 white.

$$\frac{\text{number of white marbles}}{\text{number of marbles}} = \frac{2}{6} = \frac{1}{3} \text{ or } 33\frac{1}{3}\%$$

MATH AT WORK

PAGE 223

1. **(4) 3:1** 3 cloves: 1 cup cheese

2. **(1) $\frac{6}{8} = \frac{4}{?}$**

$$\frac{\text{6 people—orig. recipe}}{\text{8 people to be served}} = \frac{\text{4 peppers—orig. recipe}}{?}$$

3. **(4) 50%** 3 cloves ÷ 6 people = .5 = 50%

UNIT 4 REVIEW

PAGES 224–225

1. **14** $8 \times 35 = 280$; $280 \div 20 = 14$

2. **324** $450\% = 4.5$

72
× 4.5
324.0

3. **2.5%**

320)8.000 = 0.025

$0.025 = 2.5\%$

4. **$45**

0.09.)$4.05. = $45.

$9\% = 0.09$

5. **40%**

$17.50
− 12.50
$5.00

$12.50.)$5.00.0 = 0.4

$0.4 = 40\%$

6. **67%**

45
−15
30

45)30.000 = 0.666
−27 0
3 00
−2 70
300

0.666 rounds to 67%

7. **17%**

$$\begin{array}{r} 60 \\ -50 \\ \hline 10 \end{array} \qquad \begin{array}{r} 0.1666 \\ 60\overline{)10.0000} \\ -6\ 0 \\ \hline 4\ 00 \\ -3\ 60 \\ \hline 400 \\ -360 \\ \hline 400 \end{array}$$

0.1666 rounds to 17%

8. **\$151.20** Subtract the down payment from the cost to find the principal.
\$520 − \$100 = \$420.
Use the formula to find the interest.

$$\begin{aligned} i &= p \times r \times t \\ &= \$420 \times 0.18 \times 2 \\ &= \$151.20 \end{aligned}$$

9. **\$23.80** Add the principal and interest (see the explanation for Question 8) to find the amount to be paid back.
\$420 + \$151.20 = \$571.20
Divide the amount to be paid back by the number of monthly payments.
(2 years = 24 months)
\$571.20 ÷ 24 = \$23.80

10. **\$7.60**

$$\frac{3}{\$2.85} = \frac{8}{?} \qquad \begin{aligned} \$2.85 \times 8 &= \$22.80 \\ \$22.80 \div 3 &= \$7.60 \end{aligned}$$

11. **\$995.94** Multiply the profit times the rate to find the commission earned.
\$4,527 × 0.22 = \$995.94

12. **(3) $\frac{3}{5}$ or 60%**

$$\frac{\text{number of nickels}}{\text{number of coins}} = \frac{3}{5} = 60\%$$

13. **(3) $\frac{1}{2}$ or 50%** After a nickel has been chosen, there are 4 coins left: 2 nickels and 2 dimes.

$$\frac{\text{number of nickels}}{\text{number of coins}} = \frac{2}{4} = \frac{1}{2} \text{ or } 50\%$$

14. **(4) \$862.50** To find the amount of the tax, multiply the cost of the car (base) by the sales tax percent (rate). \$11,500 × $7\frac{1}{2}$% = \$11,500 × 0.075 = \$862.50.

15. **(3) \$40.00 ÷ 25%** To find the regular price (base), divide the amount saved (part) by the discount (rate). \$40.00 ÷ 25%

16. **28%** To find the percent (rate) of adults who read mysteries, divide the number who read mysteries (part) by the total number in the survey (base). Add to find the total surveyed: 700 + 300 + 400 + 500 = 2,500. Divide the part by the whole: 700 ÷ 2,500 = 0.28, or 28%.

17. **$\frac{11}{25}$** To find the probability, divide the part (600 + 500 = 1,100) by the base (2,500). 1,100 ÷ 2,500 = $\frac{11}{25}$

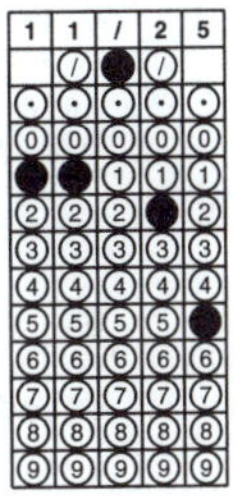

UNIT 4 MINI-TEST

PAGES 226–227

1. **(4) 25** Set up a proportion. Cross multiply:

$$\frac{15}{600} = \frac{?}{1{,}000}$$
$$15 \times 1{,}000 = 15{,}000$$
$$\frac{15{,}000}{600} = 25$$

2. **(2) 70%** To find the percent (rate), divide the actual snowfall (part) by the predicted snowfall (base). $\frac{154}{220}$ = 0.7 or 70%

3. **(1) \$1,050** To find the amount of interest, use the interest formula, $i = prt$. \$2,800 × 12.5% × 3 = \$2,800 × 0.125 × 3 = \$1,050

4. **(4) 62.5%** The dollar amount of 1-hour sales for the 1st quarter is \$8,000 and the amount for the 4th quarter is \$13,000. To find the percent of increase, subtract the two amounts: \$13,000 – \$8,000 = \$5,000. Divide by the dollar amount for the 1st quarter: \$5,000 ÷ \$8,000 = 0.625, or 62.5%

5. **599** To find the cost of the computer (base), divide the amount of tax (part) by the sales tax rate (rate):
$\$47.92 \div 8\% = \$47.92 \div 0.08 = \$599$

6. **(5) \$210** To find the amount of savings (part), multiply the original price (base) by the sale percent (rate):
$\$350 \times 60\% = \$350 \times 0.6 = \$210$

7. **(4) 20%** First find the total number of moviegoers: 50 + 40 + 30 + 20 + 10 = 150. Then divide the number of moviegoers ages 20 to 29 by the total:
30 ÷ 150 = 0.2, or 20%

8. **(3) 60%** First find the total number of moviegoers in the age 10–19 group and the under 10 age group: 40 + 50 = 90. Then find the total number of moviegoers: 50 + 40 + 30 + 20 + 10 = 150. Divide the number of moviegoers in the two groups (part) by the total (base):
90 ÷ 150 = 0.6, or 60%

9. **(4)** $\frac{2}{3}$ There are 6 sections in all, and there are 4 sections with numbers less than 5. The probability of getting a number less than 5 is $\frac{4}{6}$, or $\frac{2}{3}$.

10. **5** Set up a proportion. Cross multiply.
$\frac{3 \text{ nights}}{\$225} = \frac{?}{\$375}$
$3 \times \$375 = \$1,125$
$\$1,125 \div \$225 = 5$

POSTTEST

PAGES 231–239

Part I

1. **(3) 79 (*Whole Numbers: Procedural*)** Add the number of guests in each room and divide the sum by 8: 424 + 208 = 632 guests. 632 ÷ 8 = 79 tables.
2. **(2) \$10.35 (*Decimals: Application*)** Find the cost of three dress shirts at the regular price: \$18.95 × 3 = \$56.85. Subtract to find the difference in the sale price: \$56.85 − \$46.50 = \$10.35.
3. **(2) \$15.20 ÷ 0.4 (*Ratio, Proportion, and Percent: Conceptual*)** The original price is the base. To solve for the base, divide the part (\$15.20) by the rate (40% = 0.4).
4. **(5)** $7\frac{3}{5}$ **(*Fractions: Procedural*)** Add to find the total distance: $2\frac{1}{5} + 3\frac{9}{10} + 1\frac{1}{2} = 2\frac{2}{10} + 3\frac{9}{10} + 1\frac{5}{10} = 6\frac{16}{10} = 7\frac{6}{10} = 7\frac{3}{5}$

5. **1.85 (*Decimals: Application*)**
To find one-half of a decimal, either multiply by 0.5 or divide by 2.
$3.7 \div 2 = 1.85$ pounds.
Note: If you chose to work the problem using fractions, your answer grid should contain the answer $\frac{37}{20}$. This is also a correct answer.

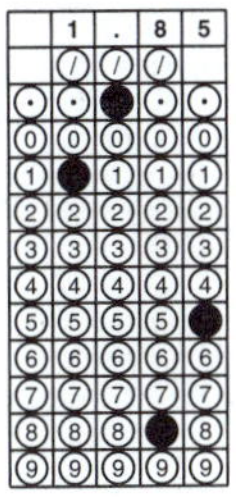

6. **(4) February, March, May, and June (*Whole Numbers: Conceptual*)** Find the line on the graph that represents 5,800. Then follow it across. The bars that are equal to or greater than this line represent months in which the goal was reached.

7. **(3) 50% (*Ratio, Proportion, and Percent: Application*)** The number sold for April is 4,400. The number sold for May is 6,600. To find the percent of change, subtract: $6{,}600 - 4{,}400 = 2{,}200$. Then divide by April's number (the original amount): $2{,}200 \div 4{,}400 = 0.5$. Change to a percent: $0.5 = 50\%$.

8. **(4) 119 (*Whole Numbers: Application*)** Use the order of operations.
$$144 + 18 \div 3 - (13 + 9 \times 2) =$$
$$144 + 18 \div 3 - (13 + 18) =$$
$$144 + 18 \div 3 - 31 =$$
$$144 + 6 - 31 =$$
$$150 - 31 = 119$$

9. **(2) $\frac{3}{10}$ (*Ratio, Proportion, and Percent: Conceptual*)** There are 3 sections out of 10 that are labeled with the number 5. The probability is $\frac{3}{10}$.

10. **$\frac{19}{6}$ (*Fractions: Procedural*)**
Multiply: $4\frac{3}{4} \times \frac{2}{3} = \frac{19}{\cancel{4}_2} \times \frac{\cancel{2}^1}{3} = \frac{19}{6}$ yards. Remember to write your answer in the grid as an improper fraction.

11. **(2) 15 (*Decimals: Procedural*)** Divide. If you do the work on a calculator, you will get a decimal remainder: $25.8 \div 1.7 = 15.17647059$. Ignore the remainder since you are trying to find how many whole pieces can be cut from the spool.

12. **(1) 1:3 (*Ratio, Proportion, and Percent: Application*)** Compare \$2,400 to \$7,200: $\frac{2{,}400}{7{,}200}$ in lowest terms is equal to $\frac{1}{3}$.

13. **(4) $\frac{3}{10}$ (*Fractions: Application*)** Write a fraction comparing the net profit from the Jazz Concert to the Total Net Profit and simplify: $\frac{5{,}400}{18{,}000} = \frac{3}{10}$.

Part II

14. **(3) 8 (*Fractions: Application*)** To estimate an answer, think: $2\frac{2}{3}$ is almost 3 and $2\frac{1}{8}$ is close to 2. Since there are 16 ounces in 1 pound, 2 pounds 2 ounces is close to 2, and 15 ounces is close to 1 pound. Add the rounded amounts: $3 + 2 + 2 + 1 = 8$ pounds.

15. **(5) $(2 \times 3) + (24 \div 12)$ (*Whole Numbers: Conceptual*)** To find the number of feet in yards, multiply by 3. To show the number of inches in feet, divide by 12. Only option (5) shows both operations.

16. **(3) 22,250 (*Decimals: Application*)** Find the data for November (10.5 = 10,500) and December (11.75 = 11,750) and add: 10,500 + 11,750 = 22,250.

17. **(5) Not enough information is given. (*Ratio, Proportion, and Percent: Conceptual*)** Each data point represents the total new subscriptions and renewals during a month. There is no way to know the number for each, so there is no way to calculate the percent.

18. **235 (*Whole Numbers: Application*)** Put the six numbers in order. Since there is an even number of values, average the two middle values: 140, 180, 210, 260, 340, 350. Average: 210 + 260 = 470, and 470 ÷ 2 = 235.

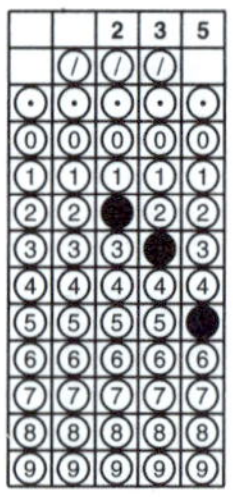

19. **(4) $\frac{2}{3}$ (*Fractions: Conceptual*)** In the figure, 10 out of 15 boxes are shaded: $\frac{10}{15} = \frac{2}{3}$

20. **(2) 3.5 × 100 (*Decimals: Conceptual*)** There are 100 centimeters, a smaller unit, in 1 meter, a larger unit. Multiply the number of meters by 100 to convert to centimeters.

21. **(3) 234 (*Whole Numbers: Application*)** Think of the room as two rectangles. Use the formula: Area = length × width. Find the area of the larger rectangle: 15 × 12 = 180 square feet. Find the area of the smaller rectangle: 9 × 6 = 54 square feet. Add: 180 + 54 = 234 square feet.

22. **(5) $\frac{3}{10} = \frac{5}{?}$ (*Ratio, Proportion, and Percent: Conceptual*)** Ron can place an ad of 3 lines for $10. Set up the proportion so that both ratios compare lines to dollars.

23. **(4) 60 (*Fractions: Conceptual*)** Think of the multiples of each denominator. The number 60 is the lowest multiple that 4, 5, and 6 have in common.

24. **2040 (*Ratio, Proportion, and Percent: Application*)** 15% of 2,400 people voted for Prow. To find that number, multiply: 2,400 × 0.15 = 360. Subtract to find the number that did not vote for Prow: 2,400 − 360 = 2,040. (You could also subtract: 100% − 15% = 85% and then find 85% of 2,400: 2,400 × 0.85 = 2,040.)

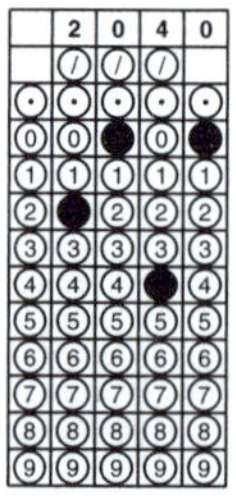

25. **(2) $255.00 (*Decimals: Procedural*)** Multiply 3,000 × $0.085. You can also solve this problem with estimation by thinking: 3,000 × $0.10 = $300, so the correct answer must be option (4), the choice nearest $300.

CALCULATOR HANDBOOK

PAGE 241

1. **$628** Add the amounts for the cost of the oven, tax, delivery and installation: $499 + $29 + $25 + $75 = $628.
2. **$1,820** Add the amounts: $780 + $345 + $290 + $85 + $320 = $1,820.

PAGE 242

1. **8** Subtract the Bluebirds' first-half score from the Redwings' first-half score: 47 − 39 = 8.
2. **27** Subtract the Bluebirds' final score from the Redwings' final score: 116 − 89 = 27.

3. **931** Subtract the paid attendance from the number of people who attended the game: 14,899 − 13,968 = 931.

PAGE 243

1. **15,914 square meters** Multiply the length times the width: 218 × 73 = 15,914.
2. **768 cubic inches** Multiply the length times the width times the height: 12 × 8 × 8 = 768.

PAGE 244

1. **$2,500,000** Divide the total lottery winnings by the number of winners: $32,500,000 ÷ 13 = $2,500,000.
2. **$15,500** Divide the state grant by the number of counties in the joint project: $93,000 ÷ 6 = $15,500.

PAGE 245

1. **3,125** 5 × 5 × 5 × 5 × 5 = 3,125
2. **6.25** 2.5 × 2.5 = 6.25
3. **25** $\sqrt{625} = 25$
4. **17** $\sqrt{289} = 17$

PAGE 246

1. **20,132.8 miles** Add the number of miles at which the oil was last changed to the number of miles after which the oil should be changed: 16,632.8 + 3,500 = 20,132.8.
2. **$10.73** Add the cost of the parts: $5.88 + $4.25 + $0.60 = $10.73.

PAGE 247

1. **350.9** Subtract the May 10 odometer reading from the May 15 odometer reading: 21,734.6 − 21,383.7 = 350.9.
2. **$5.85** Subtract the amounts from $20: $20.00 − $11.75 − $1.39 − $1.01 = $5.85

PAGE 248

1. **256.68 square meters** Multiply the length times the width: 20.7 × 12.4 = 256.68.
2. **78.26 cubic centimeters** Multiply: 5.2 × 3.5 × 4.3 = 78.26

PAGE 249

1. **2.625 feet** Divide the length of the board by the number of pieces: 10.5 ÷ 4 = 2.625.
2. **$7.50** Divide the total by the number of people splitting the cost: $22.50 ÷ 3 = $7.50.

PAGE 250

1. $2\frac{2}{3}$
2. $24\frac{1}{15}$
3. $6\frac{5}{9}$
4. $15\frac{1}{5}$
5. $\frac{42}{5}$
6. $\frac{51}{4}$
7. $\frac{25}{7}$
8. $\frac{83}{12}$

PAGE 251

Fraction	Decimal	Percent
$\frac{1}{10}$	.1	10%
$\frac{1}{8}$	.125	12.5%
$\frac{1}{5}$	.2	20%
$\frac{1}{4}$	.25	25%
$\frac{1}{3}$	.33	$33\frac{1}{3}\%$
$\frac{2}{5}$	.4	40%
$\frac{1}{2}$	.5	50%
$\frac{3}{5}$	.6	60%
$\frac{2}{3}$	.67	$66\frac{2}{3}\%$
$\frac{3}{4}$	.75	75%
$\frac{4}{5}$	.8	80%
$\frac{9}{10}$	.9	90%
1	1	100%

PAGE 252

1. **$2.19** Multiply the purchase price by the tax: $\$35 \times 6.25\% = 2.1875$
   2.1875 rounds to 2.19
2. **1%** Divide the number of defective parts (the part) by the number of parts tested (the whole): $5 \div 500 = .01 = 1\%$.

PAGE 253

1. **Subtotal = $215.93; Deposit = $145.93**
   Add the two checks to find the subtotal: $\$165.43 + \$50.50 = \$215.93$. Subtract the amount of cash received from the subtotal to find the deposit:
   $\$215.93 - \$70.00 = \$145.93$.
2. **Subtotal = $221.69; Deposit = $166.69**
   Add the three checks to find the subtotal: $\$85.25 + \$70.89 + \$65.55 = \$221.69$.
   Subtract the amount of cash received from the subtotal to find the deposit:
   $\$221.69 - \$55.00 = \$166.69$.

GLOSSARY

addition combining numbers to find a total or sum

area the amount of surface something takes up, measured in square units; for a square or rectangle, the product of the length times the width

average the usual, or typical, value that can be used to represent a group of values; the mean, the sum of a group of entries divided by the number of entries used to find the sum

axis lines the vertical and horizontal lines that create a reference point for a graph

bar graph a graph that displays data using horizontal or vertical bars to compare numbers

base (*b*) the whole amount in a percent problem; the base represents 100%; one of the terms in the percent formula; the quotient of the part divided by the rate

base the amount being multiplied by itself when working with exponents

canceling finding values that will divide evenly into the numerators and denominators of fractions to convert them into smaller fractions that are easier to work with

chance outcome; the probability that something will happen

circle graph a graph that shows a whole amount divided into parts

column information arranged in a table so that it can be read from top to bottom

common denominator a denominator that is the same for two or more fractions

convert to change from one form to another

cross-multiply multiplying the numerator of one fraction by the denominator of another fraction; used to solve a proportion and to determine if two fractions are equivalent

cross-product the result of cross-multiplication. The cross-products of equivalent fractions are equal.

cube a six-sided figure in which each side is the same-sized square

cube a number multiplied by itself three times, represented by the exponent 3; raised to the third power

cubic foot a box shape that measures one foot on each side (the length, width, and height)

data information that has been gathered to be studied and analyzed

decimal a value that shows part of a whole number; a number containing a decimal point; a fraction that uses the place value system based on the number 10

denominator the bottom number in a fraction that shows the number of parts a whole is divided into

dependent probability the chance of something happening that is affected by another outcome

difference the answer to a subtraction problem; the result of subtraction

digit one of ten numbers on which all other numbers are based. Our number system uses ten digits: 0, 1, 2, 3, 4, 5, 6, 7, 8, and 9.

digital scale a tool that measures weight using decimal amounts

distance formula the distance traveled in a specific amount of time at a specific speed; the product of the rate times the time

division finding out how many times one number goes into or divides another number

double bar graph a graph that uses two bars in each category as a means of comparison

equation a statement formed by comparing two expressions using an equals sign

equivalent equal

equivalent fraction a fraction that has the same value as another fraction; fractions that are equal

estimation a method for finding an approximate answer or value close to the actual answer or value

exponent the number that shows how many times a number (the base) is multiplied by itself

formula an algebraic equation that shows the constant relationship between certain variables

fraction a value that represents part of a whole

gram a unit of mass in the metric system (g)

graph a way to represent some mathematical relationships as a picture

higher terms finding an equivalent fraction with a greater numerator and denominator

horizontal axis the part of the graph with values that are read across

improper fraction a fraction with a value equal to or greater than one; a fraction in which the numerator is greater than or equal to the denominator

integer any positive or negative whole number or zero

interest (i) a charge paid to borrow or use someone else's money; the product of the principal times the rate times the time period

interest rate (r) the percent used to figure out how much interest a borrower has to pay

invert to switch the numerator and denominator of the fraction; to turn a fraction upside down

kilogram one thousand grams (kg)

label a word or name used to identify something

like fractions fractions that have the same, or a common, denominator

like quantities amounts that are expressed in the same, or common, units

line graph a graph that uses a line to show changes over time

lowest common denominator the smallest number that the denominators of two or more fractions will divide evenly into

lowest terms a fraction with a numerator and denominator that can only be divided evenly by the number one

mathematical expression numbers and symbols representing variables and the mathematical operations used to represent a problem

mean the average; the sum of a group of numbers divided by the number of numbers used to find the sum

median the middle number of a group or set of numbers arranged in order

mixed number a number that contains a whole number and a proper fraction

multiplication the mathematical operation that is the same as adding the same number many times

negative number an integer that is less than zero; a number to the left of zero on a number line

number line a visual that shows the relationship between positive and negative numbers and zero. On a number line, positive numbers are to the right of zero, and negative numbers are to the left of zero.

numerator the top number in a fraction that shows the number of parts of a whole being discussed

order of operations an accepted rule stating the sequence in which to perform mathematical operations such as addition, subtraction, multiplication, and division

outcome the result; the probability that something will happen

part (p) a portion of the whole amount; one of the terms in the percent formula; the product of the base times the rate

percent a way of representing part of a whole equal to 100; means "for every 100" or "out of 100"

percent formula mathematical equation that shows the relationship between the base (b) or whole, rate (r), and part (p)

percent of decrease the rate of change by which an amount went down; the quotient of the amount of change (the original amount minus the new amount) divided by the original amount

percent of increase the rate of change by which an amount went up; the quotient of the amount of change (the new amount minus the original amount) divided by the original amount

perimeter the distance around the outside of a shape; the sum of the measurements of all the sides of an object

place value the value of a digit that depends on its location in the number

population in statistics, an entire group from which information is gathered

positive number an integer that is greater than zero; a number to the right of zero on a number line

powers of ten any number that is evenly divisible by 10, such as 10, 100, 1,000, and so on

principal (p) the amount of money borrowed on which interest is paid; a factor in the interest formula

probability the study of the chance of something happening

proper fraction a fraction in which the numerator is less than the denominator

proportion an equation comparing two equal ratios or fractions; expressed as a number between 0 and 1 or as a percent; the quotient of the number of ways an outcome can occur divided by the total number of possible outcomes

rate a ratio comparing two unlike quantities (amounts that are expressed in different units); a fraction usually with a denominator of one

rate (r) one of the terms (the percent) in the percent formula; always followed by the percent sign (%) or the word *percent*; the quotient of the part divided by the base

ratio a way of comparing two like quantities (amounts expressed in the same units)

reducing finding an equivalent fraction with a smaller numerator and denominator

remainder an amount left over in a division problem

rounded number an approximate, or estimated, amount

row information arranged on a table so that it can be read across, from left to right

sample in statistics, a smaller group that represents a population

scale a key that gives values to help read a graph or map

square a rectangle that has four sides of equal length; to multiply a number times itself; the exponent 2

square root the number which, when multiplied by itself, gives a specified number

square yard a square that measures one yard on each side

statistics data that is organized and analyzed

subtotal the sum of part of a group of numbers

subtraction finding the difference between amounts; making a comparison between two numbers

sum the total; the answer to an addition problem

survey a way of gathering data

table a grid that organizes information using rows and columns

tax additional money paid based on the amount of a purchase

term one of the numbers in a proportion; the numerator or denominator of a fraction

time (t) the period or duration of something; used to compute distance and interest; often the denominator of a rate

total the sum; the complete cost of something

total cost the amount found by multiplying the number of units by the unit cost

unit cost the price of one item

unlike fractions fractions that have different denominators

unlike quantities amounts that are expressed in different units

value the number obtained by substituting a specific number or numbers in an expression

variable a letter or symbol used to represent an unknown value

vertical axis the part of the graph with values that are read from bottom to top

volume the amount of space inside a solid or three-dimensional object

whole number an amount made up of one or more digits

A

B

C

D